John D. Anderson, Jr.
Dept. of Aerospace Engineering
University of Maryland
College Park, Md, 20742
Jan, 1982

AEROTHERMODYNAMICS AND PLANETARY ENTRY

Edited by
A. L. Crosbie
Thermal Radiative Transfer Group
Department of Mechanical and Aerospace Engineering
University of Missouri-Rolla, Rolla, Missouri

Volume 77
PROGRESS IN
ASTRONAUTICS AND AERONAUTICS
Martin Summerfield, Series Editor-in-Chief
Princeton Combustion Research Laboratories, Inc.,
Princeton, New Jersey

Technical papers from the AIAA 18th Aerospace Sciences Meeting, January 1980, and the AIAA 15th Thermophysics Conference, July 1980, subsequently revised for this volume.

Published by the American Institute of Aeronautics and Astronautics,
1290 Avenue of the Americas, New York, N.Y. 10104.

American Institute of Aeronautics and Astronautics
New York, New York

Library of Congress Cataloging in Publication Data

AIAA Aerospace Sciences Meeting (18th: 1980:
Los Angeles, Calif.)
Aerothermodynamics and planetary entry.

(Progress in astronautics and aeronautics; v. 77) Bibliography: p.
1. Aerothermodynamics—Congresses. 2 Space vehicles—Atmospheric entry—Congresses. I. Crosbie, A. L. (Alfred L.) II. AIAA Thermophysics Conference (15th: 1980: Snowmass, Colo.) III. American Institute of Aeronautics and Astronautics. IV. Series.
TL1060.A36 1980 629.4'152 81-10785
ISBN 0-915928-52-3 AACR2

Table of Contents

Preface

Thermal protection of a spacecraft or probe entering a planetary atmosphere is one of the most challenging problems ever attempted. The thermal design of these vehicles requires the coupling of heat transfer and fluid mechanics to predict heating rates. The development of the Space Shuttle and the advancement of missile technology have stimulated interest in Earth entry and aerothermodynamics, in particular. The Space Shuttle's size and complex geometry, combined with the fact that it is to be a reusable spacecraft, give rise to many unique aerodynamic heating problems. By necessity much of the work in this area is experimental in nature. Exploration of the other planets in our solar system has also stimulated interest in entry technology. The high entry velocities and severe environments challenge our thermal design capabilities. For example, almost half the weight of the Galileo (Jupiter) probe is devoted to its thermal protection system. Because of the difficulty of experimentally simulating the entry conditions here on Earth, most of the thermal design work for the Galileo probe must be accomplished by numerical simulation.

This volume contains a selection of recent studies in aerothermodynamics and planetary entry. The papers for this volume and its companion volume were drawn from the AIAA 18th Aerospace Sciences Meeting in Los Angeles, California in January 1980 and the AIAA 15th Thermophysics Conference in Snowmass, Colorado in July 1980. They have been reviewed, revised, and updated especially for these volumes. The papers in Volume 77 have been grouped into three chapters: aerothermodynamics, thermal protection, and planetary entry.

Besides the obvious appeal of this volume to those working in the area of entry technology, the volume should be of interest to engineers and scientists working in related areas. Specifically, these areas include: forced convection heat transfer, gaseous radiation heat transfer, experimental fluid mechanics, turbulent and transitional boundary layers, and thermophysical properties of high temperature gases. Hopefully, the research developments reported here will stimulate Earth-based technologies which are confronted with problems of high heating rates and extreme environments.

Chapter I consists of six papers concerned with the influence of surface geometry on aerodynamic heating. The first two papers deal with supersonic flow over rectangular gaps or slots. *Bertin and Goodrich* measure heating rates in and around various transverse gap and longitudinal slot configurations in laminar and transitional boundary layers, while *Lamb* correlates heat-transfer data for a transverse gap with laminar or turbulent flow. These studies are motivated in part by the Space Shuttle. In the third paper, *McDermott* reports measurements of local heat-flux rates on a flat plate downstream of a square, subsonic, hot air jet issuing from the surface.

The next three papers deal with high-speed flow over surfaces with protuberances as opposed to gaps or slots. *Finson and Clarke* present a Reynolds stress model for studying the influence of roughness character (element shape and spacing) on skin friction heat transfer for turbulent boundary layers and high Mach numbers. *Hung and Clarke* present an extensive experimental study on interference heating caused by a three-dimensional cylindrical protuberance protruding from a flat plate. Measurements are made on both the flat plate and the protuberance. *Cassel and Jarrett* report aeroheating measurements for finite span trailing flaps in hypersonic flow. The flap deflection angles of 5, 10, and 15 deg are considered.

Various aspects of thermal protection systems are discussed in the six papers of Chapter II. *Laub* presents a thermochemical ablation model for carbon-carbon composite materials loaded with tantalum carbide. Steady-state ablation predictions are compared with available arc test data and other existing ablation models. The energy transfer catalytic recombination coefficient for nitrogen and oxygen recombination on the surface coating of high-temperature reusable surface insulation is inferred by *Scott* from stagnation point heat-flux measurements in a high-temperature dissociated arc jet flow.

The next four papers deal with experimental techniques. *Painter* uses arc-heated helium to accelerate graphite particles to hypervelocities with the objective of simulating the erosion of a thermal protection system by particles suspended in the atmosphere. *Cassaing, Lelievre, Durrengerger, and Balageas* present a novel combined ablation/temperature gage which uses the optical view of radiation from a sensor embedded in a heat shield. *Berman and Nestler* conduct a wind-tunnel study at Mach 8 to provide design information for the multiple re-entry vehicle assent heating problem. A phase-change paint technique is used to identify shock impingement hot spots. Soft-recovery techniques for moderately high ballistic coefficient re-entry vehicles are discussed by *Rigali, Sterk, and Randmaa.*

Chapter III consists of six papers concerned with vehicle heating during planetry entry. The first three papers deal with Jovian entry. The thermal design of the Galileo probe is described by *Brewer and Brant.* An approximate method is proposed by *Armaly and Sutton* for predicting the viscosity of partially ionized gas mixtures. *Kumar, Tiwari, Graves, and Weilmuenster* use a time-dependent finite-difference method for solving the problem of coupled laminar and turbulent flows over the forebody of a probe entering a nominal Jupiter atmosphere. Solutions are obtained for both a 35 deg hyperboloid and a 45 deg sphere cone.

Preliminary studies are given in the next three papers. Using methods developed for the Galileo probe, *Zoby and Moss* present a preliminary thermal analysis on Saturn entry. *French and McRonald* investigate the aerobraking entry technique to reduce spacecraft mass, while *Cruz* investigates the aerocapture technique. Aerobraking requires propulsive maneuvers, while aerocapture relies entirely on atmospheric drag. Both techniques show promise in increasing scientific payloads by reducing the weight of thermal protection systems.

This editor gratefully acknowledges the contributions of the Editorial Committee listed on page *x*, Miss Ruth F. Bryans, Associate Series Editor, Miss Brenda Hio, Managing Editor of the Series, and Dr. Martin Summerfield, Editor-in-Chief of the AIAA *Progress in Astronautics and Aeronautics* series. The efforts of Dr. Kenneth E. Harwell who organized the thermophysics sessions at the AIAA 18th Aerospace Sciences Meeting, Dr. John E. Francis who served as General Program Chairman of the AIAA 15th Thermophysics Conference and chaired the AIAA Thermophysics Technical Committee in 1979, and Mr. Jesse F. Keville who chaired the AIAA Thermophysics Technical Committee in 1980 are also greatly appreciated. Finally, the contributors to this volume are thanked for their patience, cooperation, and care in the preparation of their papers.

Alfred L. Crosbie
March 1981

Editorial Committee for Volume 77

Progress in Astronautics and Aeronautics

VOLUMES	EDITORS
*1. **Solid Propellant Rocket Research.** 1960	Martin Summerfield *Princeton University*
2. **Liquid Rockets and Propellants.** 1960	Loren E. Bollinger *The Ohio State University* Martin Goldsmith *The Rand Corporation* Alexis W. Lemmon Jr. *Battelle Memorial Institute*
3. **Energy Conversion for Space Power.** 1961	Nathan W. Snyder *Institute for Defense Analyses*
*4. **Space Power Systems.** 1961	Nathan W. Snyder *Institute for Defense Analyses*
5. **Electrostatic Propulsion.** 1961	David B. Langmuir *Space Technology Laboratories, Inc.* Ernst Stuhlinger *NASA George C. Marshall Space Flight Center* J. M. Sellen Jr. *Space Technology Laboratories, Inc.*

*Now out of print.

*6.	**Detonation and Two-Phase Flow.** 1962	S. S. Penner *California Institute of Technology* F. A. Williams *Harvard University*
7.	**Hypersonic Flow Research.** 1962	Frederick R. Riddell *AVCO Corporation*
8.	**Guidance and Control.** 1962	Robert E. Roberson *Consultant* James S. Farrior *Lockheed Missiles and Space Company*
*9.	**Electric Propulsion Development.** 1963	Ernst Stuhlinger *NASA George C. Marshall Space Flight Center*
*10.	**Technology of Lunar Exploration.** 1963	Clifford I. Cummings and Harold R. Lawrence *Jet Propulsion Laboratory*
11.	**Power Systems for Space Flight.** 1963	Morris A. Zipkin and Russell N. Edwards *General Electric Company*
12.	**Ionization in High-Temperature Gases.** 1963	Kurt E. Shuler, Editor *National Bureau of Standards* John B. Fenn, Associate Editor *Princeton University*
*13.	**Guidance and Control—II.** 1964	Robert C. Langford *General Precision Inc.* Charles J. Mundo *Institute of Naval Studies*
*14.	**Celestial Mechanics and Astrodynamics.** 1964	Victor G. Szebehely *Yale University Observatory*
*15.	**Heterogeneous Combustion.** 1964	Hans G. Wolfhard *Institute for Defense Analyses* Irvin Glassman *Princeton University* Leon Green Jr *Air Force Systems Command*

16. Space Power Systems Engineering. 1966	George C. Szego *Institute for Defense Analyses* J. Edward Taylor *TRW Inc.*
***17. Methods in Astrodynamics and Celestial Mechanics.** 1966	Raynor L. Duncombe *U. S. Naval Observatory* Victor G. Szebehely *Yale University Observatory*
18. Thermophysics and Temperature Control of Spacecraft and Entry Vehicles. 1966	Gerhard B. Heller *NASA George C. Marshall Space Flight Center*
***19. Communication Satellite Systems Technology.** 1966	Richard B. Marsten *Radio Corporation of America*
20. Thermophysics of Spacecraft and Planetary Bodies: Radiation Properties of Solids and the Electromagnetic Radiation Environment in Space. 1967	Gerhard B. Heller *NASA George C. Marshall Space Flight Center*
21. Thermal Design Principles of Spacecraft and Entry Bodies. 1969	Jerry T. Bevans *TRW Systems*
22. Stratospheric Circulation. 1969	Willis L. Webb *Atmospheric Sciences Laboratory, White Sands, and University of Texas at El Paso*
23. Thermophysics: Applications to Thermal Design of Spacecraft. 1970	Jerry T. Bevans *TRW Systems*
24. Heat Transfer and Spacecraft Thermal Control. 1971	John W. Lucas *Jet Propulsion Laboratory*

25. Communications Satellites for the 70's: Technology. 1971	Nathaniel E. Feldman *The Rand Corporation* Charles M. Kelly *The Aerospace Corporation*
26. Communications Satellites for the 70's: Systems. 1971	Nathaniel E. Feldman *The Rand Corporation* Charles M. Kelly *The Aerospace Corporation*
27. Thermospheric Circulation. 1972	Willis L. Webb *Atmospheric Sciences Laboratory, White Sands, and University of Texas at El Paso*
28. Thermal Characteristics of the Moon. 1972	John W. Lucas *Jet Propulsion Laboratory*
29. Fundamentals of Spacecraft Thermal Design. 1972	John W. Lucas *Jet Propulsion Laboratory*
30. Solar Activity Observations and Predictions. 1972	Patrick S. McIntosh and Murray Dryer *Environmental Research Laboratories, National Oceanic and Atmospheric Administration*
31. Thermal Control and Radiation. 1973	Chang-Lin Tien *University of California, Berkeley*
32. Communications Satellite Systems. 1974	P. L. Bargellini *COMSAT Laboratories*
33. Communications Satellite Technology. 1974	P. L. Bargellini *COMSAT Laboratories*
34. Instrumentation for Airbreathing Propulsion. 1974	Allen E. Fuhs *Naval Postgraduate School* Marshall Kingery *Arnold Engineering Development Center*
35. Thermophysics and Spacecraft Thermal Control. 1974	Robert G. Hering *University of Iowa*

36. Thermal Pollution Analysis. 1975	Joseph A. Schetz *Virginia Polytechnic Institute*
37. Aeroacoustics: Jet and Combustion Noise; Duct Acoustics. 1975	Henry T. Nagamatsu, Editor *General Electric Research and Development Center* Jack V. O'Keefe, Associate Editor *The Boeing Company* Ira R. Schwartz, Associate Editor *NASA Ames Research Center*
38. Aeroacoustics: Fan, STOL, and Boundary Layer Noise; Sonic Boom; Aeroacoustics Instrumentation. 1975	Henry T. Nagamatsu, Editor *General Electric Research and Development Center* Jack V. O'Keefe, Associate Editor *The Boeing Company* Ira R. Schwartz, Associate Editor *NASA Ames Research Center*
39. Heat Transfer with Thermal Control Applications. 1975	M. Michael Yovanovich *University of Waterloo*
40. Aerodynamics of Base Combustion. 1976	S. N. B. Murthy, Editor *Purdue University* J. R. Osborn, Associate Editor *Purdue University* A. W. Barrows and J. R. Ward, Associate Editors *Ballistics Research Laboratories*
41. Communication Satellite Developments: Systems. 1976	Gilbert E. LaVean *Defense Communications Engineering Center* William G. Schmidt *CML Satellite Corporation*
42. Communication Satellite Developments: Technology. 1976	William G. Schmidt *CML Satellite Corporation* Gilbert E. LaVean *Defense Communications Engineering Center*

43. Aeroacoustics: Jet Noise, Combustion and Core Engine Noise. 1976
Ira R. Schwartz, Editor
NASA Ames Research Center
Henry T. Nagamatsu, Associate Editor
General Electric Research and Development Center
Warren C. Strahle, Associate Editor
Georgia Institute of Technology

44. Aeroacoustics: Fan Noise and Control; Duct Acoustics; Rotor Noise. 1976
Ira R. Schwartz, Editor
NASA Ames Research Center
Henry T. Nagamatsu, Associate Editor
General Electric Research and Development Center
Warren C. Strahle, Associate Editor
Georgia Institute of Technology

45. Aeroacoustics: STOL Noise; Airframe and Airfoil Noise. 1976
Ira R. Schwartz, Editor
NASA Ames Research Center
Henry T. Nagamatsu, Associate Editor
General Electric Research and Development Center
Warren C. Strahle, Associate Editor
Georgia Institute of Technology

46. Aeroacoustics: Acoustic Wave Propagation; Aircraft Noise Prediction; Aeroacoustic Instrumentation. 1976
Ira R. Schwartz, Editor
NASA Ames Research Center
Henry T. Nagamatsu, Associate Editor
General Electric Research and Development Center
Warren C. Strahle, Associate Editor
Georgia Institute of Technology

47. Spacecraft Charging by Magnetospheric Plasmas. 1976
Alan Rosen
TRW Inc.

48. Scientific Investigations on the Skylab Satellite. 1976	Marion I. Kent and Ernst Stuhlinger *NASA George C. Marshall Space Flight Center* Shi-Tsan Wu *The University of Alabama*
49. Radiative Transfer and Thermal Control. 1976	Allie M. Smith *ARO Inc.*
50. Exploration of the Outer Solar System. 1977	Eugene W. Greenstadt *TRW Inc.* Murray Dryer *National Oceanic and Atmospheric Administration* Devrie S. Intriligator *University of Southern California*
51. Rarefied Gas Dynamics, Parts I and II (two volumes). 1977	J. Leith Potter *ARO Inc.*
52. Materials Sciences in Space with Application to Space Processing. 1977	Leo Steg *General Electric Company*
53. Experimental Diagnostics in Gas Phase Combustion Systems. 1977	Ben T. Zinn, Editor *Georgia Institute of Technology* Craig T. Bowman, Associate Editor *Stanford University* Daniel L. Hartley, Associate Editor *Sandia Laboratories* Edward W. Price, Associate Editor *Georgia Institute of Technology* James G. Skifstad, Associate Editor *Purdue University*
54. Satellite Communications: Future Systems. 1977	David Jarett *TRW Inc.*

55. Satellite Communications: Advanced Technologies. 1977	David Jarett *TRW Inc.*
56. Thermophysics of Spacecraft and Outer Planet Entry Probes. 1977	Allie M. Smith *ARO Inc.*
57. Space-Based Manufacturing from Nonterrestrial Materials. 1977	Gerard K. O'Neill, Editor *Princeton University* Brian O'Leary, Assistant Editor *Princeton University*
58. Turbulent Combustion. 1978	Lawrence A. Kennedy *State University of New York at Buffalo*
59. Aerodynamic Heating and Thermal Protection Systems. 1978	Leroy S. Fletcher *University of Virginia*
60. Heat Transfer and Thermal Control Systems. 1978	Leroy S. Fletcher *University of Virginia*
61. Radiation Energy Conversion in Space. 1978	Kenneth W. Billman *NASA Ames Research Center*
62. Alternative Hydrocarbon Fuels: Combustion and Chemical Kinetics. 1978	Craig T. Bowman *Stanford University* Jørgen Birkeland *Department of Energy*
63. Experimental Diagnostics in Combustion of Solids. 1978	Thomas L. Boggs *Naval Weapons Center* Ben T. Zinn *Georgia Institute of Technology*
64. Outer Planet Entry Heating and Thermal Protection. 1979	Raymond Viskanta *Purdue University*
65. Thermophysics and Thermal Control. 1979	Raymond Viskanta *Purdue University*

66. Interior Ballistics of Guns. 1979	Herman Krier *University of Illinois at Urbana-Champaign* Martin Summerfield *New York University*
67. Remote Sensing of Earth from Space: Role of "Smart Sensors." 1979	Roger A. Breckenridge *NASA Langley Research Center*
68. Injection and Mixing in Turbulent Flow. 1980	Joseph A. Schetz *Virginia Polytechnic Institute and State University*
69. Entry Heating and Thermal Protection. 1980	Walter B. Olstad *NASA Headquarters*
70. Heat Transfer, Thermal Control, and Heat Pipes. 1980	Walter B. Olstad *NASA Headquarters*
71. Space Systems and Their Interactions with Earth's Space Environment. 1980	Henry B. Garrett and Charles P. Pike *Hanscom Air Force Base*
72. Viscous Flow Drag Reduction. 1980	Gary R. Hough *Vought Advanced Technology Center*
73. Combustion Experiments in a Zero-Gravity Laboratory. 1981	Thomas H. Cochran *NASA Lewis Research Center*
74. Rarefied Gas Dynamics, Parts I and II (two volumes). 1981	Sam S. Fisher *University of Virginia at Charlottesville*

75. Gasdynamics of Detonations and Explosions. 1981	J. R. Bowen *University of Wisconsin at Madison* N. Manson *Université de Poitiers* A. K. Oppenheim *University of California at Berkeley* R. I. Soloukhin *Institute of Heat and Mass Transfer, BSSR Academy of Sciences*
76. Combustion in Reactive Systems. 1981	J. R. Bowen *University of Wisconsin at Madison* N. Manson *Université de Poitiers* A. K. Oppenheim *University of California at Berkeley* R. I. Soloukhin *Institute of Heat and Mass Transfer, BSSR Academy of Sciences*
77. Aerothermodynamics and Planetary Entry. 1981	A. L. Crosbie *University of Missouri-Rolla*
78. Heat Transfer and Thermal Control. 1981	A. L. Crosbie *University of Missouri-Rolla*

(Other volumes are planned.)

Chapter I. Aerothermodynamics

AERODYNAMIC HEATING FOR GAPS IN LAMINAR AND TRANSITIONAL BOUNDARY LAYERS

J.J. Bertin*

The University of Texas at Austin, Austin, Tex.

and

W.D. Goodrich†

NASA Johnson Space Center, Houston, Tex.

Abstract

Heating rates were measured in and around various transverse-gap and longitudinal-slot configurations over a range of Reynolds numbers from 2.5×10^6 to 8.1×10^6. The geometric variables included the number of gaps, gap width, gap depth, the spacing between gaps, and the orientation of the gaps. Although the local heating rates were relatively unperturbed by the short, deep transverse gaps for laminar flow, significant increases occurred for transitional flows. For the intersecting T gap configurations, for which the $(x_s - x_0)/d$ ratios were measured along the longitudinal slot and at the recompression surface of the intersection.

Nomenclature

d	=	depth of the gap or height of the recompression surface
L	=	total length of the flat plate
M_e	=	Mach number at the edge of the boundary layer
$\dot{q}$	=	local heat-transfer rate

Presented as Paper 80-0287 at the AIAA 18th Aerospace Sciences Meeting, Pasadena, Calif., Jan. 14-16, 1980.

*Professor, Department of Aerospace Engineering and Engineering Mechanics.

†Aerospace Technologist, Structures and Mechanics Division.

$\dot{q}_{fp}$ = local heat-transfer rate at the same station, averaged for the three y stations (used as denominator in heating-rate ratios)
$\dot{q}_{pk}$ = peak heating rate in the vicinity of a gap or slot
$\dot{q}_{pk,i}$ = peak heating rate measured on the impingement surface of a gap
Re_x = local Reynolds number based on the streamwise distance from the leading edge of the plate
Re_θ = Reynolds number based on the local momentum thickness
$Re_{\infty,L}$ = freestream Reynolds number based on the total length of the plate
w = width of the gap
x = distance from the leading edge of the plate (see Fig. 1)
x_s = x distance to the transverse gap (configurations C2 and C4)
x_0 = x distance to the beginning of the longitudinal slot (configuration C4)
y = distance from the plane of symmetry (see Fig. 1)
z* = z coordinate of the sonic point
z = distance from the horizontal surface of the flat-plate holder
δ* = displacement thickness
θ = momentum thickness

Introduction

The presence of gaps, slots, and/or steps in a surface may significantly perturb a supersonic boundary layer. The perturbations may increase the boundary-layer thickness with a reduction in the overall heating, may cause viscous/shock interactions with locally high heating rates affecting only a small fraction of the surface, or may increase the turbulence resulting in premature transition and affecting the heating rates over large surface areas. The qualitative and the quantitative effects on the flowfield due to these sudden changes in the surface slope depend on its geometry, on its dimensions relative to the boundary-layer thickness, and on the characteristics of the boundary layer.

For relatively simple two-dimensional cutouts, i.e., a single transverse gap, Charwat et al.[1,2] found that supersonic flow over a cutout (or cavity) in the surface can produce either of two distinct, stable configurations. When the length-to-height ratio of the cutouts, or the width-to-depth ratio (w/d) in the present nomenclature, is sufficiently large, the flow reattaches to the cavity floor and the cavity is "closed." If the length of the notch decreases below a critical value, the flow bridges the gap. This is called flow with an "open" cavity. For short notches, where the length of the cavity is less than one-half the critical length, the pressure rise associated with the deceleration of the inner portion of the shear layer tends to propagate upstream and to thicken the shear layer. As a result, the characteristics of the boundary-layer/freestream interaction govern the pressure field. Data presented by Charwat et al.[1] indicate that the flow that separates from an isolated, rearward-facing step impinges on the wall approximately 7d downstream of the step if the boundary layer is laminar and approximately 5d downstream for a turbulent boundary layer. Experimental investigations[3,4] have shown that laminar heating rates in the vicinity of a very short gap are only slightly different than the flat-plate values. Impingement heating-rate measurements reported by Avery[4] did not significantly exceed the flat-plate values until the gap width was 0.30 cm (0.12 in.) in a laminar boundary where δ^* was 0.57 cm (0.22 in.). That the impingement heating increases as the gap width increases has also been reported in Refs. 5 and 6 for a laminar boundary layer and in Refs. 4 and 7 for a turbulent boundary layer. Thus, the data indicate that as long as the gap width is below some critical value (which is a function of the boundary-layer character and thickness), locally high heating rates can be avoided.

In a study of the aerodynamic heating to corrugation-stiffened structures in thick, turbulent layers, Brandon et al.[8] found that although the highest peak-heating rate occurred when the corrugations were perpendicular to the flow, it was not much less when the corrugations were inclined to the freestream by a little as 15 deg.

Locally high heating rates occur on the (upstream-facing) recompression surface at the "T" formed by the intersection of the longitudinal slot and the transverse gap. The data of Dunavant and Throckmorton[3] failed to show a region of intense heating on the upstream face at the gap intersection for turbulent flows. Note that the $(x_s - x_0)/d$ for their model was 1.49. The measurements of Avery[4] indicate that the impingement region heating rates increase with increasing w, being a maximum of 2.2 times the flat-plate value for laminar flow and 4.5 times the flat-plate value for a turbulent boundary layer. Avery noted that the length of the longitudinal slot was an important parameter even though he did not have sufficient data to define the dependence. For the configurations of Ref.

4, $(x_s - x_0)/d$ was 2.40 to 4.80. Johnson[9] reported that the local heating increased by an order of magnitude. The $(x_s - x_0)/d$ ratio was 8.0 for the models of Ref. 9. From a comparison of the results of these investigators, it is clear that the heating increase in the impingement region increases dramatically with $(x_s - x_0)/d$. The phenomenon appears to be similar to that for a two-dimensional notch where the characteristics of the flowfield perturbations depend on whether the notch is "open" or "closed." Thus, the magnitude of the heating increase in the impingement region of a T depends on many parameters, including: δ^*, w, $(x_s - x_0)/d$, and the Reynolds number.

The present paper discusses heat-transfer distributions for a variety of gap configurations that were obtained by placing instrumented inserts in a flat-plate structural carrier. These data were obtained in the 3.5-ft. Hypersonic Wind Tunnel at the Ames Research Center (NASA) at a freestream Mach number of 5.10 over a range of Reynolds number (based on the length of the flat plate) from 2.57×10^6 to 8.11×10^6. The variables of the test program included the freestream Reynolds number and the gap configuration, e.g., width, depth, step height, number, and orientation.

Experimental Program

Model

Heat-transfer distributions were measured for five different geometries which were obtained by placing instrumented inserts into a structural carrier plate. The carrier plate is a flat, stainless steel plate that is 5.08 cm thick, 68.58 cm wide, and 152.40 cm long, with a sharp 5 deg leading edge. Three model-mounted total-temperature probes were used to measure T_t. The flow conditions were calculated using these values of T_t and the total pressure in the stagnation chamber.

The C1 Configuration. A smooth, flat-plate heat-transfer insert was placed in the structural carrier plate to form Configuration 1 (C1). As shown in Fig. 1a, there were three rows of thermocouples that provided heat-transfer data for axial locations in the range 35.52 cm $\leq x \leq$ 129.59 cm along the centerline of the plate (y= 0.0) and $\pm$ 20.536 cm on either side of it. The y coordinate is measured from the plane of symmetry, as shown in Fig. 1a.

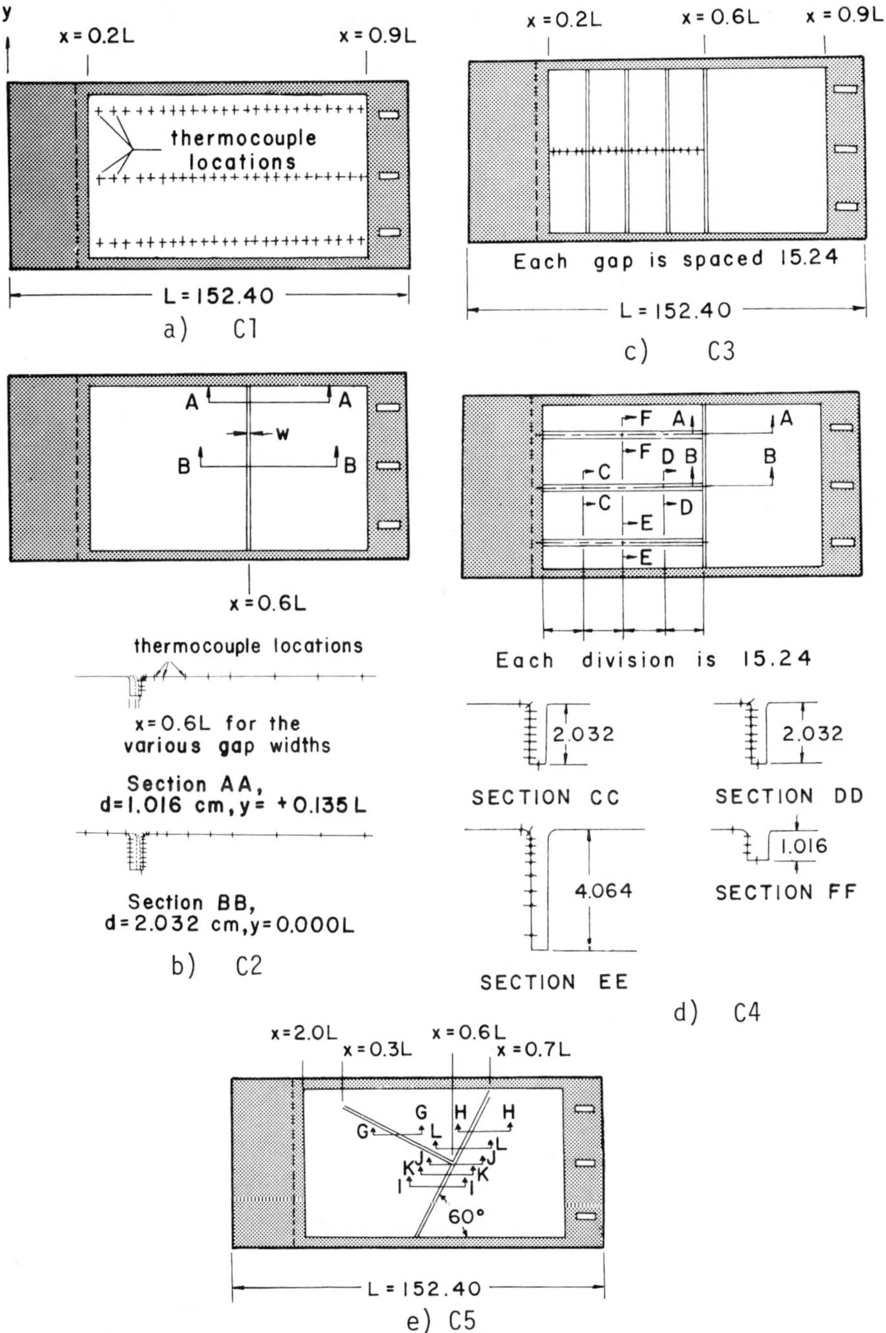

Fig. 1 Sketches of the configurations tested: C1) flat-plate insert, C2) single-transverse-gap insert, C3) multiple-transverse-gap insert, C4) longitudinal-intersecting-gap insert, and C5) diagonal-intersecting-gas insert.

The C2 Configuration. An insert containing a single transverse gap was placed in the structural carrier plate to form Configuration 2 (C2). As shown in Fig. 1b, the separation (upstream) step was at an x of 91.44 cm (0.60L). The width of the gap could be varied, so that heat-transfer distributions were measured for values of w equal to 0.127 cm, 0.254 cm, and 0.508 cm.

The insert was built so that the gap depth varied across (spanwise) the model, with instrumentation modules located at the centerline of each of three different gap-depth regions. Detailed sketches are presented to illustrate both the gap geometry and the thermocouple locations in the vicinity of the gap. Note that although the location of the recompression (downstream) step is fixed in the sketches of the section (for illustrative convenience), in reality the upstream step was fixed. Since heat-transfer distributions will be presented in this paper only for y = +0.135L (section AA), d = 1.016 cm, and for y = 0.000L (section BB), d = 2.032 cm, only these two sections are illustrated.

The corner radius was 0.152 cm for all variations for this configuration.

The C3 Configuration. An insert containing multiple, transverse gaps was placed in the structural carrier plate to form Configuration 3(C3). As shown in Fig. 1c, the gaps, which were 0.254 cm wide, were 15.24 cm apart.

The C4 Configuration. An insert containing three parallel, longitudinal gaps (or slots) that intersect a single, transverse gap was placed in the structural carrier plate to form Configuration 4(C4). As shown in Fig. 1d, the longitudinal slots were either 1.016 cm deep (which is at y = + 0.135L), 2.032 cm deep (which is at y = 0.000L), or 4.064 cm deep (which is at y = - 0.135L). Note that the thermocouples are located in-depth at one section (i.e., one x-location), FF, of the shallow slot, at two sections, CC and DD, of the slot located along the centerline, and at one section, EE, of the deepest slot.

The transverse gap and the downstream recompression surface are those for C2. Recall that the transverse gap insert was built so that its depth varied across the model (spanwise), with thermocouples located along the recompression surface at the centerline of each of the different gap-depth regions. Thus, at a given y station, the depth of the transverse gap is equal to that of the intersecting, longitudinal slot. Thermocouples located as shown in sections AA and BB of Fig. 1b provide the heat-transfer distributions on the recompression surface downstream of the T-type intersection. All the gaps (and slots) were 0.254 cm wide.

The C5 Configuration. An insert containing diagonal, intersecting gaps (or slots) was placed in the structural carrier plate to form Configuration 5(C5). The slots, which are 0.254 cm wide and 2.032 cm deep, are perpendicular to each other. As illustrated in Fig. 1e, the slots were aligned such that one was inclined 30 deg to the freestream flow and the other 60 deg to the flow. Thermocouples were located on the flat surface, upstream and downstream of the gaps, and in the gaps themselves in planes parallel to the freestream flow. The thermocouple locations for these sections (which are presented in Fig. 1e) are such that heating-rate distributions are obtained 1) in the vicinity of the 30 deg skewed gap where the boundary layer is always laminar (section GG), 2) in vicinity of the 60 deg skewed gap where the boundary layer is perturbed by the first gap and the unperturbed boundary layer is transitional at the higher Reynolds numbers (section HH), 3) near the intersection of the gaps (section JJ), and 4) at other sections along the 60 deg skewed gap. Two sets of runs were made at each of the four Reynolds numbers, using different thermocouple setups. Data will be presented for the set of runs where the heat-transfer rates were measured using the thermocouples from sections GG, II, JJ, KK, and LL.

Test Conditions. The heat-transfer data were obtained in the 3.5-ft Hypersonic Wind Tunnel at the Ames Research Center (NASA). The freestream Mach number was 5.10 and the angle-of-attack was zero for the entire run schedule. The nominal Reynolds numbers based on the model length were 2.57×10^6, 4.01×10^6, 5.90×10^6, and 8.11×10^6.

For additional information about the test program and additional analysis of the data, the reader is referred to Refs. 10 and 11.

Discussion of Results

The quantitative effect of the presence of a particular gap configuration will be defined by comparing the perturbed value of the local heat transfer to the corresponding flat-plate value measured at the same nominal flow conditions. The flat-plate heat-transfer distributions measured in the plane of symmetry (y = 0.000L) are presented in Fig. 2 for each of the four Reynolds numbers. The theoretical solutions com-

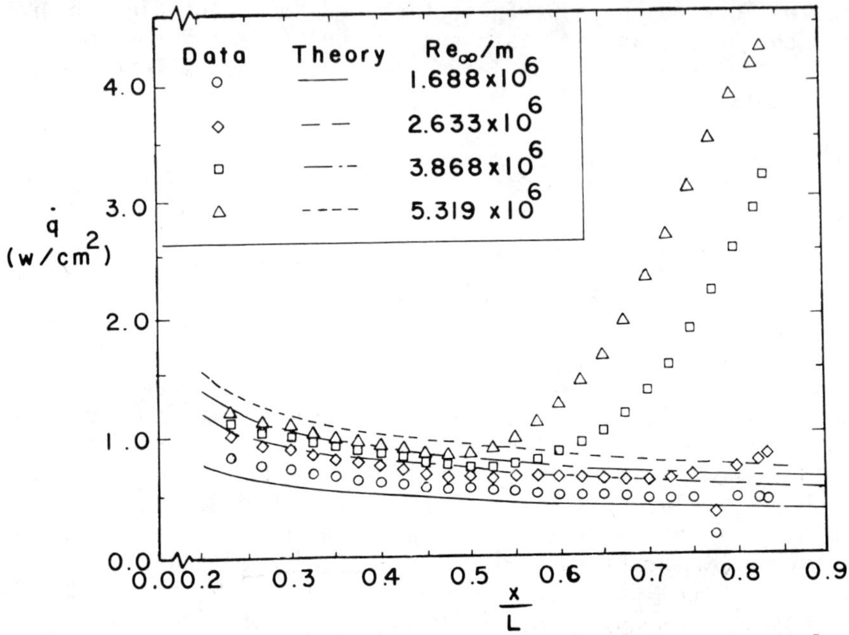

Fig. 2 A comparison of the theoretical and the experimental heat-transfer distributions for the flat-plate insert, C1.

puted using the perfect-gas relations for the thermodynamic properties in the University's nonsimilar, laminar boundary-layer code[12] are included for comparison. The perfect-gas solutions are in good agreement with the data from the laminar region. The departure of the experimental heating rates from the theoretical laminar values indicates the onset of boundary-layer transition.

Data for Configurations Having Only Transverse Gaps, i.e., C2 and C3

For two of the configurations tested, all the gaps were perpendicular to the freestream flow.

There was only one gap (located at x = 0.6L) for the C2 configuration, but its width could be varied and its depth did vary across the model, as illustrated in Fig. 1b. For the configurations studied, the width-to-depth ratio varied from 0.03125 to 0.500. None of these values even approaches the critical closure length for a two-dimensional notch.[1,2] Thus, the cavities should be open for all the C2 variations.

The data presented in Fig. 2 indicate that transition occurs upstream of the C2 gap-location for only the two highest Reynolds numbers. Although transition occurs upstream of this location when Re/m = 3.868 x 10^6, the transitional values of the boundary-layer parameters, δ^*, θ, and z^*, are virtually the same as the values computed for the same edge conditions, if one assumes the boundary layer remained laminar. For this flow condition, δ^* at the gap is 0.27 cm. However, when Re/m = 5.319 x 10^6, the transitional values are significantly different than the laminar values, such that δ^* and θ are slightly greater for the transitional boundary layer, while z^* is roughly five times less than the value for Re/m = 3.868 x 10^6. As will be discussed, the heat-transfer perturbations caused by the presence of the gap increase significantly for the transitional boundary layers. Thus, z^* appears to be a more relevant correlation parameter than either δ^* or θ.

The heat-transfer measurements for $0.583L \leq x \leq 0.700L$ are compared in Fig. 3 for C1 and for C2 over the complete Reynolds number range. Recall that, for all gap widths, the vertical face of the separation step was at x = 0.60L (91.44 cm). Thus, these data illustrate the effect of the gap on the upstream heat-transfer distribution as well as that for a distance downstream of the gap of approximately one shuttle-tile length. Data are presented only for y = 0.000L (i.e., the BB section of Fig. 1b), for which the depth of the gap was 2.032 cm.

Consider first the data for the lowest Reynolds number, which is the only flow condition for which the flat-plate measurements were significantly greater than the NSBL, perfect-gas theory (see Fig. 2). The heat-transfer rate for the C2 configuration increases slightly as the flow approaches the gap (x < 0.6L), decreases rapidly around the corner of the separation step, increases rapidly around the corner of the recompression step (reaching a maximum at the upper tangency point), and follows (at slightly lower magnitude) the heat-transfer distribution for an undisturbed laminar boundary layer. The heat-transfer data of Fig. 3a indicate that the peak values did not occur on the upstream-facing surface of the recompression step. They occurred either at the thermocouple located at the (upper) tangency point for the flat surface (for w = 0.254 cm and w = 0.508 cm) or at the first thermocouple downstream of the upper tangency point (for w = 0.127 cm). See Fig. 1b for the thermocouple locations. Thus, the peak heating does not occur where the impinging flow "stagnates" on the "recompression face." The quotes are used because the corner radius affects the character of the impinging flow for these short cavities. Because only the lower portion of the shear layer impinges on the downstream step, the "stagnation" heating on the recompression face is not the maximum value measured in this region. Instead the maximum heating occurs near the tangency point of the upper surface where the relatively high (velocity and temperature) gradients in the reattaching shear layer (which provides the "edge" conditions for the developing boundary layer) affect that boundary layer. Except at the thermocouples near the upper tangency point, the heating rates measured in the vicinity of the gap at this low Reynolds number were equal to or less than the corresponding flat-plate values. Because the boundary layer was laminar and relatively thick, the

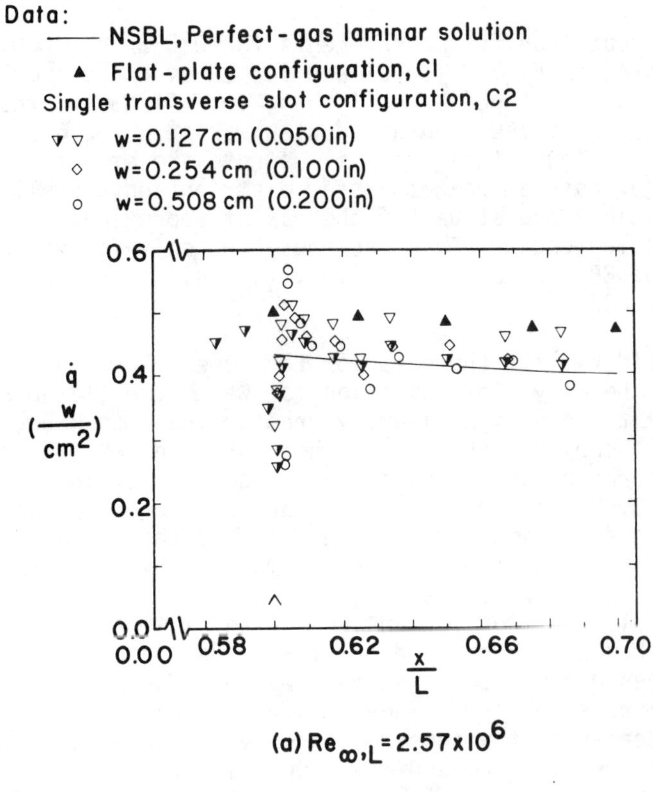

Fig. 3 A comparison of the heat-transfer distributions downstream of a single, transverse gap with that for a flat plate. Data for y = 0.000L (section BB).

adverse pressure gradient generated as the flow encountered the recompression step propagated upstream through the subsonic portion of the viscous layer and actually reduced the local heat-transfer rate to most of the surface. Dunavant and Throckmorton[3] reported no appreciable increase in the heating where the boundary layer bridged the gap. Bohon et al.[7] reported no significant spikes or hot spots for a "stacked configuration" of tiles, i.e., one in which there is no longitudinal channel intersecting (perpendicular to) the gap. Note that for these flow conditions, the peak heating in the impingement region is greatest for the widest gap.

For $Re_{\infty,L} = 4.01 \times 10^6$ (Fig. 3b), the peak heating in the impingement region is only slightly greater than the flat-plate values. This conclusion is consistent with that based on the lower Reynolds-number data. However, contrary to low-

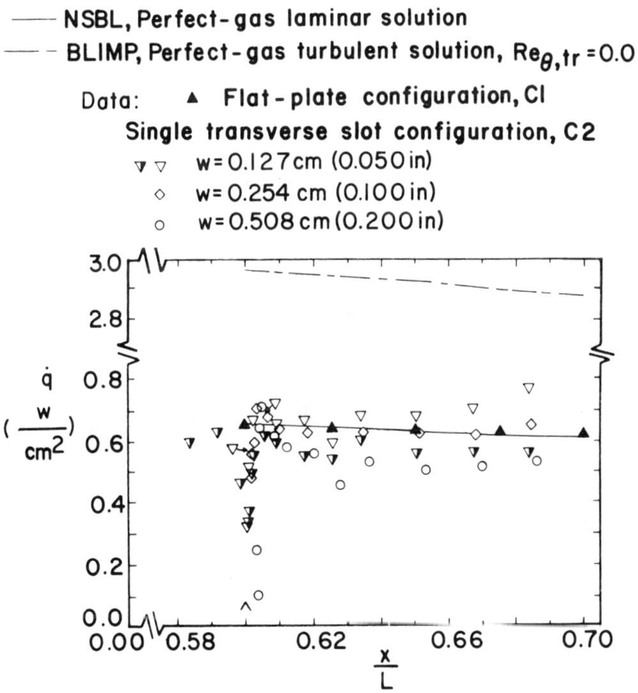

(b) $Re_{\infty,L} = 4.01 \times 10^6$

Fig. 3 (cont.)

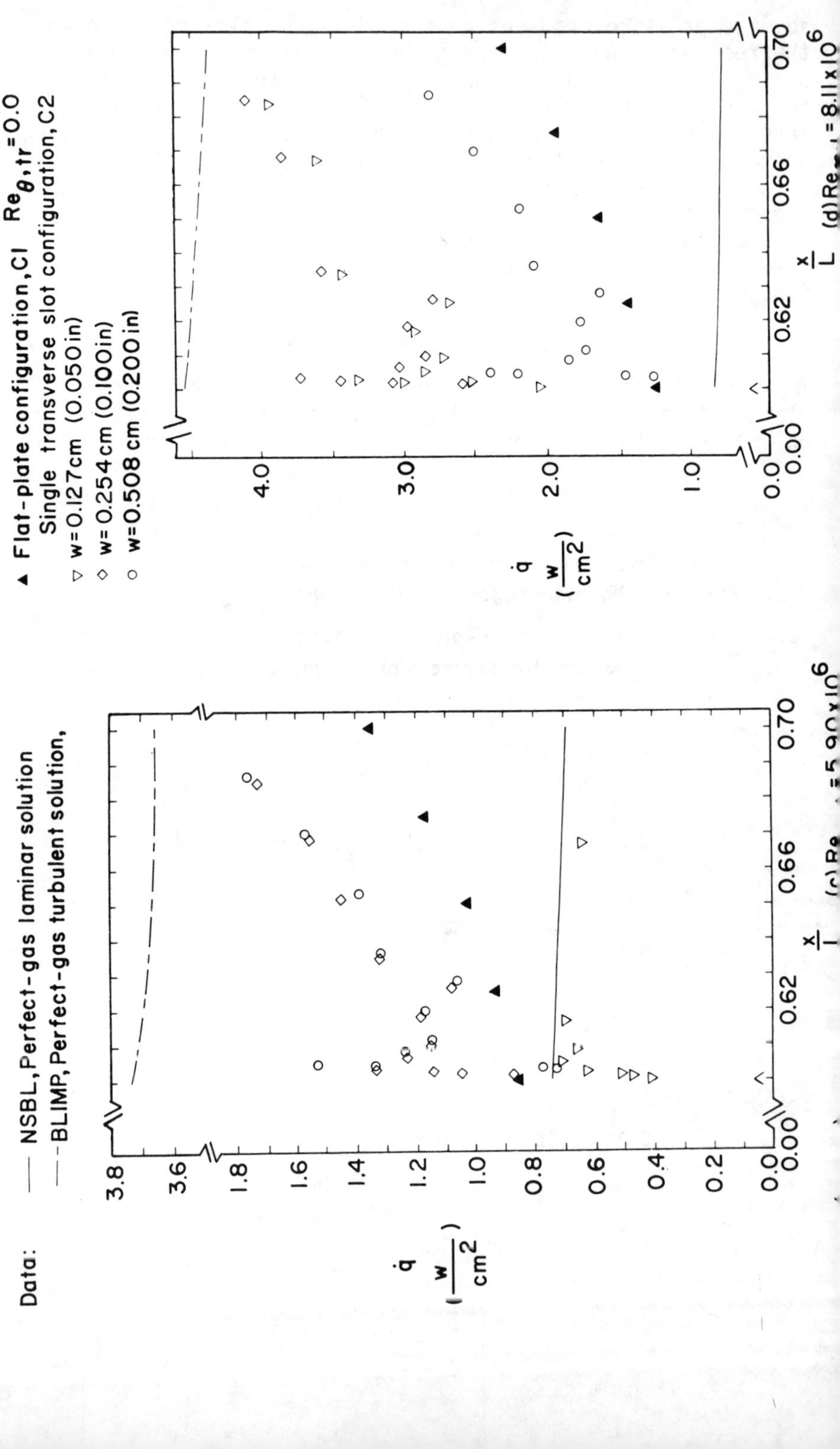
Data:
— NSBL, Perfect-gas laminar solution
-- BLIMP, Perfect-gas turbulent solution, $Re_{\theta,tr}=0.0$
▲ Flat-plate configuration, C1
Single transverse slot configuration, C2
▷ w=0.127cm (0.050in)
◇ w=0.254cm (0.100in)
○ w=0.508cm (0.200in)
$\dot{q}$ $\left(\frac{w}{cm^2}\right)$
$\frac{x}{L}$
(d) $Re_{\infty,L}=8.11\times10^6$

Reynolds-number observations, there is no consistent effect of the gap width evident in these data at these flow conditions. The streamwise variation of the heat-transfer rates measured downstream of the gap suggest the onset of transition. However, since the deviation from laminar theory is not large and since data are not available from further downstream, one cannot quantify the effect. Also included in this figure is the theoretical heat-transfer distribution for a perfect-gas boundary layer that is turbulent from the leading edge, as calculated using the University's copy of the BLIMP code.[13]

The heat-transfer distributions for the two higher Reynolds-number cases are presented in Figs. 3c and 3d. As indicated by the filled symbols, the flat-plate boundary layer was itself transitional for these two flow conditions. With the exception of the narrowest gap for $Re_{\infty,L} = 5.90 \times 10^6$, the heating rates exceed the flat-plate values at all points. At these high Reynolds numbers, the heating rates downstream of the gap are higher than the corresponding flat-plate values but do not exceed the fully turbulent values. However, the instrumentation does not extend far enough downstream to be certain that the experimental heating rates would never exceed the theoretical turbulent values. Since the upstream boundary layer is transitional for these two Reynolds numbers, the subsonic portion of the boundary layer (and the value of z^*) decreases rapidly with Reynolds number. This has two effects: the fluid in the lower portion of the shear layer moves at higher speeds and the ability of the recompression step to alter the upstream flow and spread out the recompression process is reduced. The ratio of the gap-perturbed heating rate increases rapidly as the Reynolds number increases (or as the sonic point moves near the wall). Two sets of the data produced the results contrary to the typical trends. One set is that for the shortest gap (w = 0.127 cm) for $Re_{\infty,L} = 5.90 \times 10^6$. It should be noted that although there are only a few data points from the y = 0.000L plane in Fig. 3c, the heating-rate distributions from the y = $\pm$ 0.135L planes were similar. The heating downstream of the gap closely follows the laminar theory for this test condition. Thus, this heat-transfer distribution is similar to those obtained at the two lower Reynolds numbers (Figs. 3a and 3b). However, since the flat-plate boundary layer was transitional, it is surprising that the gap-perturbed flow appears laminar. A possible explanation is that the basic flow is borderline stable and in one run transition began in this region and in another the boundary layer remained laminar. The theoretical characteristics

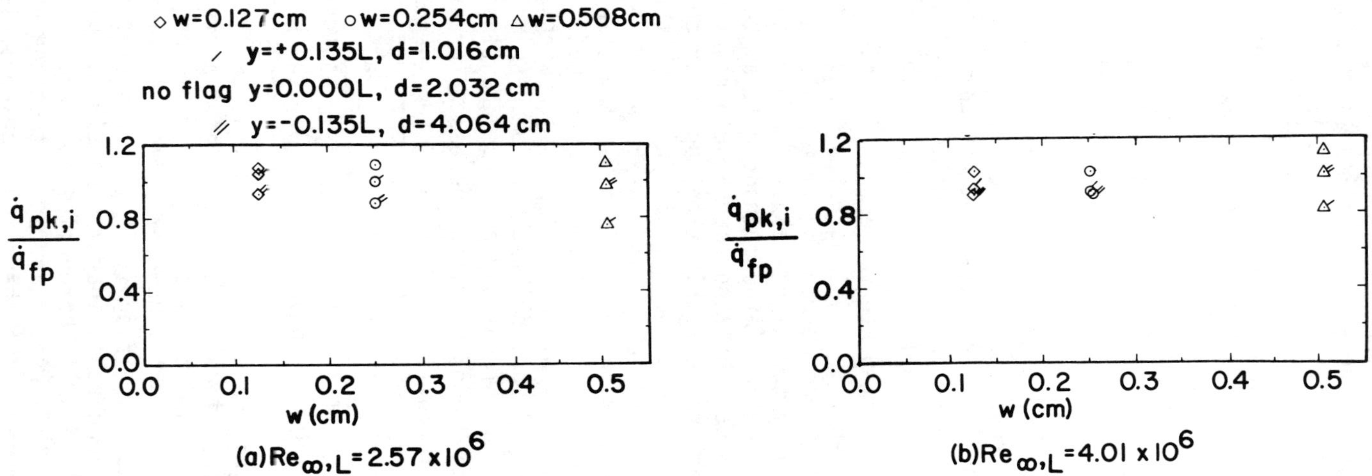

Fig. 4 The maximum heating increase in the impingement region as a function of gap width for a single transverse gap.

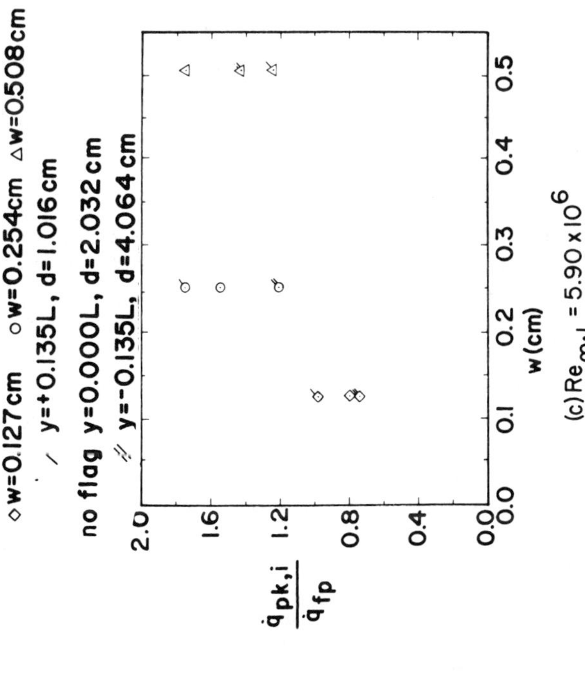
◇w=0.127cm ○w=0.254cm △w=0.508cm
/ y=+0.135L, d=1.016cm
no flag y=0.000L, d=2.032cm
// y=-0.135L, d=4.064cm
$\dot{q}_{pk,i}/\dot{q}_{fp}$
w(cm)
(c)$Re_{\infty,L}=5.90\times10^6$

Fig. 4c

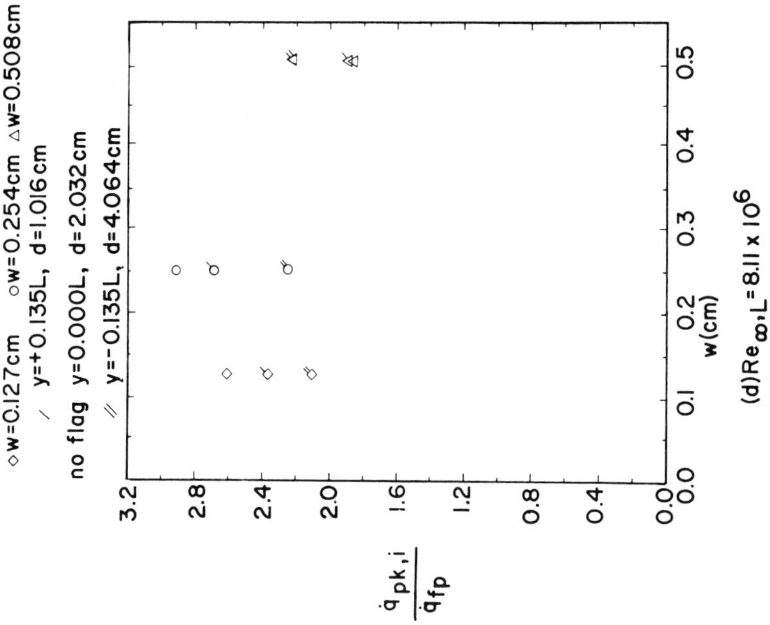
◇w=0.127cm ○w=0.254cm △w=0.508cm
/ y=+0.135L, d=1.016cm
no flag y=0.000L, d=2.032cm
// y=-0.135L, d=4.064cm
$\dot{q}_{pk,i}/\dot{q}_{fp}$
w(cm)
(d)$Re_{\infty,L}=8.11\times10^6$

Fig. 4d

calculated for this transitional boundary layer were little different from those calculated for a laminar boundary layer. Thus, since the oncoming boundary layer was relatively thick and the subsonic region relatively large, the narrow gap had little effect on the flow. For the two longer gaps, the flow perturbations were sufficiently strong to promote transition and increase the local heating rates.

The second set of anomalous data is that for the longest gap (w = 0.508 cm) for $Re_{\infty,L} = 8.11 \times 10^6$, which are presented in Fig. 3d. The reason why these heating rates appear to be relatively low is not understood. Again, the heating rates measured in the $y = \pm 0.135L$ planes exhibit the same characteristics as the data from the $y = 0.000L$ plane.

As had been discussed, the peak value of the impingement heating rate (usually) increases with gap width, once the gap width exceeds a critical value. Furthermore, the critical value is a function of the flow characteristics. This is summarized in the data presented in Fig. 4. For the two lower Reynolds number flows, the heating is relatively unaffected by presence of the gap for all three widths treated. When $Re_{\infty,L} = 5.90 \times 10^6$, the maximum values of the impingement heating rate for all three spanwise locations for the short gap, i.e., w = 0.127 cm, are essentially equal to the flat-plate value. Recall that the heat-transfer rates measured downstream of the gap follow the laminar theory even though the measured flat-plate values indicate the onset of transition. For the wider gaps at this Reynolds number, the maximum values of the impingement heating rates are significantly above the measured flat-plate values, and (with the exception of the data for section AA for w = 0.508 cm) they increase with gap width itself and with the ratio w/d. When $Re_{\infty,L} = 8.11 \times 10^6$ (the highest Reynolds number of the present test program), the maximum value of heating rate in the impingement region exceeds the flat-plate values for all gap widths and, with the exception of the data for the widest gap for all three sections, increases with gap width. However, these data do not exhibit a clear dependence on w/d.

As has been noted, the maximum value of the heating relative to the corresponding flat-plate value occurred near the tangency point of the upper surface. The relatively high (velocity and temperature) gradients in the reattaching sphear layer, which provides the "edge" conditions for the developing boundary layer, affect that boundary layer. As transition moves upstream of the gap (for the two higher Reynolds number flows), the velocity gradients near the wall increase rapidly with a corresponding decrease in z*. Furthermore, although there are numerous exceptions to the trend, the heat flux increased with increased values for the ratio w/d (which agrees with the analysis of Nestler[5]). The heating-rate ratio, $\dot{q}_{pk,i}/\dot{q}_{fp}$, is presented in Fig. 5 as a function of the parameter $(x_s/z^*)(w/d)^{0.2}$. Considering the experimental uncertainties (e.g., a finite number of thermocouples, relatively few flow conditions in the text matrix, and anomalous data), the correlation is reasonable. Note also that the streamwise distance to the step (x_s) is used to make the parameter dimensionless because it is a constant and, therefore, does not mask the significance of z*. However, it does not appear to be a significant parameter on its own. Additional data may provide insights that could identify a more relevant length.

For a limited number of runs, the flat surface of the downstream (recompression) portion of the insert (i.e., x >

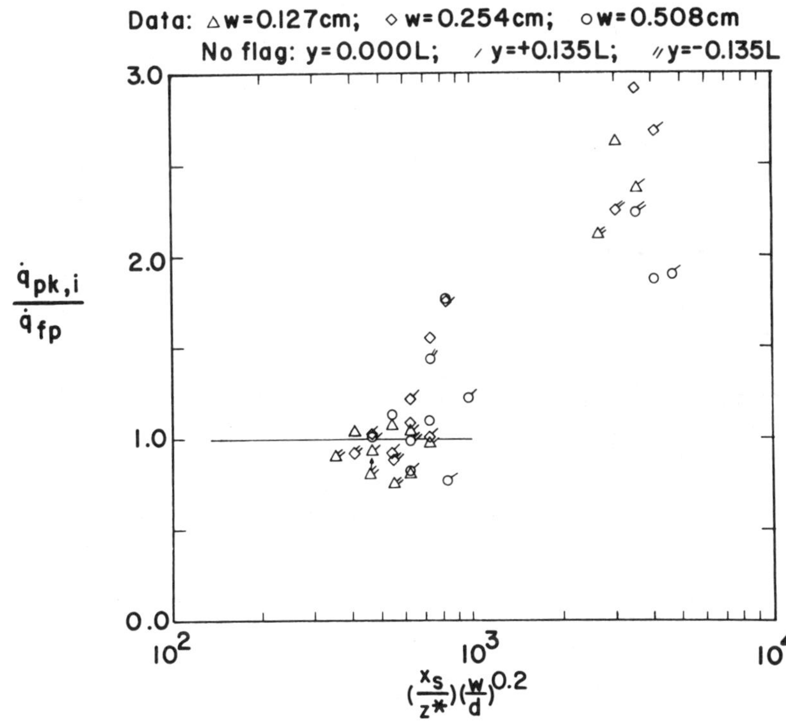

Fig. 5 Local maxima of the ratio of the perturbed heat-transfer rate (C2) to the corresponding flat-plate value (C1).

Table 1 The effect of vertical misalignment of adjacent surfaces (Δz) on the peak heat transfer measured on the impingement surface of a gap. Data for y = 0.000L

w (cm)	Re/m $\times 10^{-6}$	$\dot{q}_{pk,i}/\dot{q}_{fp}$: Δz = -0.159 (cm)	0.000 (cm)	+0.159 (cm)	+0.318 (cm)
0.127	1.688	0.67	0.93	1.37	2.10
0.127	2.633	0.59	0.96	1.32	2.32
0.127	5.319	0.55	0.74	2.26	4.43
0.254	2.633	0.62	1.09	1.45	3.13
0.508	2.633	...	1.10	2.18	3.20

0.6L) was not coplanar with the upstream portion. Thus, the two surfaces were vertically "misaligned," such that the recompression surface represented either a step up to the flow ($\Delta z > 0$) or a step down to the flow ($\Delta z < 0$). The misalignment had a significant effect on the peak heating rates measured on the recompression surface. This is illustrated in the data presented in Table 1. Unfortunately, the heating rates measured at the highest Reynolds number during these runs differed significantly from those measured during runs from other phases of the program. This experimental uncertainty can be seen by comparing the measurements of Table 1 with those presented in Fig. 4d. Nevertheless, these data illustrate that the local heating increases become more severe for configurations having both a gap and vertically misaligned adjacent surfaces.

The experimental heat-transfer distributions for Configuration 3, which has four gaps perpendicular to the free-stream flow, are presented in Fig. 6. The gap locations and the flat-plate transition location for each Reynolds number are illustrated in these figures. When studying the heat-transfer distributions, the reader should note that, with the exception of the thermocouples located near x = 0.6L, they are on the flat surface, parallel to the freestream.

At the lowest Reynolds number ($Re_{\infty,L} = 2.57 \times 10^6$), the C3 measurements closely follow the laminar theory. Recall

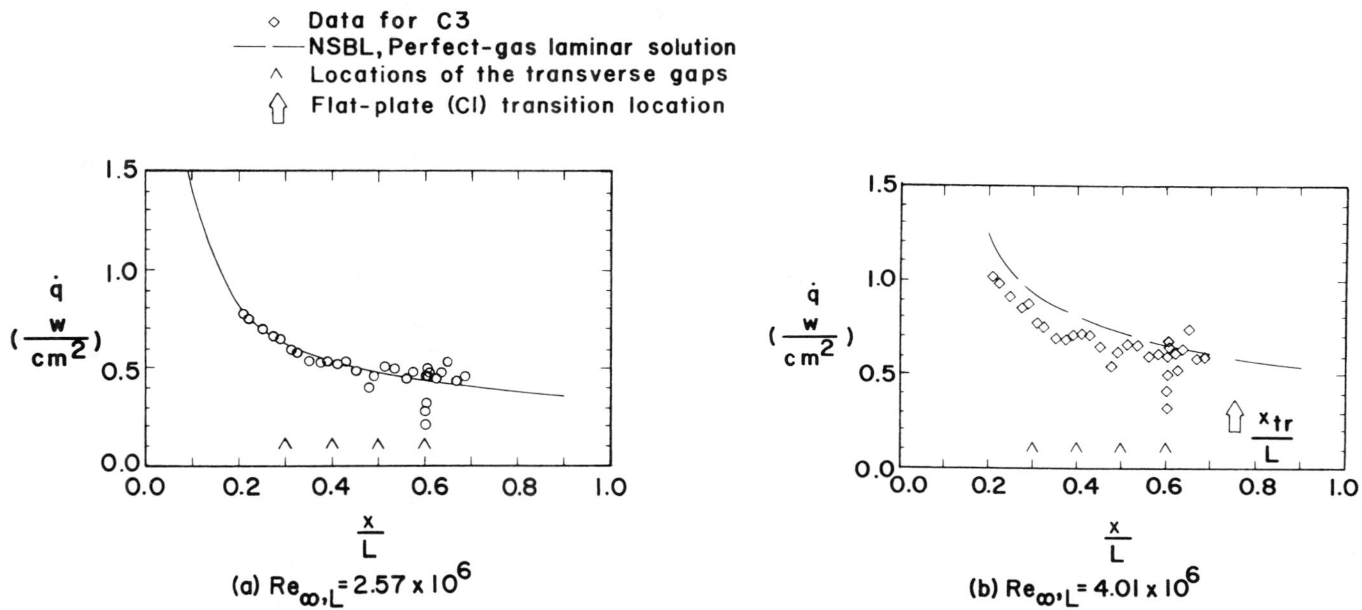

Fig. 6 The streamwise heat-transfer distribution for the multiple-transverse-gap configuration, C3.

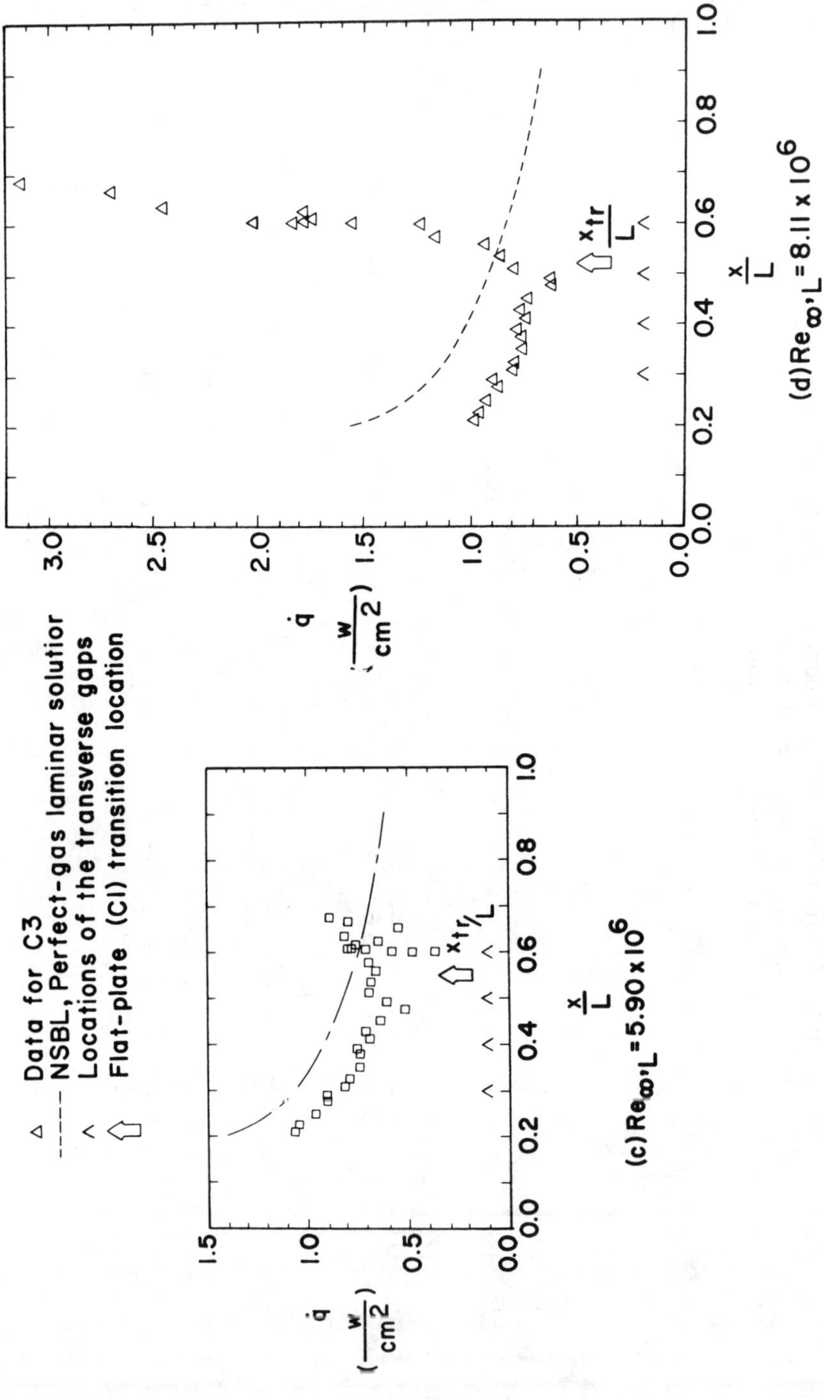

Data for C3
NSBL, Perfect-gas laminar solutior
Locations of the transverse gaps
Flat-plate (CI) transition location
q̇ (w/cm²)
x/L
x_{tr}/L
(c) $Re_{\infty,L} = 5.90 \times 10^6$
(d) $Re_{\infty,L} = 8.11 \times 10^6$

Fig. 6 (cont.)

that, for this particular flow condition, the flat-plate data are significantly above the laminar theory. Thus, since the data follow the theory very closely for $x < 0.6L$, the presence of the multiple transverse gaps causes a general reduction in the surface heating for this low Reynolds number flow. For $x > 0.4L$, the roughness-induced perturbations produce an erose heating distribution. The distribution for $Re_{\infty,L} = 4.01 \times 10^6$ (Fig. 6b) is similar to that measured at the lower Reynolds number. Note that the heat-transfer rates measured in the region $0.6L < x < 0.7L$ are essentially equal to the theoretical values for a laminar flat-plate boundary layer. Since transition of the flat-plate boundary layer was observed at 0.75L, these data indicate that the presence of the multiple transverse gaps did not dramatically affect the transition location.

For the two higher Reynolds number flows, the gap-induced perturbations to the flow cause transition to occur slightly earlier than for the flat-plate boundary layer. Furthermore, as evident in Fig. 6d, the streamwise increase in heating in the transitional region is much greater for these C3 measurements than that which occurred for the "natural" transition on the flat-plate configuration. Thus, it seems that tripping the boundary layer causes the transition process to proceed much more rapidly. Despite the rapid streamwise increase in the local heating rates, they do not exceed the values for a turbulent, flat-plate boundary layer (see Fig. 6d) over the limited length for which data are available.

Thus, over the entire Reynolds number range, the C3 heat-transfer rates were lower than the corresponding flat-plate values as long as the boundary layer remained laminar. This is consistent with the results of previous investigators. Charwat[14] cites supersonic data for a series of notches with w/d between 1 and 2 that are spaced one cavity length apart. The notched surface "shielded" the body somewhat, so that the total heat flux was about 70 to 80% of the heat flux to an equal length of a smooth-surfaced body. Charwat notes that the downstream heating will usually be decreased unless the cavity induces transition. For the present tests, the multiple transverse gaps, which were relatively narrow (0.254 cm), moved the onset of transition only slightly upstream. When considering transition, note that the flow is supersonic ($M_e = 5.10$) and that $T_w \approx 0.25\ T_t$.

Data for Configurations Having Intersecting Gaps, i.e., C4 and C5

Two of the configurations tested had perpendicular, intersecting gaps. For one configuration (the C4) gaps were aligned parallel to the freestream flow; in the other (the C5), they were skewed.

Locally high heating rates were measured at various thermocouples for Configuration 4. They were due to two distinctively different mechanisms. Locally high heating rates were measured along the length of the longitudinal slots parallel to the freestream. Because the flow spills into the slot, the boundary layer on the flat surface thins with a corresponding increase in the local heating. The data presented in Figs. 7 and 8 illustrate the effects of this mechanism. Locally high heating rates were also measured at the recompression surface at the T formed by the intersection of the longitudinal slot and the transverse gap. A comparison of the present data with the results of other investigators shows clearly that the heating in the impingement region increases dramatically with $(x_s - x_0)/d$. The results appear to be similar to those for a two-dimensional notch where the characteristics of the flowfield perturbations depend on whether the notch is "open" or "closed." These data are presented in Figs. 9 and 10.

Presented in Fig. 7 are the heating rates recorded by the three thermocouples that are nearest the flat surface for the four stations along the longitudinal slot (refer to Fig. 1d). Included for comparison are the perfect-gas solutions for a laminar, flat-plate boundary layer as computed using the University's NSBL code. Heating rates were measured at all four stations, as indicated by the open symbols of Fig. 7. A second set of runs were made in which data were obtained at section DD only. These measurements are represented by the filled symbols. Thus, the measurements presented in Fig. 7 indicate the effect of the unit Reynolds number, of the distance from the beginning of the slot, and of the depth of the slot. Recall that the width of the slot was 0.254 cm for all C4 runs.

At both Reynolds numbers, the heating rates measured at the two sections located at x = 0.4L are higher for the deeper slot. Note also that for the shallow slot (d = 1.016 cm), the heating is greatest at the thermocouple at the upper tangency point (i.e., on the flat surface). However, for the deeper slot (d = 4.064 cm), the heating is greatest at the thermo-

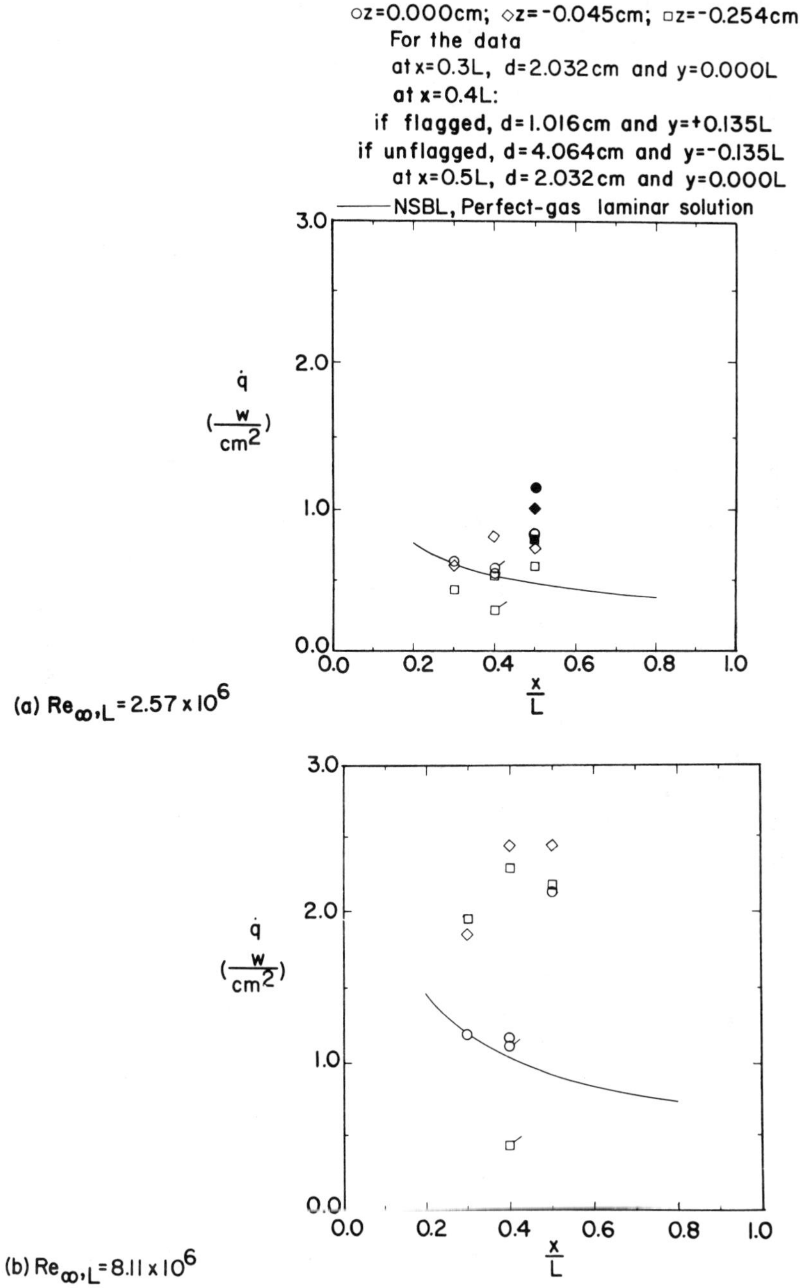

Fig. 7 The heat-transfer rates measured along the longitudinal slots of C4.

couple located on the radius. These observations are consistent with the assumption that the increased heating results because the boundary layer thins as the flow spills into the slot. The level of heating tends to increase with the distance from the beginning of the slot.

The ratio of the C4 heating rates to the experimental flat-plate value at the same x location are presented as a function of $Re_{(x-x_0)}$ in Fig. 8. The peak value ($\dot{q}_{pk}$) is the numerator. The ratio $\dot{q}_{pk}/\dot{q}_{fp}$ correlates reasonably well as a function of $Re_{(x-x_0)}$, increasing "linearly" with the Reynolds number until it is approximately constant, just less than 3.0. Note that the ratio is smallest for the slot with the smallest

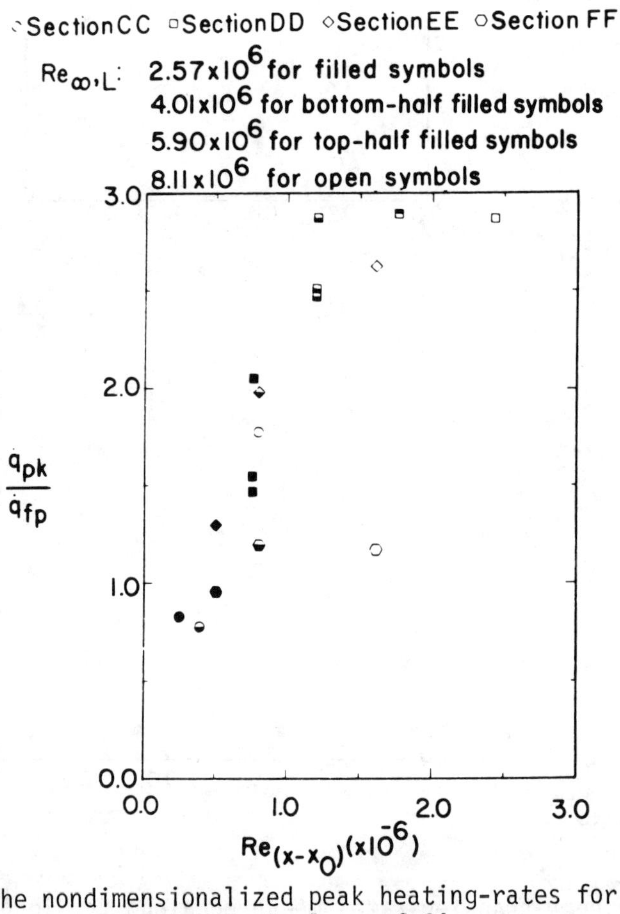

Fig. 8 The nondimensionalized peak heating-rates for locations along the longitudinal slots of C4.

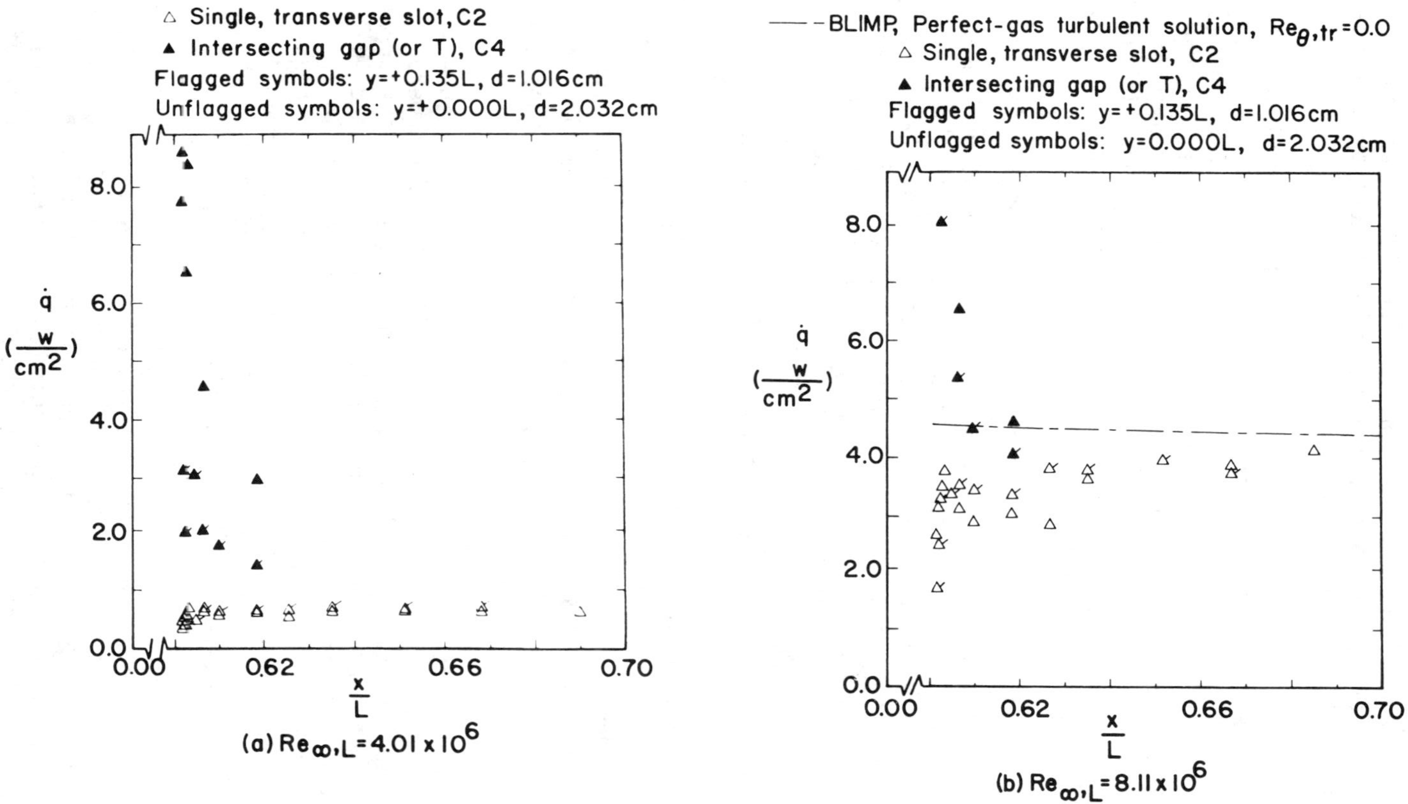

Fig. 9 A comparison of the heat-transfer rate distributions downstream of a single transverse gap (C2) and of a T (C4).

d (i.e., the most shallow section, section FF). Thus, some of the scatter in the correlation could be reduced by accounting for the (nonlinear) effect of d.

The streamwise heat-transfer distributions downstream of the T gap are presented in Fig. 9. The measurements are from the y = 0.000L plane, which is downstream of the 2.032 cm deep slot, and from the y = +0.135L plane, which is downstream of the 1.016 cm deep slot. Thus, these heating rates represent the measurements from the thermocouple locations shown in Fig. 1b, starting with the thermocouple at the lower tangency point in the gap and proceeding downstream onto the plate.

Note that the $(x_s - x_0)/d$ ratios for the longitudinal slots are 30 and 60, respectively. Thus, the relative lengths of the longitudinal portion (slot) of the T gaps in the present study are significantly greater than those of the previous investigations. The heat transfer is markedly greater downstream of the deep slot, being an order of magnitude greater than the flat-plate values that are presented in Fig. 2. Note that these peak values for the ratio $\dot{q}/\dot{q}_{fp}$ are significantly greater than those of the previous investigations where $(x_s - x_0)/d$ was 5.0, or less. However, since the heating rates are greatest downstream of the deeper slot (which has the lower $(x_s - x_0)/d$ ratio of two), these data indicate that the depth itself is the more significant parameter once $(x_s - x_0)/d$ is sufficiently large. The heating remains significantly greater than the corresponding flat-plate values well downstream of the corner.

Included for comparison are the heating rates measured downstream of the single, transverse slot (C2), which were discussed earlier, e.g., Fig. 3. The heating rates downstream of the T gap (C4) are much greater than those measured downstream of the single transverse gap both for the shallow gap and for the deeper gap. Note that, at the highest Reynolds number, the heating rates to the flat surface downstream of the T gap exceed the theoretical values for a flat-plate boundary layer that is turbulent from the leading edge.

Rapid variations in the heating rate that occur over very short distances on the corner radius of the recompression surface are evident in the streamwise heat-transfer distributions presented in Figs. 9. Examination of the vertical heat-transfer distributions (not presented in the present paper) reveal multiple local maxima. Because of the long length of the longitudinal slot, the flow attaches to the floor and the recompression surface serves as a forward-facing step. The side-wall boundary layers and the cross flow into transverse gap contribute to the three-dimensional character of the viscous interactions at the higher Reynolds number.

Based on the discussions of the basic data presented thus far, the correlations developed using the data from investigations where the $(x_s - x_0)/d$ of the longitudinal slot was relatively small would not be expected to work with the data from the present program. This lack of correlation is discussed in Ref. 11. Based on the comparison between the present data and those of other investigators, it appears that $(x_s - x_0)/d$ is an important parameter until it becomes sufficiently large. Then its numerical value does not matter and d (or d/δ^*) becomes more important. The heating-rate ratios $(\dot{q}/\dot{q}_{fp})$ for the recompression surface as measured in the present program are presented as a function of

$$\left[\frac{\delta^*}{w}\right]^2 \quad \left[\frac{z}{d}\right]$$

which is a modification of parameters developed by previous investigators, e.g., Refs. 4 and 7. Although this parameter correlates the data better than the parameters used by other investigators, it does not uniquely correlate the data, since there are considerable variations in the heating-rate ratios near the surface, i.e., low values of z/d. Considerable differences exist between the heating rates near the surface for the deeper slot (the filled symbols) and those for the more shallow slot (the open symbols), as shown in Fig. 10.

The heat-transfer distributions for configuration C5 are presented in Fig. 11. It can be seen that the gaps that are swept either 30 deg or 60 deg to the flow have roughly the same effect on the heating rates as does the transverse gap of C2. Heat-transfer distributions are presented in Fig. 11 for Reynolds numbers based on the total plate length $(Re_{\infty,L})$ of 2.57×10^6 and 8.11×10^6. The x locations of the vertical recompression surface for each of the instrumented sections are also shown. Included for comparison are the heat-transfer measurements for the flat-plate configuration (C1) at the same flow conditions, which are represented by the filled symbols. The heat-transfer rates are only slightly affected by the presence of the gaps at the lower Reynolds number, i.e.,

2.57 x 10^6. Locally high heating rates were recorded by the thermocouples at, and just downstream of, the upper tangency point of the "recompression" surface. This is true for all sections, even the one at the intersection of the gaps, i.e., section JJ. These trends are similar to those observed for the single, transverse gap (C2) in the lowest Reynolds number stream. In a study of the aerodynamic heating to corrugation-stiffened structures in thick, turbulent boundary layers, Brandon et al.[8] found that although the highest peak heating occurred when the corrugations were perpendicular to the flow, it was not much less when the corrugations were inclined to the freestream by as little as 15 deg.

Note that the heat-transfer rates measured at the thermocouples just upstream of the separation surface decrease continuously as the flow proceeds toward the gap and are consistently below the flat-plate values. Thus, the C5 measure-

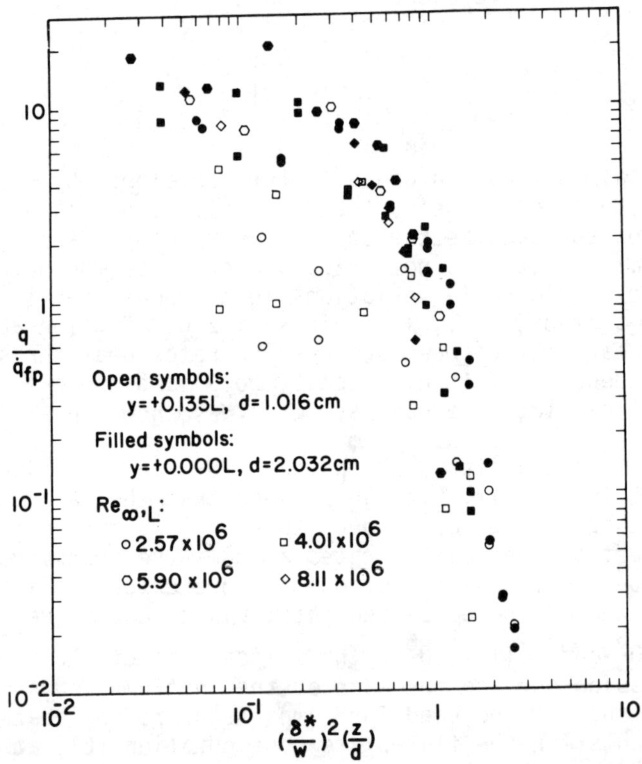

Fig. 10 Correlation of the heating rates on the vertical (recompression) surface of the T gap (C4).

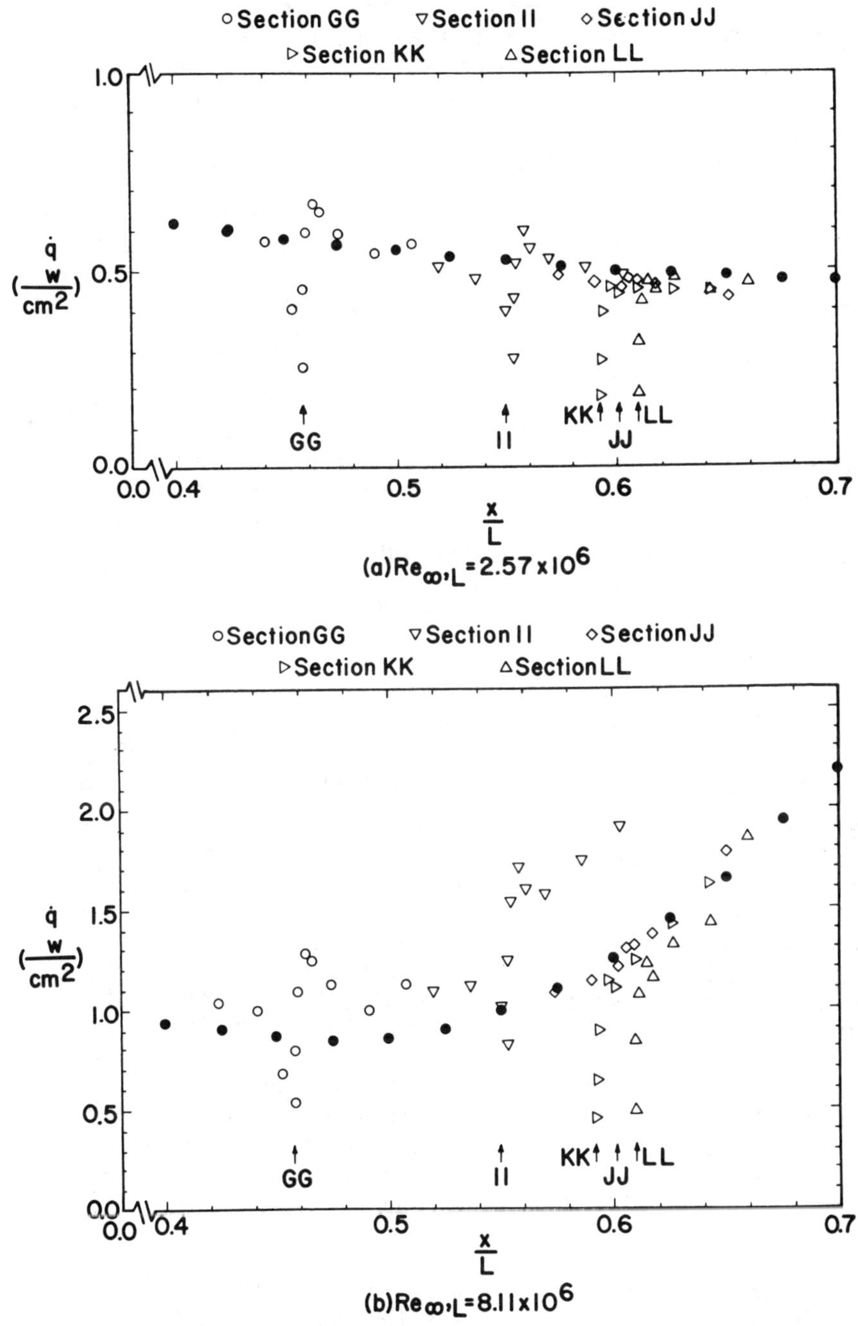

Fig. 11 The heat-transfer distributions in the vicinity of the diagonal intersecting gaps of C5.

ments from this region differ from the C2 results where the heat transfer increased slightly as the flow approached the gap and then decreased rapidly around the corner of the separation step (see Fig. 3 and the discussion of the data presented therein). The heating increase for the upstream surface of the C2 was attributed to the pressure rise created by the impingement process propagating through the relatively thick subsonic portion of the shear layer. Unless the thermocouples for the C5 are spaced such that all the local peaks were missed (which seems unlikely), the C5 data indicate that the cross flow associated with a skewed gap provided an alternative path for the oncoming flow and minimized the upstream effects of the impingement process.

At the higher Reynolds number, i.e., $Re_{\infty,L} = 8.11 \times 10^6$, these relatively narrow, oblique gaps promote transition as indicated by the data presented in Fig. 11b for section II. However, the perturbations for the heating are relatively small when compared to the increases recorded in the vicinity of the intersection of the C4.

Conclusion

Heat-transfer distributions were measured in the vicinity of gap inserts located in a flat-plate structural carrier. The data were obtained in the 3.5-ft Hypersonic Wind Tunnel at the Ames Research Center at a Mach number of 5.1 over a Reynolds number range of 2.57×10^6 to 8.11×10^6. Because of significant differences in the configuration geometries, the correlations developed by previous investigators did not work with the present data. Furthermore, because of the experimental uncertainties, it was not possible to develop general correlations incorporating local boundary-layer parameters using only the present data. However, for the range of conditions studied several trends were clearly evident.

1. The narrow transverse gaps of the present configurations had relatively little effect on the laminar boundary layer. This is true for the multiple-gap configuration (C3) as well as for the single-gap configuration (C2). The local heating increases were relatively small and the transition location changed only slightly.

When the boundary layer was transitional, locally high heating rates were measured near the upper tangency point of the recompression surface once the gap width exceeded a minimum value. The perturbed heating increased as z* decreased and as w/d increased, i.e., when $(x_s/z^*)(w/d)^{0.2}$ is greater than 500 (approximately). However, none of the measured heat-transfer rates exceeded the values calculated for a fully turbulent boundary layer. The local heating increases become more severe for configurations having both a gap and vertically misaligned adjacent surfaces.

2. For the C4 configuration whose intersecting gaps are aligned either parallel or perpendicular to the freestream, locally high heating rates were measured along the length of the longitudinal slot and at the recompression surface at the T formed by the intersection of the longitudinal slot and the transverse gap. The heating-rate ratio at points along the longitudinal slot increased with the unit Reynolds number, with the distance from the separation step $(x - x_0)$, and with the depth of the slot.

The extremely high heating rates were measured on the recompression surface downstream of the T gap. The heating rates increased with the z dimension of the recompression step, d. Furthermore, at the highest Reynolds number, the heating rates to the flat surface downstream of the T gap exceeded the theoretical values for the flat-plate boundary layer that is turbulent from the leading edge.

3. For the configuration with intersecting gaps, where the gaps are swept either 30 deg or 60 deg, i.e., the C5 configuration, the heating rates were only slightly affected by the presence of the gaps. The swept gaps did promote transition at the higher Reynolds number.

Acknowledgment

This work was supported by the Johnson Space Center (NASA) through Grant No. NAS 9-13680.

References

[1]Charwat, A.F., Roos, J.N., Dewey, C.F., Jr., and Hitz, J.A., "An Investigation of Separated Flows - Part I: The Pressure Field," Journal of the Aerospace Sciences, Vol. 28, No. 6, June 1961, pp. 457-470.

[2]Charwat, A.F., Dewey, C.F., Jr., Roos, J.N., and Hitz, J.A., "An Investigation of Separated Flows - Part II: Flow in the Cavity and Heat Transfer," Journal of the Aerospace Sciences, Vol. 28, No. 7, July 1961, pp. 513-527.

[3]Dunavant, J.C., and Throckmorton, D.A., "Aerodynamic Heat Transfer to RSI Tile Surfaces and Gap Intersections," Journal of Spacecraft and Rockets, Vol. 11, No. 6, June 1974, pp. 437-440.

[4]Avery, D.E., "Aerodynamic Heating in Gaps of Thermal Protection System Tile Arrays in Laminar and Turbulent Boundary Layers," NASA Technical Paper 1187, July 1978.

[5]Nestler, D.E., "An Engineering Analysis of Reattaching Shear Layer Heat Transfer," AIAA Paper 72-717, presented at the 5th Fluid and Plasma Dynamics Conference, Boston, June 1972.

[6]Brewer, R.A., Saydah, A.R., Nestler, D.E., and Florence, D.E., "Thermal Performance Evaluation of RSI Panel Gaps for Space Shuttle Orbiter," Journal of Spacecraft and Rockets, Vol. 10, No. 1, Jan. 1973, pp. 23-28.

[7]Bohon, H.L., Sawyer, J.W., Hunt, L.R., and Weinstein, J., "Performance of Thermal Systems in a Mach 7 Environment," Journal of Spacecraft and Rockets, Vol. 12, No. 12, Dec. 1975, pp. 744-749.

[8]Brandon, H.J., Masek, R.V., and Dunavant, J.C., "Aerodynamic Heating to Corrugation Stiffened Structures in Thick Turbulent Boundary Layers," AIAA Journal, Vol. 13, No. 11, Nov. 1975, pp. 1460-1466.

[9]Johnson, C.B., "Heat Transfer Data to Cavities Between Simulated RSI Tiles at Mach 8," NASA CR 128,770, June 1973.

[10]Foster, R.G., Lockman, W.K., and Grifall, W.J., "Thermal Protection System Gap Heating Rates of the Rockwell International Flat Plate Heat Transfer Model (OH2B/OH2N)," NASA CR-134,077, Nov. 1973, Data Management Services, Chrysler Corporation Space Division.

[11]Bertin, J.J., Burley, P.J., and Hayden, T.E., "Convective Heating in and Around Gaps in Laminar and Transitional Boundary Layers," Aerospace Engineering Report 79004, The University of Texas at Austin, Nov. 1979.

[12]Stalmach, D.D., and Bertin, J.J., "The Analysis of a Nonsimilar Laminar Boundary Layer," Aerospace Engineering Report 78001, The University of Texas at Austin, Feb. 1978.

[13]Tong, H., Buckingham, A.C., and Morse, H.L., "Nonequilibrium Chemistry Boundary Layer Integral Matrix Procedure," Aerotherm Final Report No. 73-67, The Acurex Corporation, July 1973.

[14]Charwat, A.F., "Supersonic Flows with Imbedded Separated Regions," from Advances in Heat Transfer, Vol. 6, Academic Press, New York, 1970, pp. 1-132.

CORRELATION OF CONVECTION HEAT TRANSFER FOR OPEN CAVITIES IN SUPERSONIC FLOW

J. P. Lamb*
The University of Texas at Austin, Austin, Tex.

Abstract

Heat-transfer rates at various locations in planar rectangular cavities are correlated with appropriate flow and geometric variables. It is shown that the effects of cavity depth and cavity length on heat-transfer distributions are entirely different. The primary scaling parameter for cavity geometry in the laminar regime is the upstream boundary-layer thickness. Correlations are illustrated for the downstream recompression corner as well as for the average heating rate over the entire cavity surface; no Mach number effects were included in the present data base. For turbulent flows, more complex scaling parameters are required because of the complex relationship between shear layer and upstream boundary layer. A new parameter, the characteristic cavity size, is shown to produce satisfactory correlations of both maximum and minimum floor heat-transfer rates. The average floor heating rate is shown to be directly correlated to the foregoing maximum and minimum values.

Nomenclature

b = thickness of free shear layer (Fig. 1)
H = depth of cavity (Fig. 1)
L = length of cavity (Fig. 1)
M = freestream Mach number
q = convective heat-transfer rate
Q = average heat-transfer parameter, Eq. (3)
S = characteristic cavity size, Eq. (1)
$\tilde{S}$ = length parameter (Fig. 6)

Presented as Paper 80-1526 at the AIAA 15th Thermophysics Conference, Snowmass, Colo., July 14-16, 1980.

*Professor, Department of Mechanical Engineering.

St = Stanton number
URe = unit Reynolds number (per meter)
δ = thickness of boundary layer (Fig. 1)
σ = turbulent free layer parameter, Eq. (2)

Subscripts

a = average value
i = incompressible flow
max = maximum value
min = minimum value
rc = recompression corner (Fig. 1)
1 = approach flow conditions
0 = stagnation conditions (freestream)

Introduction

The convective transport characteristics of a separated flow region are inherently more complex than those of an attached flow due to the occurrence of a secondary vortical flow zone immediately adjacent to the solid surface. This additional complexity arises as a result of the coupling between the secondary flow and the external freestream; the coupling mechanisms are functions of Reynolds number and Mach number as well as the shape and size of the separated region. These added levels of interaction have prohibited the development of any but highly complicated prediction methods.[1] Moreover, despite the existence of an extensive body of experimental data[2] there has been only limited success in developing heat-transfer correlations for separated regions produced by specific geometries, e.g., the extensive development of semiempirical models by Nestler.[3] More generalized correlations furnish two primary benefits: they can provide the basis for making rapid estimates of heat loads for preliminary design studies as well as for planning detailed experimental programs. Moreover, through the identification of characteristic dimensionless parameters one can gain added insight into the governing physical mechanisms.

The present paper is an extension of a recent study[4] by the author in which the minimum level of separated region heat transfer for a wide range of body geometries (backsteps, cavities, circular and noncircular cylinders) was correlated using a parameter which is proportional to the ratio of overall transverse temperature gradients in the separated zone and in the upstream approach flow. Current discussion will be concerned with the extensive parametric measurements on turbulent cavities by Charwat et al.[5] along with similar data of Nestler.[6] Analysis of the turbulent regime is supplemented with laminar cavity results of Hahn,[7] Nicoll,[8] and Wieting.[9] The primary objective is to consider heat-transfer rates at various locations within the cavity, to determine the extent to which these rates are correlatable, and to describe the governing flow features which are inferred by the measured heat-transfer distributions.

Present emphasis is on the so-called "open" cavity configuration which occurs when the external flow completely spans the surface discontinuity. The comprehensive experimental study of Charwat et al.[10] showed that, if the ratio of length to depth is larger than a critical value, the flow will reattach to the cavity floor and immediately separate in front of the downstream cavity face, thus forming two separated zones which are collectively described as a "closed" cavity. In addition, the current data base is composed primarily of cavity flows in which there was a well-developed vortex structure so that convective transport was the dominant mechanism. However, no attempt was made in the current study to include influences relating to the upstream history of the thermal boundary layer or to the oscillating pressure field which often exists in cavity flows.

The major features of a planar, open cavity flowfield are depicted schematically in Fig. 1. An approaching flow,

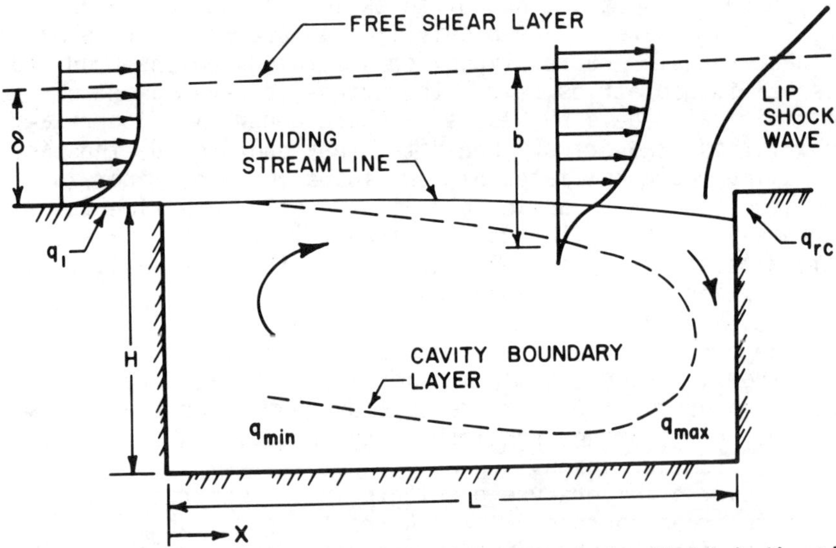

Fig. 1 Schematic representation of cavity flowfield indicating heat-transfer levels at various locations.

with boundary-layer thickness, δ, spans the cavity and develops a free shear layer of thickness, b. A portion of the free layer must be recirculated to satisfy continuity requirements; therefore all flow below the dividing streamline is identified with the cavity vortex. The circulating flow is driven through shear stresses along the dividing streamline which subsequently terminates on the recompression face at a point slightly below the corner.

Energy is transferred between the external stream and the cavity through transverse derivatives of temperature and through shear work along the dividing streamline, whereas cavity surface convection is governed by the vortex boundary layer, the thickness of which is dependent upon the dimensions L and H. In general the stagnation region on the recompression face produces the highest heating rate over the cavity surface with the absolute maximum value appearing at the recompression corner, q_{rc}, and decreasing slightly to the maximum value on the floor, q_{max}. The increased thickness of the lower surface viscous layer reduces the heating rate to its minimum value near the upstream end of the cavity.

Experimental heat-transfer distributions for cavities of constant depth and increasing length display increasing values of q_{max} but decreasing values of q_{min}. Such an inverse relation can be explained through reference to Fig. 1 where it is seen that the velocity along the dividing streamline will increase with cavity length; hence the impingement point heating will be larger due to the higher impact velocity. However

Table 1 Test conditions for cavity experiments

Investigator	M_1	$URe_1 \times 10^{-6}$	H, cm	L/H
Hahn[7]	6.3[a]	1.0, 2.1[d]	1.9	5.3, 8.0, 10.7
Nicoll[8]	6.5[b]	23[d]	0.24	3.3, 6.7, 13.3
Wieting[9]	6.9[c]	5.7, 7.4[d]	2.0	0.063, 0.21 0.38, 0.52
Charwat[5]	2.1[a]	7.6[e]	0.64, 1.27	1, 3, 6, 9, 12
	2.9	4.7	Same	Same as above
	3.5	3.1	Same	Same as above
Nestler[6]	6.3[a]	13.5[e]	2.5	5
	8.5	12.1	2.5	5

[a]Air
[b]Helium
[c]Combustion gas
[d]Laminar flow
[e]Turbulent flow

the larger dividing streamline velocity also implies a larger amount of entrained flow (i.e., flow benealth the dividing streamline) which is captured by the vortex boundary layer. A thicker cavity boundary layer therefore is produced and consequently a lower value of q_{min} will occur at the upstream wall.

Flow conditions for the five experimental investigations of present interest are given in Table 1. It is seen that the Charwat study included independent variations of unit Reynolds number, URe, Mach number, cavity length, and cavity depth whereas the Nestler data at hypersonic speeds allows one to assess effects of freestream compressibility. It is also observed that, while the three laminar flow cases involve hypersonic Mach numbers, there is a wide range of unit Reynolds numbers, cavity depths, and L/H values. It is significant that the data of Wieting include extremely small values of L/H and that the results of Nicoll were obtained for a conical surface. (Equivalent planar parameters were employed in the present correlations.)

Laminar Flows

For ease of correlation it is desirable to employ approach flow data and geometry parameters where possible and to minimize the use of structural details of the free layer or cavity vortex which, if known, would obviate the need for correlations of the type proposed here. For the laminar regime it is recognized that, unlike turbulent flow, thickness parameters for both free layers and wall layers are proportional to the half-power of development distance.[11] Hence one may use the ratio L/δ to indicate the relative thickness of the free layer b(L). In turn, this relative thickness serves as a scale factor for the transverse gradients of velocity and temperature which determine transport rates across the dividing streamline. In contrast, the value of δ/H provides scaling for cavity endwall convection in both laminar and turbulent flows.

Figure 2 (upper) indicates the variation of heat transfer at the recompression corner, q_{rc}, with L/δ. Despite the relative values of unit Reynolds numbers in the three cases, the values of δ/H for the Wieting tests are much smaller than for the other cases, which have approximately equal values of δ/H. Consequently, one finds that the data collapse about a single correlation line when q_{rc}/q_1 is scaled with δ/H as in Fig. 2 (lower).

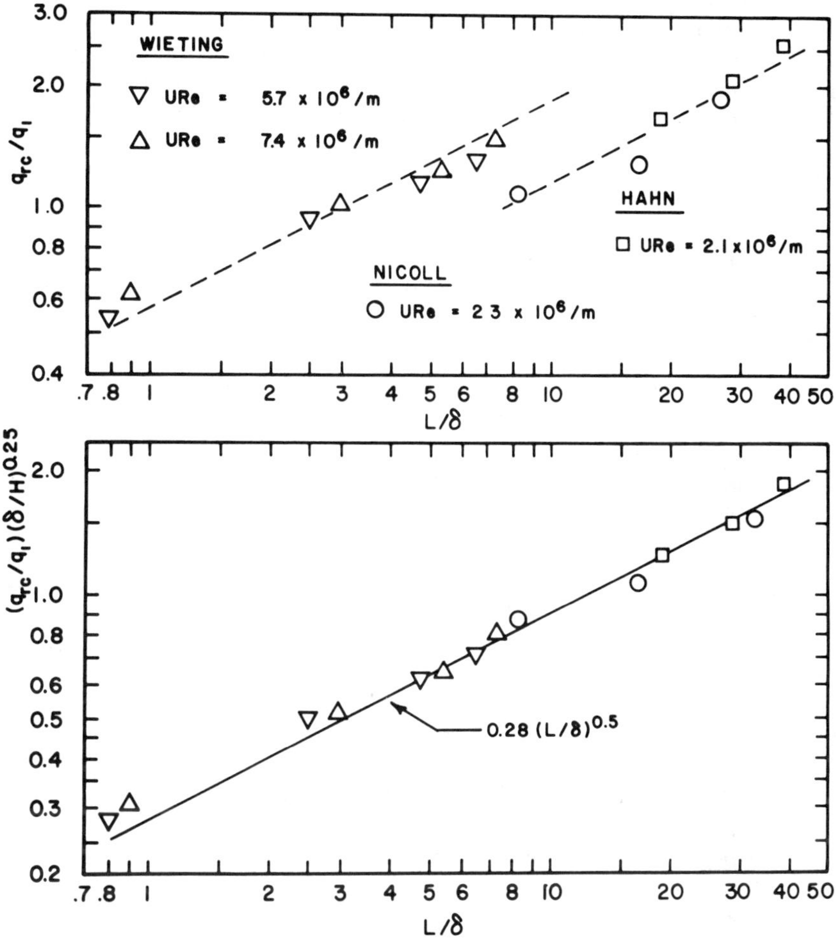

Fig. 2 Correlation of heating rate at downstream recompression corner in laminar flow.

Corresponding values of average heat-transfer rate, q_a, for the entire cavity surface are shown in Fig. 3 (upper). The values of average heat transfer denoted as "Nicoll" were obtained by integrating the reported heat-transfer distributions for an equivalent planar, rectangular cavity (same L and H). It is of interest that the actual average heat-transfer rates were substantially larger than for the equivalent cavity; this was due largely to the use of a cavity cross section which was not rectangular but contained rounded corners on the lower surface. Such a smooth lower surface, while not affecting the q_{rc} value, produced thinner boundary layers than would exist in a rectangular cavity and hence higher average heating rates.

As in Fig. 2 the general half-power variation of heat transfer with development distance is observed in Fig. 3 but, unlike the foregoing variation of q_{rc}, the effect of δ/H is in the opposite sense. This difference is consistent with the earlier discussion, viz., the larger impingement heat transfer is associated with thicker cavity boundary layers which, in turn, reduce the overall (or average) cavity

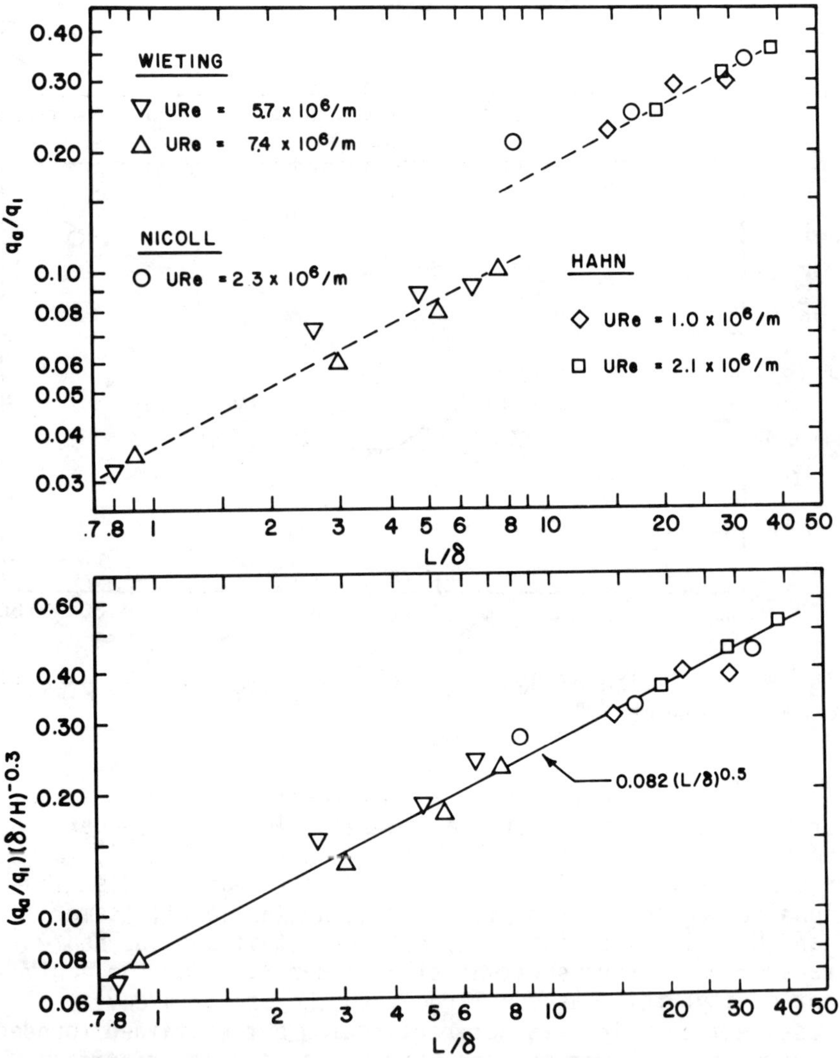

Fig. 3 Correlation of average heat-transfer (over entire cavity surface) in laminar flow.

transport. Thus it is observed in the lower part of Fig. 3 that a negative exponent is required to scale q_a/q_1 with δ/H. It should be remembered that the present laminar correlation results in Figs. 2 and 3 do not contain any Mach number influence since all three data sets were obtained at nearly identical values of M_1; thus further consideration of free-stream compressibility influences must await the availability of additional data.

Turbulent Flows

Figure 4 displays experimental variations[5] of q_{max}/q_1 and q_{min}/q_1 in terms of the conventional geometric parameter L/H. These distributions can be used as a basis for developing specific correlations which do not involve free layer parameters such as b(L) employed previously.[4] It is clearly evident that the family of curves for q_{max}/q_1 cannot be correlated with simple functions. In particular, the decrease in q_{max} for L/H = 9 (deeper cavity) is a result of an increasing thickness of the boundary layer on the downstream face and, hence, represents the occurrence of additional mechanisms not present in the remainder of the tests. These three data points can, therefore, be excluded from initial consideration.

As in laminar flow one can choose L/δ as the basic length parameter for turbulent flows while recognizing that the relationship between $b(L)/\delta$ and L/δ is more complex than in the laminar regime. Although a plot of q_{max}/q_1 vs L/δ will tend to correlate data for each cavity depth, it fails to provide any scaling for effects of the cavity boundary layer which develops on the recompression face and which largely governs q_{max}.

To provide the missing information one requires a parameter which, for constant L/δ, reflects the increase in angular velocity of the cavity vortex as the value of H decreases. One can rationalize that, with constant external flow conditions and constant L/δ, the driving energy into the vortex will be essentially constant but the retarding effect of boundary-layer drag will decrease as the depth decreases and hence the vortex will rotate faster, thus leading to higher wall convection. A characteristic cavity size, S, which displays the proper variation with H, can be defined in a manner analogous to the hydraulic diameter as

$$S = \frac{LH}{L + H} \sim \frac{\text{cross-sectional area of vortex}}{\text{total perimeter of vortex}} \tag{1}$$

The ratio $S/L = (1 + L/H)^{-1}$ can then be used as a correlation parameter.

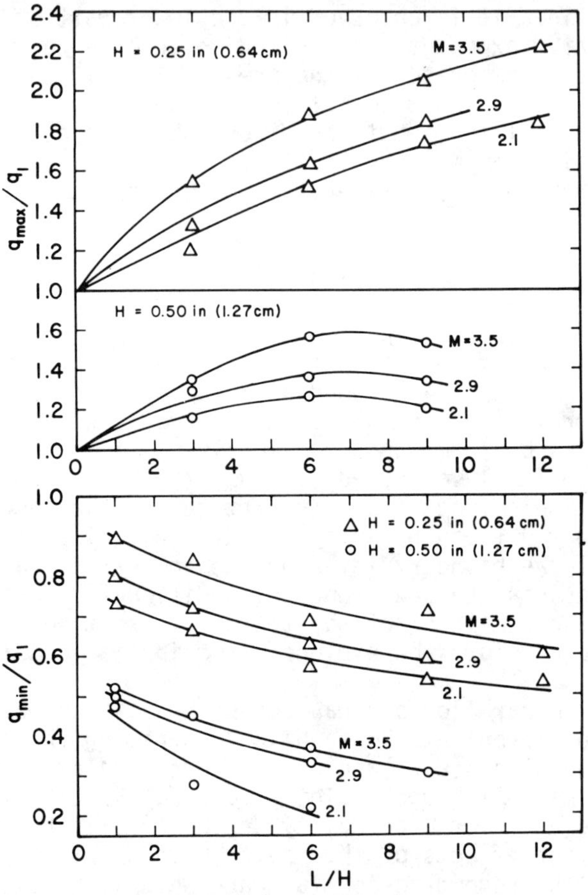

Fig. 4 Variation of maximum and minimum heat-transfer on cavity floor in turbulent flow. Data of Ref. 5.

The successful use of S/L to correlate q_{max}/q_1 for turbulent cavities is shown in Fig. 5 (upper). Although the deep cavity data of Charwat are not strongly indicative, the corresponding shallow cavity information along with the Nestler data suggest a slight Mach number influence which has been included in the lower part of Fig. 5 through use of the temperature ratio T/T_0.

Also shown in Fig. 5 (lower) are the three points at L/H = 9 (for deep cavity) which were excluded from initial consideration. It is encouraging to note that these points are also correlated by the proposed parameters but exhibit,

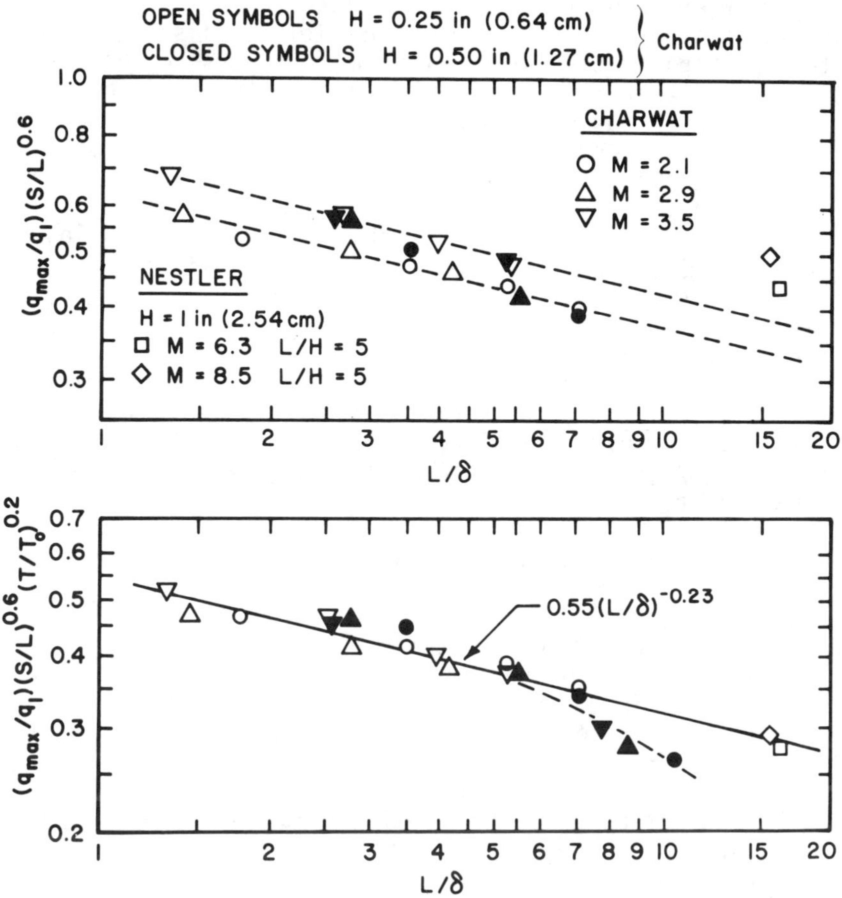

Fig. 5 Correlation of maximum floor heat-transfer rate for turbulent flow.

as would be expected, a different power law relationship with L/δ. Due to a lack of additional data which would more clearly indicate the proper trend, no correlation line is proposed.

Considering the variation of q_{min} with L/H shown in Fig. 4 (lower), one notes that there is some scatter in the data for M_1 = 3.5 (shallow cavity) and for M_1 = 2.1 (deep cavity) which is attributed to the difficulty in measuring small heat-transfer rates. Moreover, it is noted that the general shape of the curves reflects an exponential decrease with L/H. Thus one obtains linear variations in q_{min}/q_1 when plotted vs L/δ on semilogarithmic coordinates.

In order to eliminate the specific use of $b(L)$ as a correlation variable[4] it is necessary to establish an alternate length parameter which can be done by considering the classical Görtler linearized solution (see Ref. 11) for a fully developed turbulent shear layer. The velocity profile is expressed in terms of $\mathrm{erf}(\sigma y/x)$ where $\sigma = \sigma_i \simeq 12$ for incompressible flow. This early analysis also showed that the thickness of the layer is inversely proportional to σ. The author's later analysis[12] of experiments for compressible layers indicated the nature of the direct relationship between σ and Mach number (i.e., the growth rate decreases as M increases); a companion study[13] showed that, even for developing or preasymptotic regions, the layer thickness can be expressed as $b(x)/\delta = f[x/(\delta\sigma)]$. Hence, for the characteristic cavity size, S, the corresponding thickness is $b(S)/\delta = f[S/(\delta\sigma)]$. For proper scaling, however, one must use the parameter $(S/\delta)(\sigma_i/\sigma)$ where the relative growth σ/σ_i can be estimated[13] as

$$\sigma/\sigma_i \simeq 1 + 0.056\ \exp(3.4C^2) \quad \text{(for } M < 8) \qquad (2)$$

It will be remembered that the Crocco number, C^2, is related to the Mach number through the temperature ratio, i.e., $T/T_o = 1 - C^2$. The aforementioned study[13] also showed that σ/σ_i is a weak function of the stagnation temperature ratio across the free layer but that influence has been ignored in the present correlation.

The variation of q_{min}/q_1 with the foregoing length parameter (denoted as $\tilde{S}$) is displayed in Fig. 6 (upper) which includes both Charwat and Nestler data. A generally satisfactory correlation is obtained for each of the cavity depths tested by Charwat. Thus the final correlation must also include a weak function δ/H as shown in Fig. 6 (lower) along with the proposed correlation line.

The variation of average floor heat transfer shown in Fig. 7 (upper) appears to be exceedingly complex because of its dependence on the distribution of floor heat transfer as well as on both maximum and minimum values. It is clear that the relatively large values of q_{max} for the shallow cavity (Fig. 4, upper) cause the value of q_a to increase for all L/H when $M_1 = 3.5$ whereas, for the lower Mach numbers, the increase in q_{max} does not affect the value of q_a until L/H is larger than 3. Thus, since the variation of average floor

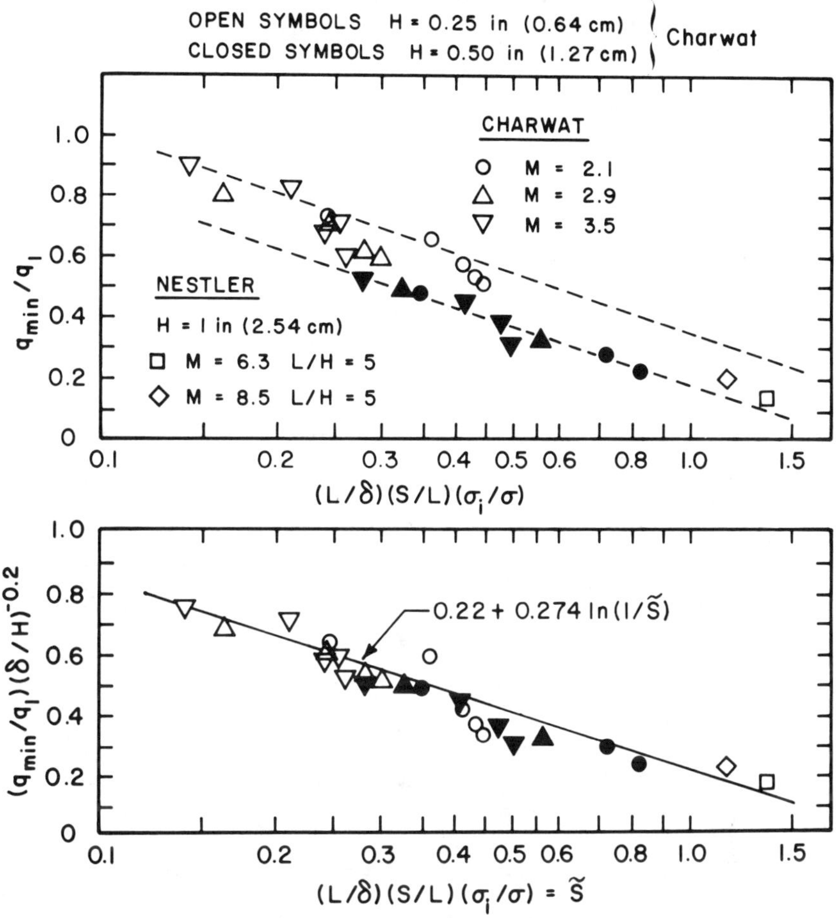

Fig. 6 Correlation of minimum floor heat-transfer rate for turbulent flow.

heat transfer is coupled strongly to the corresponding distributions of q_{max} and q_{min}, it is convenient to define a correlation variable for q_a in the form

$$Q = (q_a - q_{min})/(q_{max} - q_{min}) \quad (3)$$

The variation of this parameter with L/H is depicted in Fig. 7 (center) where it is seen that there is reasonable correlation of results for each cavity depth. The scatter in data for the deeper cavity is attributed to the uncertainty in defining the variation of q_{min} (Fig. 4, lower).

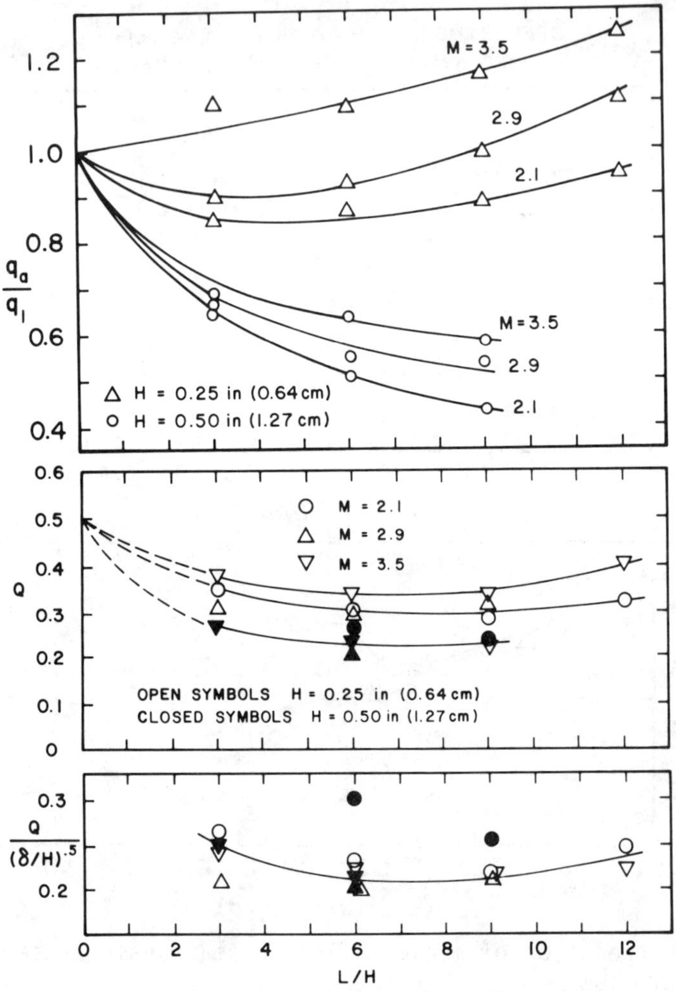

Fig. 7 Correlation of average floor heating rate for turbulent flow. Data of Ref. 5.

The final correlation of Q, using the half-power of δ/H as a depth scaling factor, is displayed in the lower part of Fig. 7; the apparently poor agreement for the data at M - 2.1 is exaggerated somewhat by the expanded scale. However, the fact that the composite variable, Q, is easily correlated with L/H indicates a weak dependence of Q on the detailed structure of the cavity flow pattern.

Conclusions

The present study of convection heat-transfer distributions for planar, rectangular, open cavities in supersonic flow has shown that effective correlation of heating rates with appropriate flow and geometric variables is possible. It was shown that the effects of cavity depth and cavity length on the flowfield and consequent heat-transfer distributions are entirely different.

For the laminar regime relatively simple correlation expressions were obtained for the heat-transfer rate at the downstream recompression corner and as well as for the average heating rate over the entire cavity surface. The primary scaling parameter for cavity geometry was the upstream boundary-layer thickness. This relative simplicity was attributed to the similar growth rates for both wall and free layers in the laminar regime. No Mach number effects were included in the present data base.

In turbulent flow cases, more complex scaling parameters were required for correlation as a consequence of the relationship between the shear layer and upstream boundary layer. In particular, a new length parameter, the characteristic cavity size, was shown to produce satisfactory correlations of both maximum and minimum floor heat-transfer rates for a wide range of unit Reynolds numbers, Mach numbers, and cavity dimensions. The average floor heating rate was shown to be directly related to the foregoing limiting values of floor heat transfer and not correlatable independently.

References

[1]Charwat, A. F., "Supersonic Flows with Imbedded Separated Regions," Advances in Heat Transfer, Vol. 6, Academic Press, New York, 1970, pp. 1-132.

[2]Fletcher, L. S., Briggs, D. G., and Page, R. H., "A Review of Heat-Transfer in Separated and Reattached Flows," AIAA Paper 70-767, June 1970.

[3]Nestler, D. E., "An Engineering Analysis of Reattaching Shear Layer Heat-Transfer." AIAA Paper 72-717, June 1972.

[4]Lamb, J. P., "Convective Heat-Transfer Correlations for Planar, Supersonic, Separated Flows," Transactions of ASME, Journal of Heat Transfer, Vol. 102, No. 2, May 1980, pp. 351-356.

[5]Charwat, A. F., Roos, J. N., Dewey, C. F., and Hitz, J. A., "An Investigation of Separated Flows - Part II: Flow in the Cavity and Heat-Transfer," Journal of Aerospace Sciences, Vol. 28, No. 7, July 1961, pp. 513-527.

[6]Nestler, D. E., Saydah, A. R., and Auxer, W. L., "Heat-Transfer to Steps and Cavities in Hypersonic Turbulent Flow," AIAA Journal, Vol. 7, No. 7, July 1969, pp. 1368-1370.

[7]Hahn, M.,"Experimental Investigation of Separated Flow Over a Cavity at Hypersonic Speed," AIAA Journal, Vol. 7, No. 6, June 1969, pp. 1092-1098.

[8]Nicoll, K. M., "A Study of Laminar Hypersonic Cavity Flows," AIAA Journal, Vol. 2, No. 9, Sept. 1964, pp. 1535-1541.

[9]Wieting, A. R., "Experimental Investigation of Heat-Transfer Distributions in Deep Cavities in Hypersonic Separated Flow," NASA TN D-5908, Sept. 1970.

[10]Charwat, A. F., Roos, J. N., Dewey, C. F., and Hitz, J. A., "An Investigation of Separated Flows - Part I: The Pressure Field," Journal of Aerospace Sciences, Vol. 28, No. 6, June 1961, pp. 457-470.

[11]White, F. M., Viscous Fluid Flow, McGraw-Hill, New York, 1974.

[12]Lamb, J. P. and Bass, R. L., "Some Correlations of Theory and Experiment for Developing Turbulent Free Shear Layers," Transactions of ASME, Journal of Basic Engineering, Vol. 90, No. 4, 1968, pp. 572-580.

[13]Lamb, J. P., "An Approximate Theory for Developing Turbulent Free Shear Layers," Transactions of ASME, Journal of Basic Engineering, Vol. 89, No. 3, 1967, pp. 633-642.

HEAT TRANSFER AND PRESSURE ON A FLAT PLATE DOWNSTREAM OF A HEATED SQUARE JET IN A MACH 0.4 TO 0.8 CROSSFLOW

D. J. McDermott*

Wright-Patterson Air Force Base, Ohio

Abstract

Local mean heat flux rates and static pressures were measured on a flat plate (wind tunnel nozzle wall) downstream of a square, subsonic, heated air jet issuing from the surface. Freestream Mach numbers at a constant unit Reynolds number of 1.9×10^6/ft ranged from M = 0.40 to M = 0.80. The jet total temperature was 300^oR above the freestream value and the jet to freestream mass flux ratios were varied from MR = 0.30 to MR = 0.70. The centerline (z/D = 0) heat-transfer data could be expressed as a power function of x/D. For z/D > 1.3, jet effects on the heat-transfer data were negligible. In the separated region of the jet (2.9 < x/D < 5) at z = 0, the heat-transfer data correlated as an exponential function of the pressure coefficient. Two prediction methods were formulated that show good correlation with measured centerline heat-transfer data.

Nomenclature

D = width of a square jet or diameter of a circular jet
k = thermal conductivity
M = Mach number
MR = mass flux ratio = $\rho_J V_J/\rho_\infty V_\infty$ or $\rho_2 V_2/\rho_\infty V_\infty$
P = pressure
Pr = Prandtl number
$\dot{q}$ = heat flux
Re = Reynolds number

Presented as Paper 80-0290 at the AIAA 18th Aerospace Sciences Meeting, Pasadena, Calif., Jan. 14-16, 1980.

*Aerospace Engineer, Flight Dynamics Laboratory.

T = temperature
u = velocity in the x direction
x = distance along the model centerline in the freestream direction
X = distance along the model centerline in the freestream direction measured in jet diameters D
y = distance from the model centerline normal to the freestream
z = spanwise distance as measured from the jet centerline
Z = spanwise distance from the jet centerline measured in jet diameters D
μ = dynamic viscosity
ρ = mass density

Subscripts

∞ = freestream condition
2 = secondary or injected fluid
aw = adiabatic wall
J = conditions at the jet exit
x = local conditions in a streamwise direction
w = conditions at the wall

Superscripts

o = stagnation condition

Introduction

Various studies have shown some important characteristics of an exhaust jet in a subsonic crosswind. These studies have been primarily directed toward determining the aerodynamic interference effects of a circular jet downwash.[1] The mixing process of a jet with a crossflow has also been studied with emphasis on the associated vortex structure.[2-4] Recently, the turbulent mixing of a square, heated exhaust jet issuing from the lower fuselage of an advanced aircraft configuration and the associated plume impingement on the surface have been under investigation by the Air Force. The purpose of the investigation has been to predict the surface heating effects from a hot-exhaust-flow device to determine whether a protective covering is necessary. Preliminary wind tunnel tests were performed to define the plume impingement boundaries. However, heat-transfer data were not obtained. The objective of this study was to measure local heat-transfer and pressure data on a flat-plate configuration (illustrated in Fig. 1) for various operating conditions and jet to freestream mass flux ratios representative of those for flight.

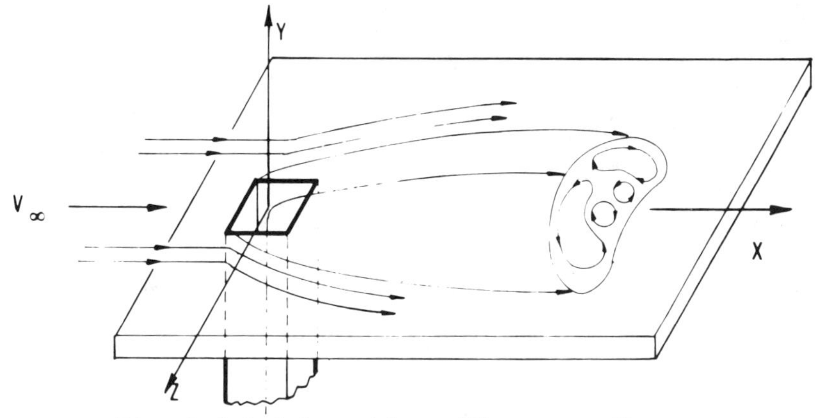

Fig. 1 Jet interaction with a crossflow.

Protection of a surface exposed to a high-temperature freestream flow has received considerable study. Insulating the surface by injecting a secondary fluid into the boundary layer has been investigated for the past 30 years. These studies are commonly referred to as film-cooling. The aircraft surface-heating problem above is very similar to the film-cooling problem. In fact, since similar heat-transfer relations govern whether the temperature of the injected fluid is lower or higher than the freestream fluid (assuming constant properties), many experiments are performed with a higher temperature injected fluid. The majority of film-cooling experiments have been concerned with the prediction and measurement of adiabatic wall temperatures of a surface downstream from a two-dimensional injection slot.[5] In several three-dimensional film-cooling experiments, adiabatic wall temperatures have also been measured on a flat plate downstream of a circular hole.[6-8] However, in contrast to the two-dimensional studies, no unified prediction theory has been developed for the three-dimensional film-cooling problem.

Although most of the film-cooling studies have been concerned with adiabatic wall temperature measurements, heat-transfer data have also been measured. Wilson and Goldstein[9] investigated the effect of slot film-cooling on the downstream heat-transfer coefficients in a Mach 3 mainstream flow. A simple theoretical model based on the thickening of the boundary layer owing to the injected fluid was also developed to predict the heat-transfer coefficients. The authors state that, considering its simplicity, the theory does remarkably well. However, both the theory and measurements are limited to small mass velocities of the injected fluid (<2% of the

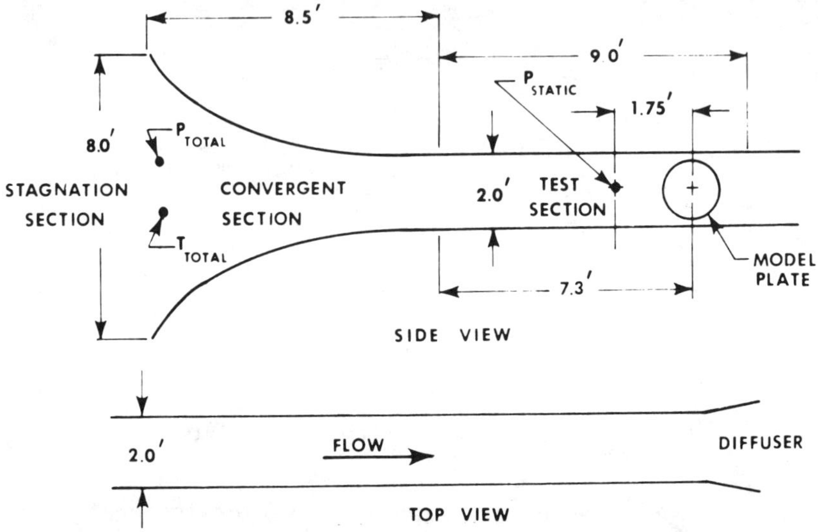

Fig. 2 Wind tunnel flow channel.

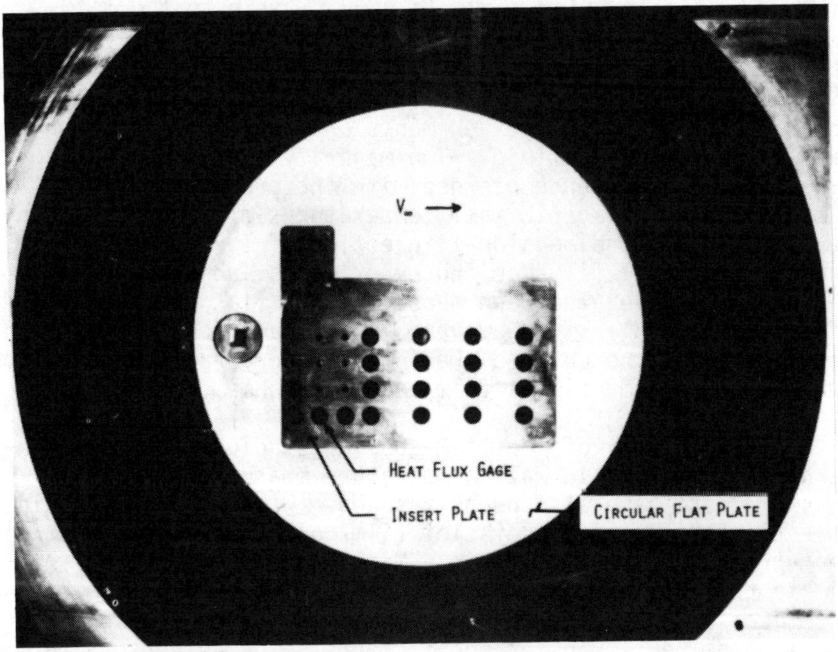

Fig. 3 Wind tunnel installation photograph of the model test plate with the heat-transfer insert plate.

mainstream value). Hartnett, Birkebak, and Eckert[10] measured heat-transfer on a surface downstream from a 30 deg tangential slot for both an accelerated and decelerated freestream flow. In this experiment, mass flux ratios MR were varied over a large range (0.28 to 1.23), and the freestream velocity was kept constant at 137 ft/s. The ratio of heat-transfer coefficients with and without film-cooling was only slightly affected by the imposed pressure gradients, but significant increases could be obtained by increasing the mass flux ratio except for zero pressure gradient. In this case, differences were found only for MR > 0.50. In addition, for MR > 1.0, the heat-transfer coefficients with and without film-cooling agreed within 10% at distances downstream from the slot greater than 30 slot widths. Foster and Haji-Sheikh[11] found similar agreement for a perpendicular slot geometry for distances greater than approximately 40 slot widths.

Bouchez and Goldstein[12] investigated the effect of film-cooling and also measured the local heat-transfer coefficients due to an air jet impinging on a wall and subjected to a low-speed crossflow (40 ft/s). The coolant air was injected through a circular tube that could be mounted either flush with the top wall of the wind tunnel or extended into the freestream flow. In this study, however, the heat-transfer coefficients and adiabatic wall temperatures were measured on the wind tunnel wall opposite the injection tube.

The measurement of local surface heat-transfer and pressure data downstream of a heated jet injected into a high subsonic crossflow has to date received little attention. In addition, previous studies have been restricted to either a two-dimensional injection slot or a circular hole jet. The measurements of this study provide some insight into the effects of compressibility and square jet geometry on surface heat-transfer and pressure distributions.

Theoretical Models

Two-Dimensional Film-Cooling Theory

Consider a heated mainstream of air flowing over a flat surface. If a cooler secondary fluid is injected into the boundary layer of the surface through a continuous spanwise slot, then the problem is referred to as two-dimensional film-cooling. The temperature of the secondary fluid could also be higher than the freestream fluid (film heating). As previously mentioned, film-cooling studies have been mainly concerned with the measurement and prediction of adiabatic wall temperatures downstream of an injection slot. The adiabatic wall temperatures are usually presented as a film-cooling effectiveness η defined by

$$\eta = (T_{aw}-T_{\infty})/(T_2-T_{\infty}) \tag{1}$$

Once the adiabatic wall temperature with injection is known, then the heat transfer to the surface is computed using the adiabatic wall temperature instead of the freestream temperature in the convective heat-transfer equation. The heat transfer to the surface is then

$$\dot{q}_{wall} = h_x(T_{aw}-T_{wall}) \tag{2}$$

where the heat-transfer coefficient h is computed using the equation for flow over the surface without injection. Fluid properties are determined using Eckert's reference temperature.[13]

Several existing theories are presented in Ref. 5 that predict adiabatic wall temperatures downstream of a two-dimensional slot. The results of one of these theories is outlined below using the present notation. Basically all the theories are derived from a mass and energy balance in the boundary layer. The sum of the mass of the injected fluid $\dot{m}_2$ and the freestream mass entrained into the boundary layer $\dot{m}_{\infty}$ is considered to be the total mass flowing within the boundary layer. In all the theories the main correlating parameters are the slot width S, the slot Reynolds number Re_2, the mass flux ratio MR, and the distance x from the origin of the boundary layer.

The two-dimensional theory presented here was developed by Goldstein and Haji-Sheikh.[5] Similar to their predecessors, their analysis begins with an overall mass and energy balance. In addition, their analysis accounts for temperature variations throughout the boundary layer and the effect of injection on thickening the boundary layer. When the injected and mainstream fluids are the same and the slot is oriented perpendicular to the freestream flow, the effectiveness equation was found to be

$$\eta = \frac{1.9\ Pr^{2/3}}{1 + 0.329\ \xi^{0.8}\ \beta} \tag{3}$$

where

$$\xi = (1/MR)(x/S)\left[(\mu_2/\mu_{\infty})Re_2\right]^{-0.25} \tag{4}$$

and

$$\beta = 1 + 1.5 \times 10^{-4}(\mu_2/\mu_\infty)Re_2 \tag{5}$$

It should be noted that all the film-cooling correlations rely on a considerable amount of empirical data. The term β in the effectiveness Eq. (3) is of particular importance. This term represents an experimental measurement of the entrained mass flow with injection as a function of the secondary injected flow. Specifically, $\beta = \dot{m}_\infty/\dot{m}_0$ where $\dot{m}_0$ is the freestream mass flow entrained into the boundary layer for zero injection and $\dot{m}_\infty$ is the freestream mass flow entrained into the boundary layer with injection. Although Eq. (3) was developed for the two-dimensional problem, it is hypothesized that a relationship between the two- and three-dimensional film-cooling problems may be found by considering only the β term. At least directly downstream of a three-dimensional jet (a square hole, for example) such a connection seems plausible. For a three-dimensional flow pattern the entrainment of the freestream flow $\dot{m}_\infty$ into the boundary layer increases with both jet velocity and the distance downstream from the jet. Theories of Braun and McAllister[1] have shown that for a jet directed perpendicular to the freestream

$$\frac{\dot{m}_\infty}{\dot{m}_0} = 1 + \frac{V_J}{V_\infty}\frac{dx}{dy} \tag{6}$$

where dx/dy describes the trajectory of the jet centerline. Using the jet trajectory equations of Ivanov[1], the mass entrainment ratio derived by the author is

$$\beta' = \dot{m}_\infty/\dot{m}_0 = 1 + 3(x/D)^{2/3}(V_J/V_\infty)^{2/15} \tag{7}$$

Substituting β' for β in Eq. (3), the effectiveness equation for the line directly downstream of the jet can be written as

$$\eta' = \frac{1.9\ Pr^{2/3}}{1 + 0.329\ \xi^{0.8}\ \beta'} \tag{8}$$

Three-Dimensional Film-Cooling Theory

Perhaps the most extensive experimental study investigating the effects of film-cooling downstream of a three-dimensional jet was conducted by Goldstein, Eckert, and Ramsey.[7] In a later study by these authors a point source model was developed (information concerning this study was extracted from Ref. 14). According to Mayle and Camarata[14], the "model is based on the solution of the energy equation for a point sink (or source) of heat moving at a constant velocity in an infinite conducting medium." The resulting solution for the effectiveness η was

$$\eta = \frac{1}{8} \frac{MR\ V_\infty D}{\varepsilon_T \sqrt{x^2 + z^2}} \exp\left[- \frac{V_\infty D}{2\varepsilon_T} \left[\sqrt{x^2 + z^2} - x \right] \right] \tag{9}$$

where the turbulent thermal eddy diffusivity ε_T must be determined experimentally.

To compare Eq. (9) with the experimental results of this investigation, the following approach was used to estimate ε_T for the case z = 0. Assume that in the turbulent core of the boundary layer the momentum and thermal eddy diffusivities are equal, that is

$$\varepsilon_T = \varepsilon_M \tag{10}$$

Introducing the free jet relationship[15] for ε_M, one obtains

$$\varepsilon_T = \varepsilon_M = E\ V_J\ D \tag{11}$$

where the constant E is determined from the experimental data. For z = 0, Eq. (9) can be written as

$$\eta = (MR\ V_\infty\ D^2)/(8\ \varepsilon_T\ x) \tag{12}$$

Combining Eqs. (11) and (12), one obtains

$$\eta = (MR\ V_\infty\ D)/(8\ E\ V_J\ x) \tag{13}$$

Combining Eqs. (1), (2), and (13) and noting that $T_W \simeq T_\infty$, one obtains

$$\dot{q}_W = h_x\ (T_2 - T_W)(MR\ V_\infty\ D)/(8\ E\ V_J\ x) \tag{14}$$

Using the experimental data for $\dot{q}_W$, Eq. (14) was used to calculate values of E and obtain an average $\overline{E}$ value. In these calculations, $T_2 = T^o{}_J$ and h_x was computed using the equation for flow over a flat plate without injection. Substituting the average $\overline{E}$ value for E in Eq. (11), one obtains

$$\varepsilon_T = \overline{E}\ V_J\ D = 0.177\ V_J\ D \tag{15}$$

Equation (15) can be used to calculate values of ε_T and the effectiveness η can be obtained from Eq. (9). Heat transfer to the wall is then calculated using Eq. (2).

Description of Experiments

Wind Tunnel Facility and Test Flow Conditions

The test program was conducted in the subsonic test section of the Air Force Flight Dynamics Trisonic Gasdynamics Facility, which is a closed-circuit, variable density, continuous flow wind tunnel. Air dryer and cooled settling chamber sections are used to maintain predetermined air dew point and temperature. Figure 2 shows the wind tunnel internal configuration, the placement of the model circular flat plate, and the position of the wind tunnel temperature and pressure instrumentation. Throughout the test the wind tunnel and jet total temperatures were 100°F and 400°F respectively, and the jet to freestream mass flux ratios MR were 0.30, 0.50, and 0.70. The wind tunnel was operated at Mach numbers of 0.40, 0.60, 0.70, and 0.80 with corresponding total pressures (psfa) of 1730, 1215, 1100, and 1020. Consequently, all test runs were at a constant freestream unit Reynolds number of 1.9 x 10^6/ft.

Model Description and Test Variables

Figure 3 is a wind tunnel model installation photo that shows the square jet on the model circular flat plate with a copper insert plate instrumented with heat-transfer gages. The circular flat plate that mounts flush with the wind tunnel side wall is the model's main structural housing. Two insert plates that mount into and flush with the circular plate were used in this experiment. One insert is a 1/4 in. copper plate that mounts on a 1/2 in. aluminum frame. The copper plate was instrumented with both heat flux gages and thermocouples. The other insert is a 3/4 in. aluminum plate instrumented with static pressure ports. The dimensions of the circular flat plate and insert plates are shown in Fig. 4.

Nine jet configurations of different geometries as shown in Fig. 5 and Table 1 were investigated. When mounted in the circular flat plate, the exit plane of each jet is the same as that of the circular flat plate. The varied parameters were the width of the jet square channel and the angle α of the jet exhaust flow with respect to the tunnel airflow. The angle β indicated in Fig. 5 is 90 deg for all the tests.

Before the heated air enters the square channel of the jet, it passes through a plenum chamber designed to produce smooth flow at the jet entrance region. The plenum chamber is shown in Fig. 6. Heated air enters through a circular pipe having an i.d. of 1.043 in. The air then flows radially out the sides of a porous canister and expands into the plenum chamber area. The canister, which was approximately 30% porous, had an open area 1.5 times as great as the cross-sectional area of the circular pipe. The ratio of the plenum chamber inside cross-sectional area to that of the jet channel cross-sectional area varied from 9 to 25 depending on the width of the square jet channel.

Model Instrumentation

The copper insert plate was instrumented with 28 heat flux gages. Wall temperatures, used in the calculation of the local heat-transfer coefficients, were measured by 14 iron-constantan thermocouples imbedded in the copper plate along two rows. The center location of the heat flux gages and thermocouples with respect to the insert plate leading edge and jet center is given in Fig. 7. Eight of the gages were 1/4 in. o.d. and the other twenty gages were 11/16 in. o.d. As a result, the local heat-transfer measurements in this experiment are limited by the spatial resolution of the gages.

The aluminum insert plate that was used to make the flat-plate pressure measurements was instrumented with 28 static pressure taps. The locations of the pressure taps were the same as the center locations of the heat flux gages (see Fig. 7).

Each jet was instrumented with four static pressure ports and one total pressure probe (see Fig. 5). The four static pressures were used to calculate an average jet static pressure. The total temperature of the jet exhaust was measured by an iron-constantan thermocouple located in the plenum chamber stagnation region (see Fig. 6). The jet static pressure and jet total pressure and temperature were used to calculate all the jet flow properties such as mass flow, Mach number, density, and velocity.

Results and Discussion

In the following, local heat-transfer coefficients are based on the difference between the jet total temperature and the local wall temperature, that is

$$h_x = (\dot{q}_{wall})/(T_J^0 - T_{w_x}) \tag{16}$$

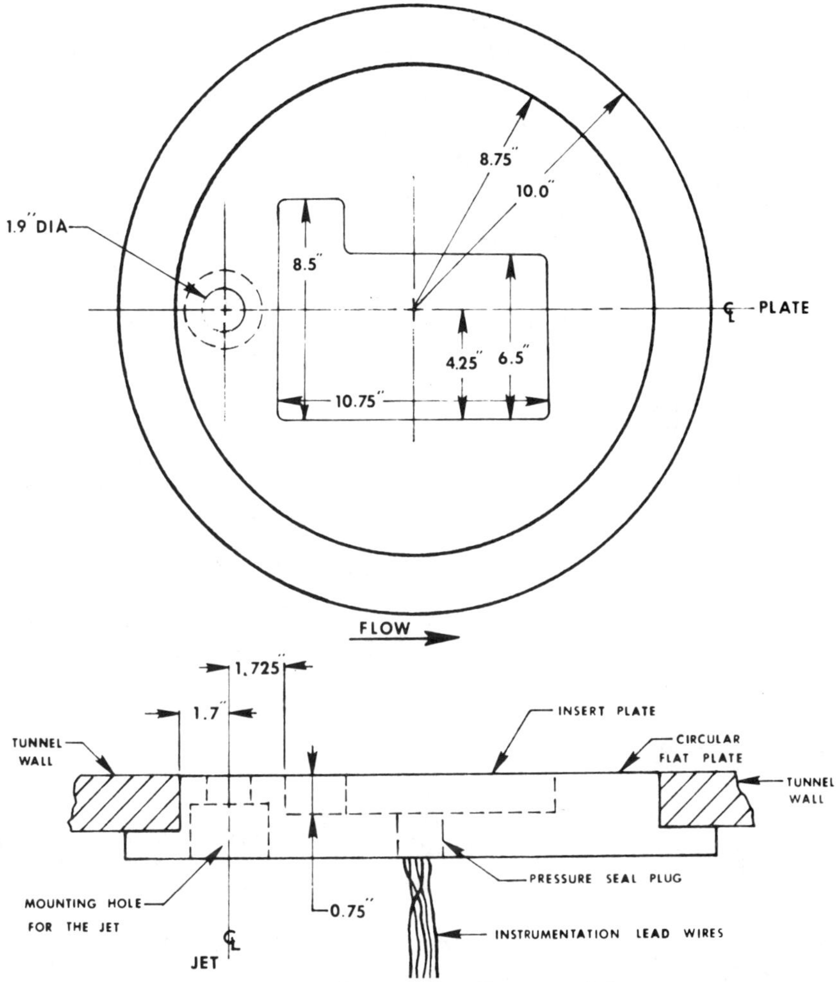

Fig. 4 Circular flat-plate model.

The wall temperatures were 100°F for all test points and the jet total temperature was 400°F. Thus, the driving potential for the heat-transfer coefficients was the same for all test points. Pressure data are presented as a local pressure coefficient defined by the equation

$$\frac{C_{P_x}}{MR} = \frac{P_{w_x} - P_\infty}{MR\ q_\infty} = \frac{P_{w_x} - P_\infty}{\rho_J V_J V_\infty / 2} \qquad (17)$$

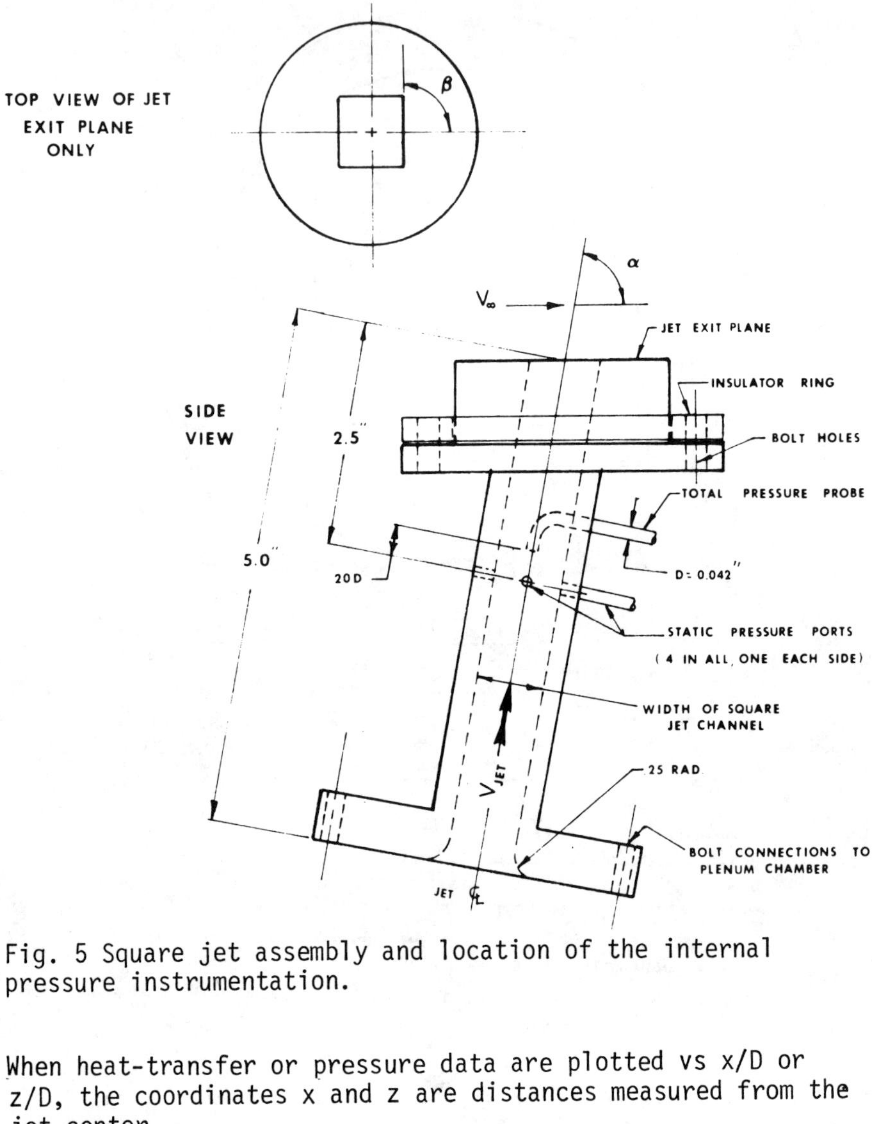

Fig. 5 Square jet assembly and location of the internal pressure instrumentation.

When heat-transfer or pressure data are plotted vs x/D or z/D, the coordinates x and z are distances measured from the jet center.

Test Plate Boundary Layer

The boundary-layer velocity profile ahead of the jet was measured with a total pressure rake located at 0.95 in. upstream of the jet center. The total pressures were converted to velocities and are plotted in Fig. 8. The boundary-layer thickness δ (value of y when $u = 0.99\ V_\infty$), as indicated by these data, was approximately 1.37 in. For $\delta = 1.37$ in. a boundary-layer velocity profile was plotted using the 1/7th power-law equation. The agreement between this profile and the experimental data indicates that the boundary layer ahead of the jet is indeed turbulent.

Characteristic Heat-Transfer Results

Figures 9 through 13 illustrate the following similarity of results obtained over the entire range of test variables.

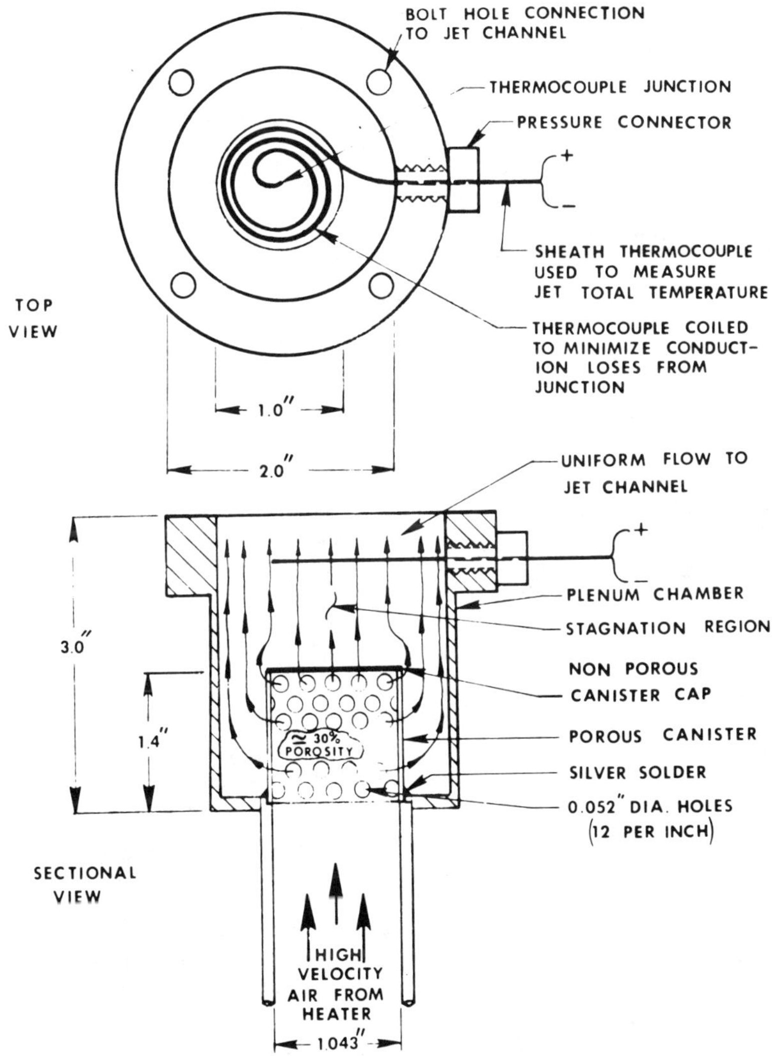

Fig. 6 Cross-sectional view of the plenum chamber design.

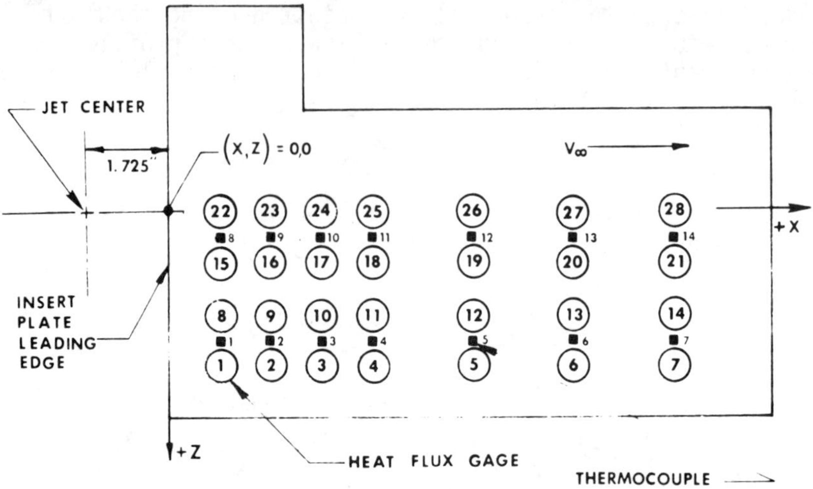

HEAT FLUX GAGE NUMBER	THERMOCOUPLE TO MEASURE LOCAL WALL TEMP.	HEAT GAGE LOCATION (IN.) X	HEAT GAGE LOCATION (IN.) Z	THERMOCOUPLE LOCATION (IN.) X	THERMOCOUPLE LOCATION (IN.) Z
1	1	0.5	3.0	0.5	2.5
2	2	1.5		1.5	
3	3	2.5		2.5	
4	4	3.5		3.5	
5	5	5.5		5.5	
6	6	7.5		7.5	
7	7	9.5		9.5	
8	1	0.5	2.0	0.5	
* 9	2	1.5		1.5	
* 10	3	2.5		2.5	
11	4	3.5		3.5	
12	5	5.5		5.5	
13	6	7.5		7.5	
14	7	9.5		9.5	
* 15	8	0.5	1.0	0.5	0.5
* 16	9	1.5		1.5	
* 17	10	2.5		2.5	
18	11	3.5		3.5	
19	12	5.5		5.5	
20	13	7.5		7.5	
21	14	9.5		9.5	
* 22	8	0.5	0	0.5	
* 23	9	1.5		1.5	
* 24	10	2.5		2.5	
25	11	3.5		3.5	
26	12	5.5		5.5	
27	13	7.5		7.5	
28	14	9.5		9.5	

* SMALLER HEAT FLUX GAGE

Fig. 7 Location of instrumentation on insert plate.

Directly downstream from the jet (z = 0) the heat-transfer coefficients decay from values of about 5 Btu/h-ft^2-^{o}F at x/D = 5 to values of about 1 Btu/h-ft^2-^{o}F at x/D = 26. Note that for all spanwise stations of z = 1 in., which corresponds to z/D = 1.3 for the largest jet, the heat-transfer coefficient drops off rapidly (as compared to data at z = 0), thus indicating very little spreading of the jet. For this station the heat-transfer coefficients are on the order of 1 Btu/h-ft^2-^{o}F at 2.9 < x/D < 15. At the next station out (z = 2 in.) heat-transfer coefficients of approximately 1 Btu/hr-ft^2 - ^{o}F are also observed but are confined between 5 < x/D < 10. Still further out at z = 3 in. the heat-transfer coefficients are extremely small and any effect of the jet is buried in the error band of the instrumentation.

It was found that the centerline local heat-transfer coefficients for all freestream Mach numbers, mass flux ratios, and jet flow angles, could be expressed by the power function

$$h_x = 42.2\ (x/D)^{-1.2} \tag{18}$$

for z = 0 and 5 < x/D < 26.

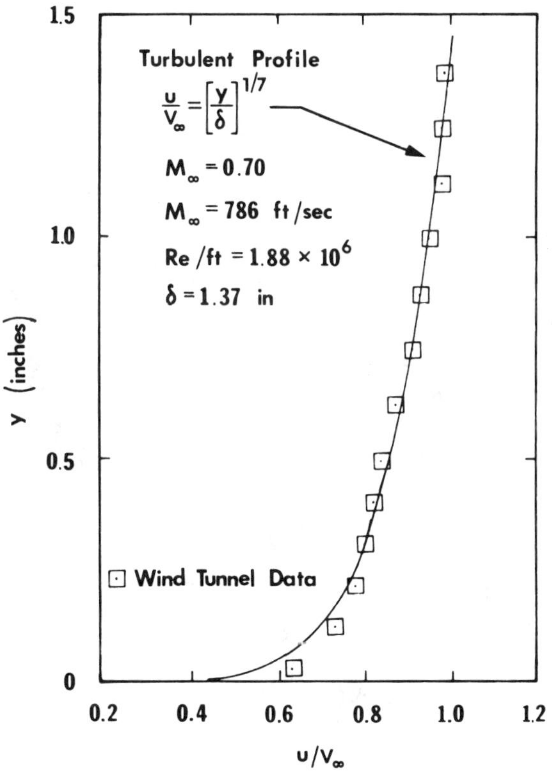

Fig. 8 Boundary-layer profile at 0.95 in. upstream of the jet center.

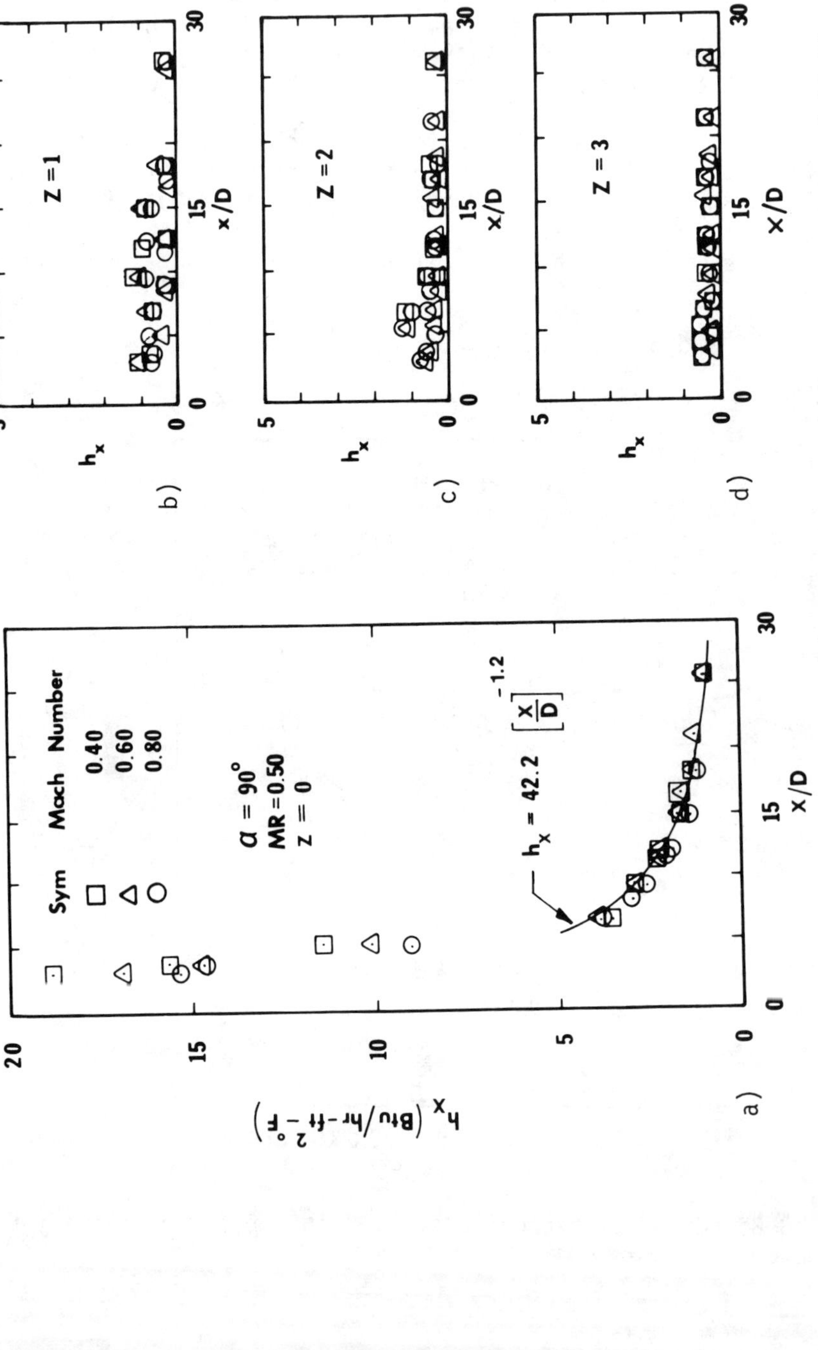

Fig. 9 Effect of Mach number on local heat-transfer coefficients (all D).

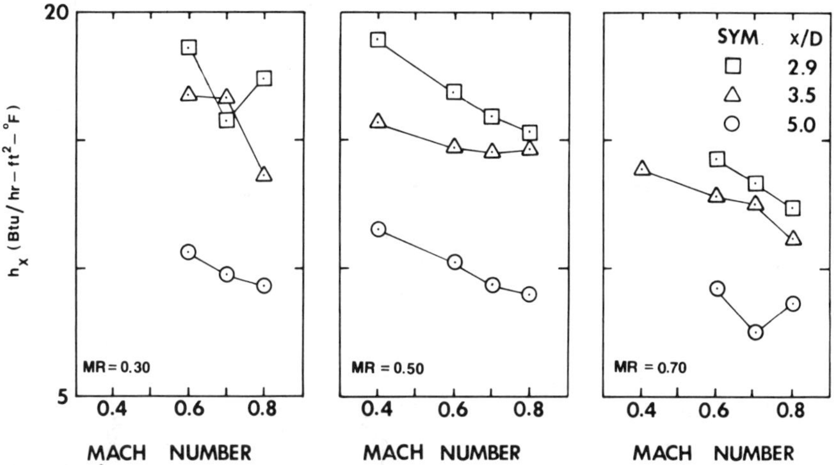

Fig. 10 Effect of Mach number on local heat-transfer coefficients in the near region of the jet at z = 0 (all D).

Table 1 Jet geometry

Jet no.	Angle α jet flow (deg)	Width of jet channel (in.)	Jet area (in.^{2})	% Area of largest jet
1	90	0.750	0.563	100
2	86	0.750	0.563	100
3	82	0.750	0.563	100
4	90	0.612	0.375	66
5	86	0.612	0.375	66
6	82	0.612	0.375	66
7	90	0.433	0.188	33
8	86	0.433	0.188	33
9	82	0.433	0.188	33

Flow and Geometry Heat-Transfer Effects

The effect of freestream Mach number on heat-transfer, for the largest Mach number range tested, is presented in Figs. 9a through 9d. A trend of decreasing heat-transfer coefficient with increasing Mach number exists. Although the trend exists, its effect is negligible at all spanwise stations of $z \geq 1$ in. for all x/D. This trend is more clearly seen in

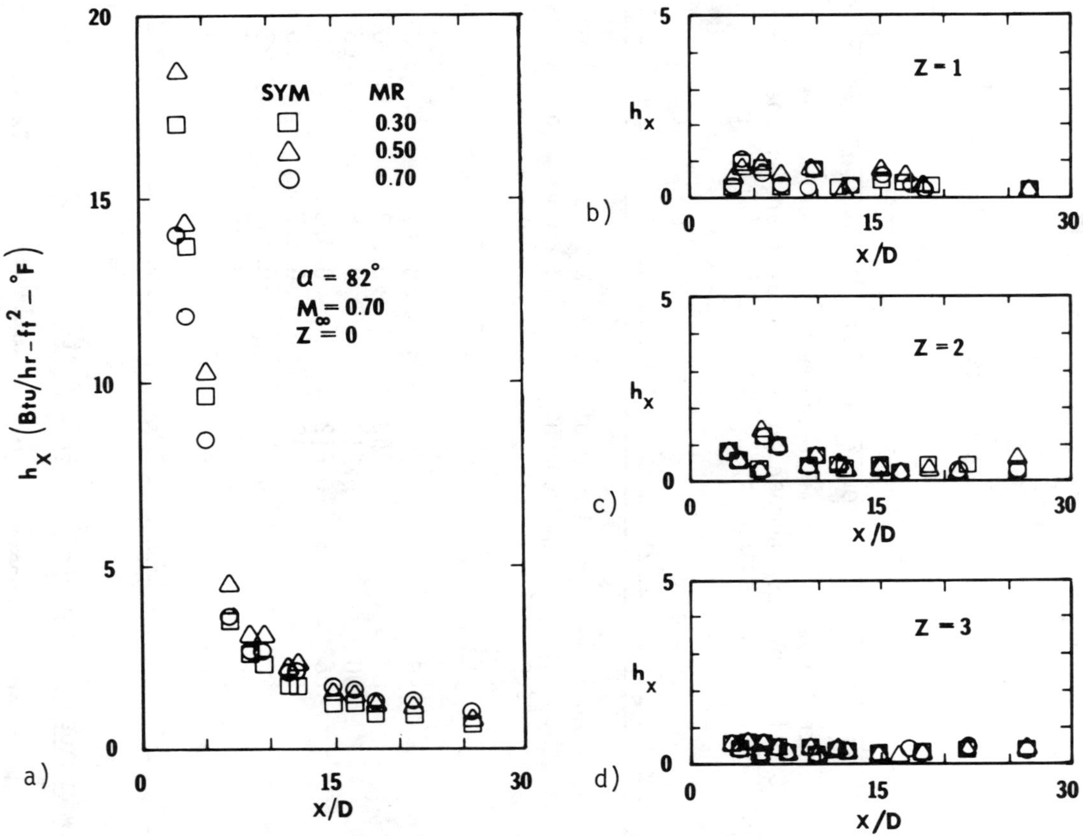

Fig. 11 Effect of MR on local heat-transfer coefficients (all D).

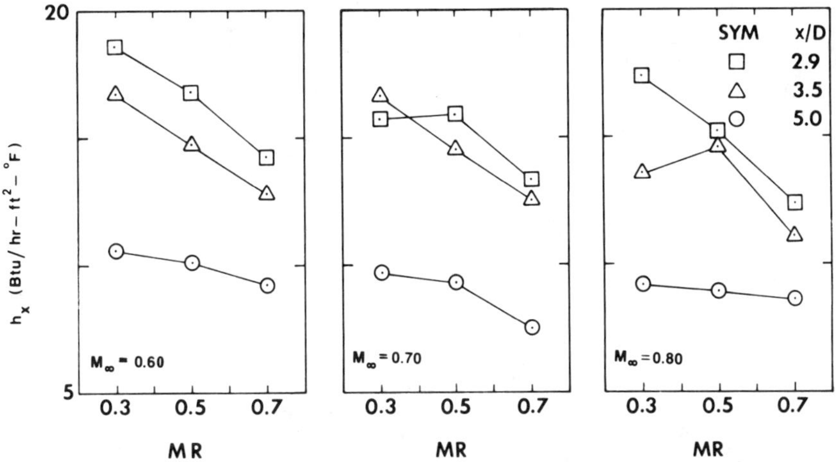

Fig. 12 Effect of MR on local heat-transfer coefficients in the near region of the jet at z = 0 (all D).

Fig. 10 where the centerline heat-transfer coefficients are also plotted vs Mach, but only at $2.9 < x/D < 5$. Although some exceptions in the data exist, the general trend is that for almost all jet flow angles and mass flux ratios the local heat transfer coefficient decreases with increasing Mach number.

As shown in Figs. 11a through 11d, mass flux ratio has no significant affect on heat transfer for spanwise stations $z > 1$ in. However, mass flux ratios do affect the centerline heat-transfer. Figure 12 shows that for all freestream Mach numbers heat-transfer coefficients are lower at mass flux ratios of MR = 0.7 as compared to MR = 0.3. Although this decrease is most predominant at $2.9 < x/D < 5$ (Fig. 12), the same effect is found for the entire range $2.9 < x/D < 26$ (Figs. 11a through 11d). These findings seem reasonable. One would expect that for higher mass flux ratios, and consequently higher dynamic pressure ratios, the jet would penetrate further into the freestream flow before being deflected downstream. Consequently, the hot jet would mix with more of the cooler freestream air resulting in a continual decrease in heat transfer with increasing mass flux ratio. This effect was observed for the 90 deg jet at all Mach numbers ($2.9 < x/D < 5$).

A typical plot of heat transfer coefficient for various jet flow angles is shown in Figs. 13a through 13d. No effect of jet flow angle on heat transfer was observed for spanwise

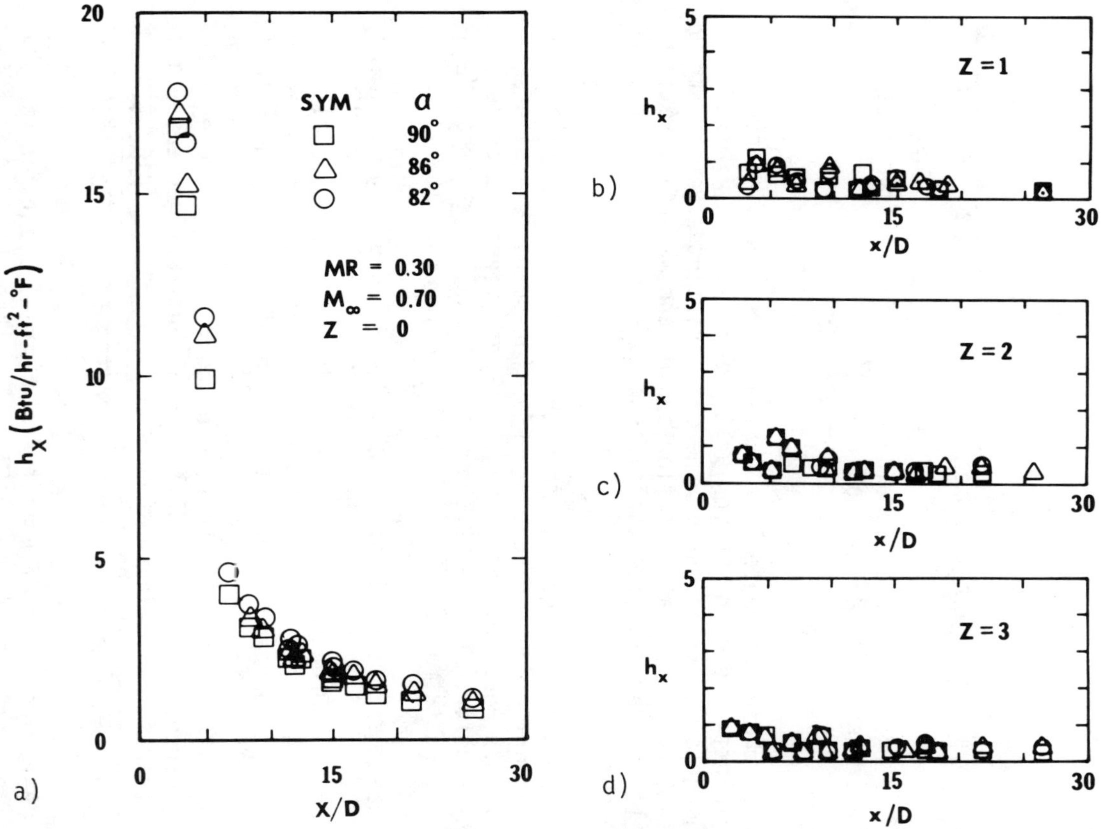

a) z = 0 in.

b) z = 1 in.

c) z = 2 in.

d) z = 3 in.

Fig. 13 Effect of jet flow angle on local heat-transfer coefficients (all D).

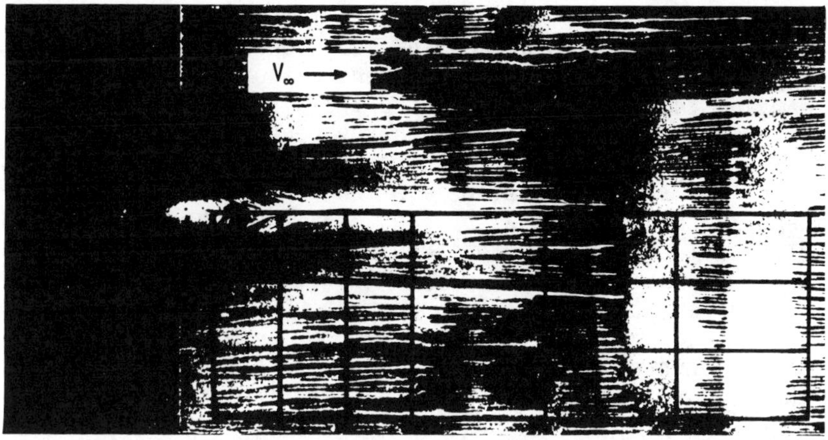

Fig. 14 Oil flow visualization indicating the maximum spreading of the jet. Grid intersection lines correspond to locations of the pressure and heat flux gage instrumentation (M_∞ = 0.80, MR = 0.70, D = 0.75 in.).

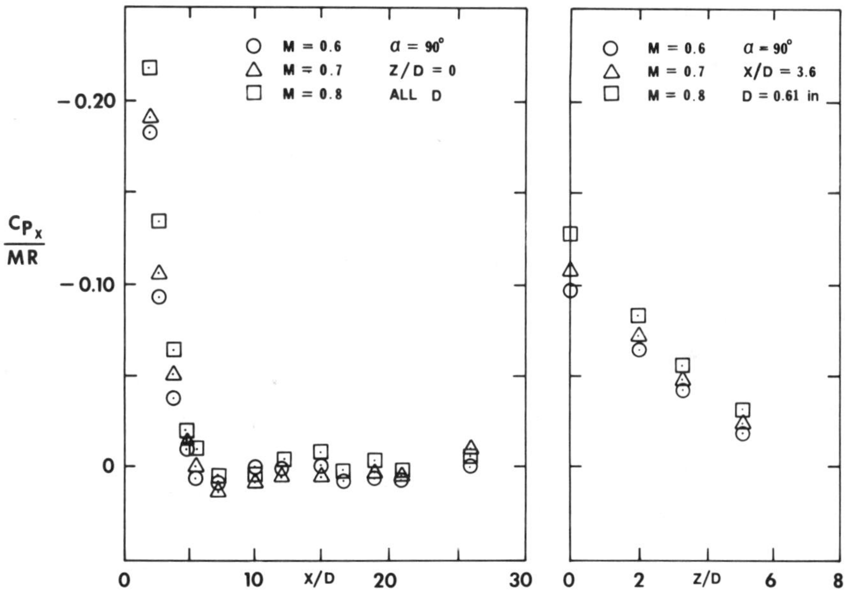

Fig. 15 Effect of Mach number on streamwise and lateral static pressure distributions.

stations of z > 1 in. However, centerline heat-transfer coefficients increase with decreasing jet flow angles. Since smaller jet angles mean less jet penetration into the freestream and thus less mixing of the two air streams, such an effect was to be expected. Along the centerline of the jet at $2.9 < x/D < 5$ the above results (increasing heat-transfer coefficients with decreasing jet flow angle) were not as consistent as those indicated by Figs. 13a through 13d. In fact, plots of all the data in this region indicate a very erratic pattern. Thus, the effect of jet flow angle on centerline heat transfer at $2.9 < x/D < 5$ was not predictable.

The downstream heating is also a function of jet diameter. At the same x location, increasing the jet diameter increases the heat transfer to the wall. In fact, the centerline data represented by Eq. (18) show the heat-transfer coefficients increase by a factor of 2 for the largest vs smallest jet tested. This result is primarily due to the increase in mass flow rate with increasing jet area. Thus, more energy is being added to the boundary layer at the point of injection. One can argue that the energy injected into the boundary layer is also more diffused for the larger jet diameters and should have an opposite effect. Apparently this effect is secondary. Indeed, oil flow studies (see Fig. 14) indicate very little

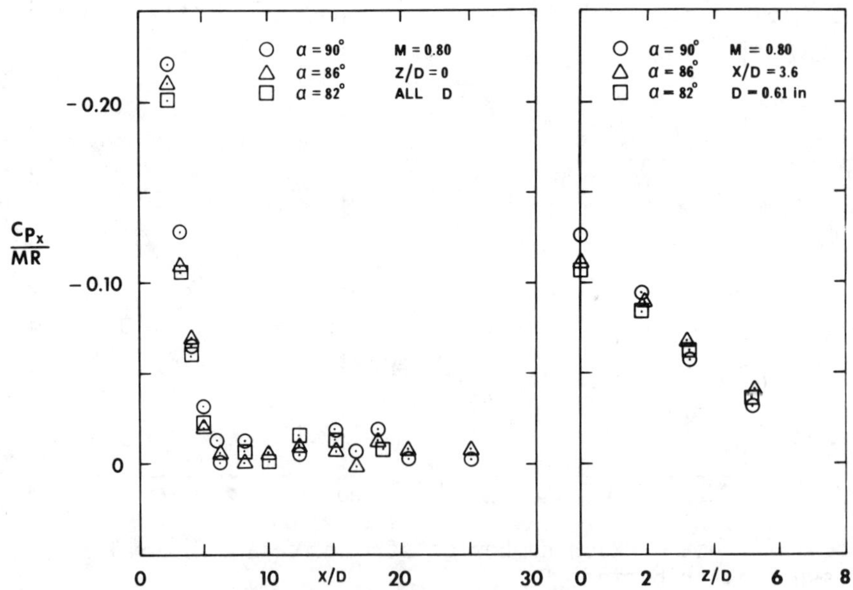

Fig. 16 Effect of jet flow angle on streamwise and lateral static pressure distributions.

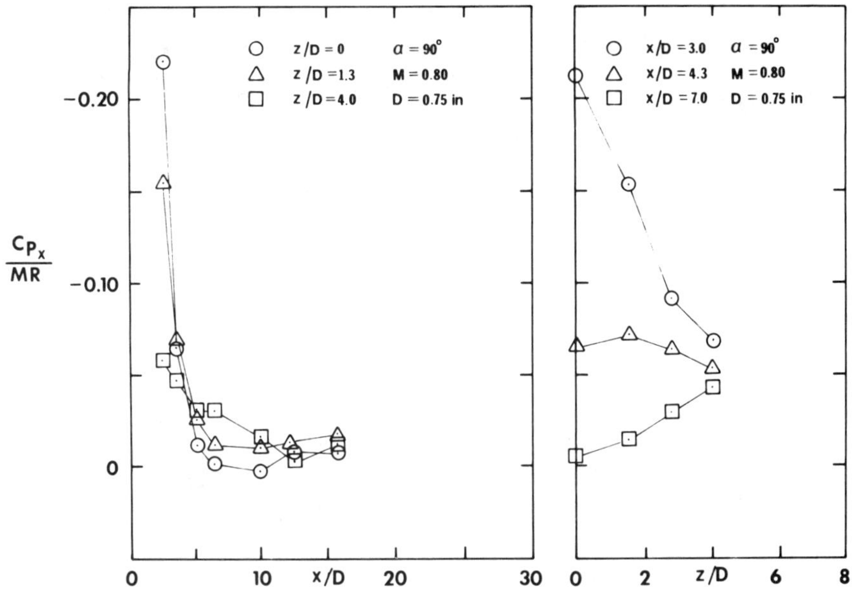

a) D = 0.75 in.

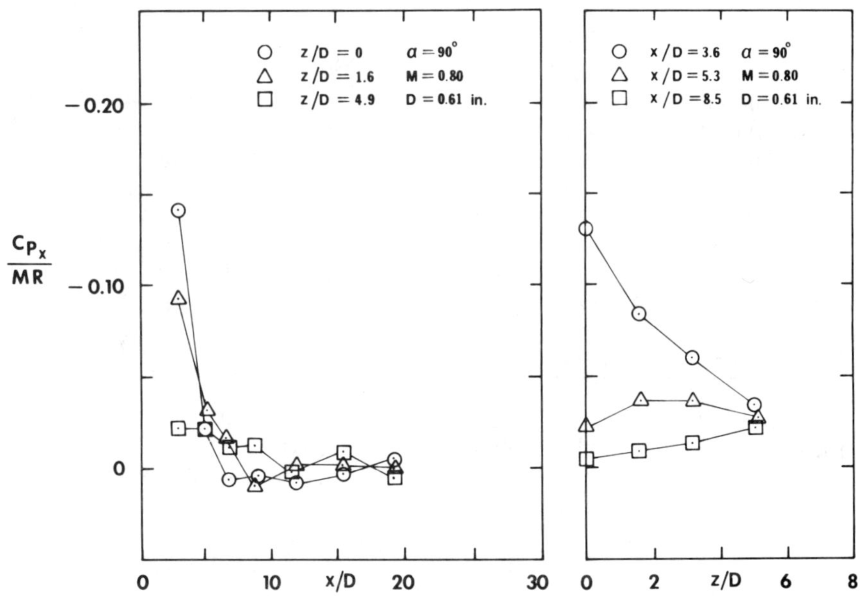

b) D = 0.61 in.

Fig. 17 Effect of jet diameter on streamwise and lateral static pressure distributions.

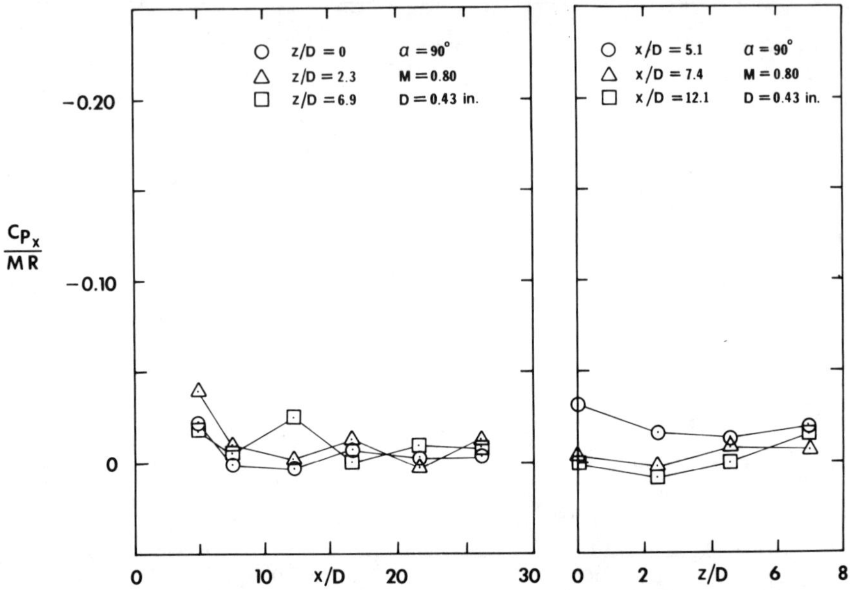

c) D = 0.43 in.

Fig. 17 (cont.) Effect of jet diameter on streamwise and lateral pressure distributions.

spreading of the jet as it is swept downstream. Although the oil flow pattern is not necessarily indicative of the flow-field above the plate, the minimum and maximum disturbance distances for the largest jet were found to be approximately z = 1 in. and z = 1.5 in. for 2.9 < x/D < 26. Perhaps a more significant observation from the oil flow pattern is the strong interaction between the jet flow and the mainstream flow which occurs directly behind the jet. It is in this region where the highest heat-transfer coefficients were measured.

Static Pressure Distributions

The result of increasing the Mach number or jet flow angle is to lower the pressure field to the side and in the wake region of the jet as shown in Figs. 15 and 16. The effect, however, is small. Nevertheless, these data were found to be consistent for all Mach numbers, jet flow angles, and jet sizes. Notice that the influence of these parameters on the pressure field is also restricted to the near region of the jet as was found for the heat-transfer data. As expected, the magnitude of the pressure field increases and the region of influence decreases with decreasing jet size as shown in Figs. 17a through 17c. It should be noted that these pressure data were observed to be independent of the mass flux ratio when normalized by this parameter.

Generally speaking, the hydrodynamic character of the jet cannot be completely defined from surface measurements alone. The complex flowfield behind the jet requires detailed measurements of the velocity fields off the plate before blockage, mass entrainment, and wake effects can be completely defined. In addition, for the case of the heated jet, temperature profiles behind the jet need to be measured. From the velocity and temperature profiles, the wall shear stress and heat transfer could then be calculated (and possibly correlated) as a function of the jet geometry and the jet flow characteristics. Nevertheless, in this experiment where only local heat transfer and static pressure data were measured, a certain similarity between the pressure distributions and the heat-transfer distributions exist. The effect of increasing the freestream Mach number was a decrease in the heat-transfer and pressure fields behind the jet (near region). The similarity between the two results appears to be associated with the same flow phenomena. Increasing the freestream Mach number accelerates the flow around the jet, thus creating the low-pressure region. As the flow accelerates around the jet, the turbulent mixing between the freestream flow and jet flow is increased. This greater mixing process between the cooler freestream air and the heated jet flow lowers the mean temperature of the two fluids in the boundary layer and lowers the heat transfer to the surface. Increasing the jet flow angle has the same effect. In this case, both the penetration of the jet further from the surface and the increase in the turbulent mixing of the two flows results in a decrease in the heat transfer to the surface.

Correlation of Heat-Transfer and Pressure Data

The region of maximum surface heating on the aircraft's fuselage is of particular importance. In this experiment, the region of maximum surface heating located at $(x,z) = (2.175,0)$ in. from the jet center $(2.9 < x/D < 5)$ was also the region of maximum negative pressure coefficient. In this separated region behind the jet a simple correlation between the maximum heat-transfer coefficient h_m and the pressure coefficient was obtained. It should be pointed out that these values are probably not the maximum values downstream of the jet but merely the maximum values measured in this experiment.

Figure 18 is a plot of the maximum heat-transfer coefficient as a function of pressure coefficient and jet area for

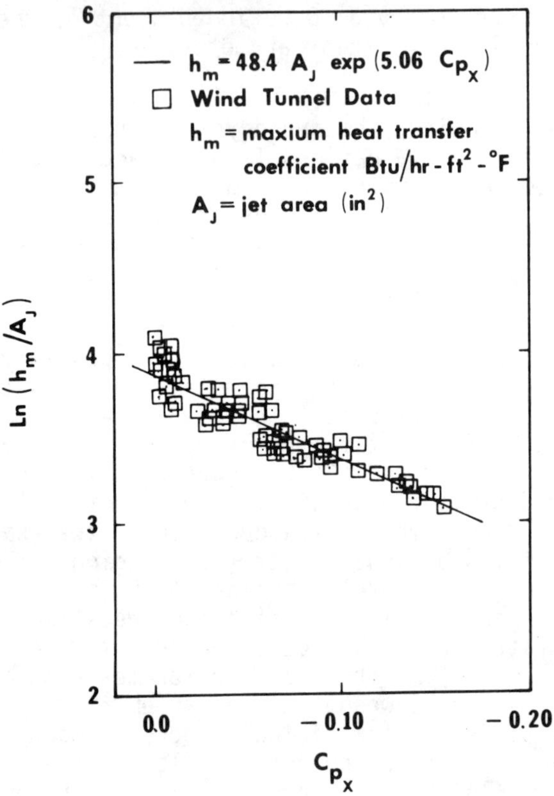

Fig. 18 Correlation of local heat-transfer and pressure coefficients at (x,z) = (2.175,0) in. from the jet center.

all mass flux ratios, jet flow angles, and freestream Mach numbers. The correlation for the maximum heat-transfer coefficient h_m was found to be

$$h_m = 48.4\ A_J\ \exp(5.06\ C_{p_x}) \qquad (19)$$

where

$$A_J = \text{jet area (in.}^2)$$

Experimental and Calculated Heat-Transfer Results

Figures 19, 20 and 21 compare the experimental heat-transfer data (at z = 0) with the various theories developed in the section on theoretical models. From these equations the adiabatic wall temperatures were computed. The heat-transfer

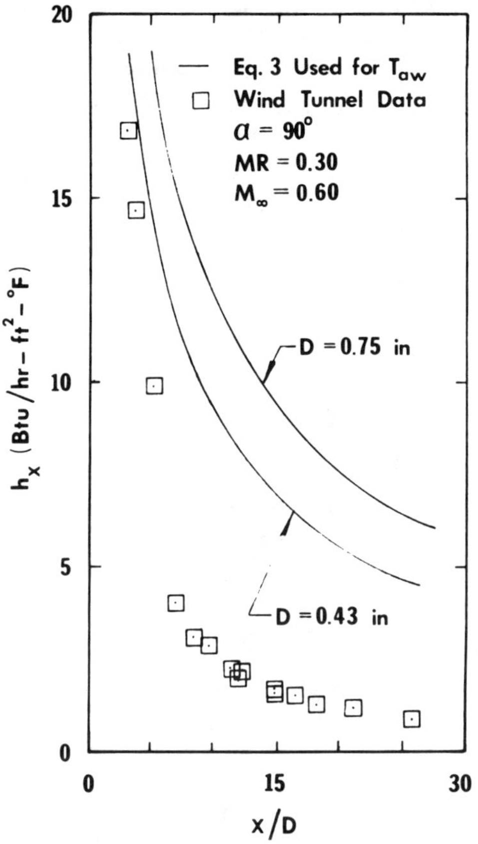

Fig. 19 Comparison of experimental data with local heat-transfer coefficients computed from adiabatic wall temperatures using two-dimensional film-cooling theory.

coefficients were then computed using the equation for turbulent flow over a flat plate

$$h_x = 0.0288\ (k/x)Pr^{1/2}Re_x^{4/5} \tag{20}$$

To compare the theory with the experimental data the heat flux to the wall was then computed from Eq. (2) and the heat-transfer coefficients were then recomputed using the jet total temperature in Eq. (16).

Figure 19 clearly illustrates that the prediction of local heat-transfer coefficients if one assumes a two-dimensional slot (Eq. 3) of width S equal to the jet diameter D

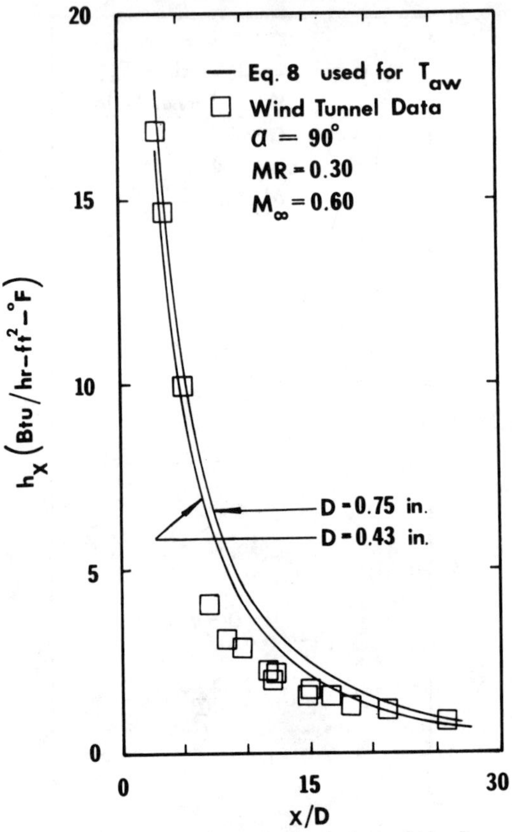

Fig. 20 Comparison of experimental data with local heat-transfer coefficients computed from adiabatic wall temperatures using a modified two-dimensional film-cooling theory.

can result in extremely high errors. In contrast, Fig. 20 illustrates that Eq. (8), which accounts for more entrained mass flow into the boundary layer owing to the three-dimensional effects of the jet, provides a much more accurate model for predicting the experimental data. However, the errors are still significant in the region $2.9 < x/D < 15$. It is interesting to note that the results using Eq. (8) appear to be insensitive to the jet diameter D. Such an effect can probably be attributed to the fact that the mass entrainment (Eq. 7) is not constant for a particular diameter D, but rather a function of the diameter and the distance away from the jet, that is, x/D. Figure 21 compares the experimental data with the three-dimensional film-cooling equation (Eq. 9). Surprisingly, the results are remarkably good considering the rather crude

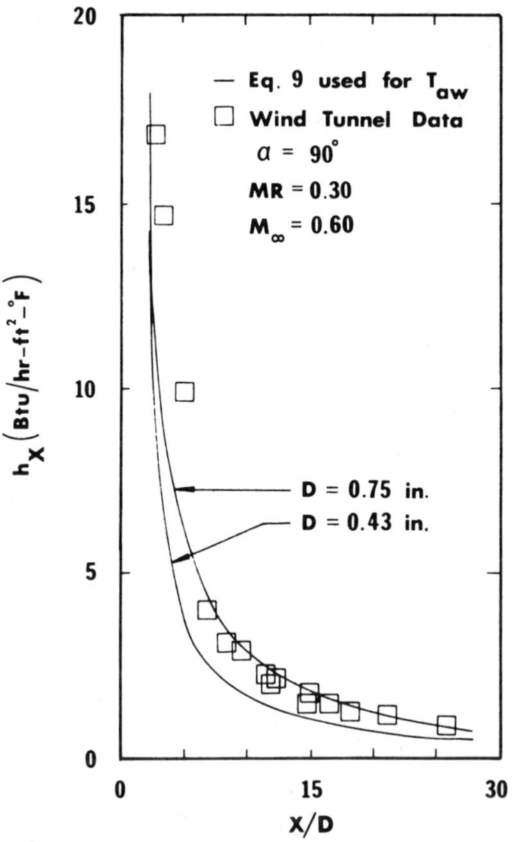

Fig. 21 Comparison of experimental data with local heat-transfer coefficients computed from adiabatic wall temperatures using three-dimensional film-cooling theory.

method used to estimate the thermal eddy diffusivity ε_T. Since the results using the three-dimensional equation (Eq. 9) vary significantly with the hole diameter D, the accuracy obtained is comparable to Eq. (8) where the three-dimensional mass entrainment effects were considered.

A few general remarks concerning the three equations appear to be in order. First, all three equations display similarly shaped profiles. Second, none of the equations considered can predict the heat transfer to the surface in the near region of the jet with accuracy ($2.9 < x/D < 5$). Indeed, any agreement between predictions and data in this separated region of the jet is clearly fortuitous. Third, all the theories presented require a considerable amount of empirical input which is evident in the attempt to estimate such flow dependent properties as mass entrainment and turbulent diffusivities. In fact, in an attempt to compare the experimental data with the three-dimensional theory at other spanwise stations ($z > 0$), the results were not consistent. For example, higher heat-transfer coefficients were obtained at $z = 3$ in. as compared to $z = 1$ in. Nevertheless, the approach taken here illustrates that, lacking experimental data, a rough estimate of the heat transfer to the surface could be made. More importantly, it illustrates that by modifying the existing two-dimensional film-cooling theories, there exists a strong possibility that a new theory could be developed that would accurately predict the heat transfer to the surface downstream of a three-dimensional jet. Development of such a theory would require measurements of both mass flow entrainment into the boundary layer and also jet trajectory data in a high-speed crossflow. It should be emphasized that such a theory would apply only to the region directly downstream of the jet ($z = 0$). It is in this region, however, where the heat-transfer to the surface is a maximum and therefore of greatest importance.

V. Summary

Both heat-transfer and pressure data were measured on a flat plate downstream of a heated, subsonic square air jet subjected to a high subsonic crossflow. Freestream Mach numbers at a constant unit Reynolds numbers of 1.9×10^{0}/ft ranged from $M = 0.40$ to $M = 0.80$. Mass flux ratios ranged from $MR = 0.30$ to $MR_o = 0.70$. The jet total temperature was held constant at 400°F. Configuration variables included the jet cross-sectional area and the jet flow angle. The analysis of the test results leads to the following conclusions:

1. The centerline local heat-transfer coefficients ($5 < x/D < 26$) for all freestream Mach numbers, mass flux ratios, and jet flow angles could be expressed by the power function

$$h_x = 42.2\ (x/D)^{-1.2}$$

2. Local heat-transfer coefficients were small ($h_x \leq 1$) at spanwise stations of $z > 1$ in. For all the test variables no significant effect on heat transfer was found at these spanwise stations.

3. In the separated region behind the jet ($2.9 < x/D < 5$), the heat-transfer data were correlated with the local

pressure coefficient by the equation

$$h_m = 48.4\ A_J \exp(5.06\ C_{p_x})$$

4. Increasing the freestream Mach number or the mass flux ratio decreases the local centerline heat-transfer coefficients. The most significant decreases occur at $2.9 < x/D < 5$.

5. Decreasing the jet flow angle from 90 deg to 82 deg increases the local centerline heat-transfer coefficients at $5 < x/D < 26$.

6. The local centerline heat-transfer coefficients downstream of a three-dimensional jet can be calculated by modifying the existing two-dimensional film-cooling theories to account for the mass entrained into the boundary layer because of the three-dimensional effects of the jet.

Acknowledgments

This work was conducted while a graduate aeronautical engineering student at the Air Force Institute of Technology and was sponsored by the Flight Dynamics Laboratory.

References

[1]"Analysis of a Jet in a Subsonic Crosswind," NASA SF-218, 1969.

[2]Thames, F. C. and Weston, R. P., "Properties of Aspect-Ratio 4.0 Rectangular Jets in a Subsonic Crossflow," Paper 78-1508, presented at AIAA Aircraft Systems & Technology Conference, Los Angeles, Calif., 21-23 Aug. 1978.

[3]Moussa, Z. M., Trischka, J. W., and Eskinazi, S., "The Near Field in the Mixing of a Round Jet With a Cross-Stream," Journal of Fluid Mechanics, Vol. 80, Part 1, 1977, pp. 49-80.

[4]Le Grives, E., "Mixing Process Induced by the Vorticity Associated with the Penetration of a Jet Into a Cross Flow," Journal of Engineering for Power, Vol. 100, July 1978, pp. 465-475.

[5]Goldstein, R. J., Advances in Heat Transfer, Academic Press, Inc., Vol. 7, 1971, pp. 321-379.

[6]Ramsey, J. W., Goldstein, R. J., "Interaction of a Heated Jet With a Deflecting Stream," NASA CR-72613, HTL TR No. 92, April 1970.

[7]Goldstein, R. J., Eckert, E. R. G., and Ramsey, J. W., "Film Cooling With Injection Through a Circular Hole," NASA CR-54604, HTL TR No. 82, 14 May 1968.

[8]Bergeles, G., Gosman, A. D., and Launder, B. E., "Near-Field Character of a Jet Discharged Through a Wall at 30° to a Mainstream," AIAA Journal, Vol. 15, No. 4, April 1977, pp. 499-504.

[9]Wilson, D. J. and Goldstein, R. J., "Effect of Film Cooling Injection on Downstream Heat Transfer Coefficients in High Speed Flow," Journal of Heat Transfer, ASME Transactions, Series C, Vol. 87, Nov. 1973, pp. 505-509.

[10]Hartnett, J. P., Birkebak, R. C. and Eckert, E. R. G., "Velocity Distributions, Temperature Distributions, Effectiveness and Heat Transfer in Cooling of a Surface with a Pressure Gradient," Paper 81, presented at the 1961 International Heat Transfer Conference, University of Colorado, Boulder, 28 Aug. - 1 Sept. 1961.

[11]Foster, R. C. and Haji-Sheikh, A., "An Experimental Investigation of Boundary Layer and Heat Transfer in the Region of Separated Flow Downstream of Normal Injection Slots," Journal of Heat Transfer, May 1975, pp. 260-266.

[12]Bouchez, J. P. and Goldstein, R. J., "Impingement Cooling From a Circular Jet in a Cross Flow," International Journal of Heat and Mass Transfer, Vol. 18, 1975, pp. 719-730.

[13]Kays, W. M., Convective Heat and Mass Transfer, Mc-Graw-Hill, New York, 1966.

[14]Mayle, R. E. and Camarata, F. J., "Multihole Cooling Film Effectiveness and Heat Transfer," Journal of Heat Transfer, ASME Transactions, Series C, Vol. 97, Nov. 1975, pp. 534-538.

[15]Schlichting, H., Boundary-Layer Theory, McGraw-Hill, New York, 1968.

THE EFFECT OF SURFACE ROUGHNESS CHARACTER ON TURBULENT RE-ENTRY HEATING

Michael L. Finson* and Andrew S. Clarke+

Physical Sciences Inc., Woburn, Mass.

Abstract

A Reynolds stress model for turbulent boundary layers is used to study surface roughness effects on skin friction and heat transfer. The issues of primary interest are the influence of roughness character (element shape and spacing) and the nature of roughness effects at high Mach numbers. Computations based on the model compare satisfactorily with measurements from experiments involving variations in roughness character, in low-speed and modestly supersonic conditions. The more limited data base at hypersonic Mach numbers is also examined with reasonable success, although no quantitative explanation is offered for the reduction of heat transfer with increasing roughness observed by Holden at $M_e = 9.4$. The present calculations indicate that the mean velocity is approximately uniform over much of the height range below the tops of the elements, $y \leq k$. With this constant "roughness velocity," it is simple to estimate the form drag on the elements. This roughness velocity has been investigated by systematically exercising the present model over ranges of potential parameters. The roughness velocity is found to be primarily a function of the projected element frontal area per unit surface area, thus providing a new and simple method for predicting roughness character effects. The model further suggests that increased boundary-layer temperatures should be generated by roughness at high edge Mach numbers, which would tend to reduce skin friction and heat transfer, perhaps below smooth wall levels.

Introduction

Surface roughness plays an important role in turbulent boundary-layer skin friction and heat transfer for many high-

---

Presented as Paper 80-1549 at the AIAA 15th Thermophysics Conference, Snowmass, Colo., July 14-16, 1980.

*Principal Research Scientist, Executive Vice President.

+Scientist/Engineer.

speed flight applications. Although the general nature of roughness effects for typical types of "sand grain" roughness has been known for many years, dating back to the classic study by Nikuradse,[1] modern composite materials introduce a different character of roughness. By roughness character, we mean the shape, spacing, and perhaps the distribution of roughness heights. With a woven composite material, for example, the exposed fibers would be approximately cylindrical in shape, in contrast to the more nearly hemispherical or pyramidal shape of conventional roughness elements.

Previous studies of roughness character have been somewhat limited experimentally, and the only theoretical investigations have been quite empirical. Schlichting[2] measured the drag due to various element shapes (spheres, spherical segments, cones) at several relative spacings on the side wall of a water channel. There are several reported experiments involving two-dimensional roughness elements (machined grooves normal to the flow direction). Some years ago, Bettermann[3] correlated the available data to obtain k_s/k, the ratio of the effective sand grain roughness height to the actual roughness height, as a function of roughness shape and spacing. Of course, the effective sand grain roughness can be used in Nikuradse's results to predict the skin friction increase. Dvorak,[4] Simpson,[5] and Dirling[6] have applied correlations similar to that of Bettermann to practical heating applications. Unfortunately, with the exception of Schichting's water channel measurements, nearly all of the data upon which these correlations are based were obtained on two-dimensional roughness patterns. One might expect a difference in the nature of the flow over two-dimensional vs three-dimensional roughness. For example, with two-dimensional roughness the flow would be more likely to separate, resulting in a cavity flow in the grooves between the elements. The three-dimensional case is of far greater pratical importance, and the two-dimensional type of roughness will not be considered further here.

A second important issue concerns the nature of roughness effects in supersonic flow conditions. Density variations could alter the extent of roughness-induced augmentation of friction or heating, and the character of the flow about the elements might change if supersonic conditions prevail locally. Only recently have measurements started to become available on supersonic and hypersonic rough wall boundary layers, offering an opportunity to examine the Mach number issue in a preliminary fashion.

In this study we employ a Reynolds stress turbulent boundary-layer model which specifically accounts for roughness effects. Roughness is represented by distributed sources and sinks in the various governing equations. The most important term is a sink term in the mean momentum equation representing form drag on the roughness elements. In previous studies,[7] the approach was developed and compared against subsonic rough wall boundary-layer measurements. The present objective is to apply the theory to variations in roughness character, as well as to supersonic conditions. A model extension to treat closely packed roughness elements will be described. Calculations will be compared with a variety of data to establish confidence in the model; then the model will be exercised systematically over a range of parameters to develop a scaling law for roughness character that is simple and quite different from existing approaches.

Rough Wall Turbulence Model

The turbulence model used here is one in which closure approximations are applied at second order. With the exception of the treatment of roughness, the formulation is somewhat standard at this time, and has been successfully applied to a variety of smooth wall boundary-layer and free shear flows. The model accounts for both mean and fluctuating velocities and temperatures. The dependent velocity variables are the mean velocity vector U_i, the Reynolds stress tensor $\overline{u_i' u_j'}$, and the isotropic dissipation rate Φ. The analogous thermal variables (temperature or, more precisely, enthalpy h) are the mean enthalpy $\bar{h}$, the mean square fluctuating enthalpy $\overline{h'^2}$, and the Reynolds heat flux vector $\overline{u_i'h'}$. Under the boundary-layer approximation, this set of variables reduces to U, V, $\overline{u'^2}$, $\overline{v'^2}$, $\overline{w'^2}$, $\overline{u'v'}$, Φ, $\bar{h}$, $\overline{h'^2}$, $\overline{u'h'}$, and $\overline{v'h'}$. The development of the governing equations and the required closure approximations are described in Ref. 7 and will not be repeated here. The actual equations are given in the Appendix.

The effect of roughness is described by distributed source or sink terms in the appropriate equations. As already noted, only distributed roughness is considered here, and we make the fundamental assumption that the flow around individual elements is attached to the elements. For two-dimensional roughness, the flow might be treated more appropriately as cavity flows between the elements. In the present model, roughness elements provide a distributed sink (due to drag) for mean momentum, and distributed sources for mean turbulent kinetic energy and dissipation. We idealize the rough surface as being made up of identical elements (although the extension to a size distribution is feasible). The bottom of the elements corresponds to y = 0. Let k be the element height, D(y) be the element diameter at height y (for $0 \leq y \leq k$), and ℓ be the average center-to-center element spacing. The functional form of the diameter D(y) is easily prescribed for simple shapes such as cones or hemispheres. As discussed in Ref. 7 and in the Appendix, the drag on the elements per unit volume is the appropriate sink term in the equation for the mean velocity:

$$R_u = - (1/2)\ \rho\ U^2\ C_D\ D(y)/\ell^2 \tag{1}$$

A drag coefficient value of $C_D = 0.6$ is roughly appropriate for elements such as cones or hemispheres. The source terms for kinetic energy and dissipation, which are less important numerically, are given in the Appendix. Except in the Stokes flow regime, heat transfer to an element should be small. Therefore, the only roughness term appearing in the thermal equations is a source for the mean static enthalpy. This term is simply constructed so that, in combination with Eq. (1), form drag does not alter total enthalpy:

$$R_h = + (1/2)\ \rho\ U^3\ C_D\ D(y)/\ell^2 \tag{2}$$

In the approach that we have just outlined, roughness elements are assumed to occupy no physical space. This assumption becomes progressively worse as the roughness density increases. Accordingly, the model has been extended to account for the blockage effect of the roughness elements. This is done in a simple manner. At a given height y, the fraction of the flow area in the x direction that is open to the flow is $1 - D(y)/\ell$; terms that act in the streamwise direction, such as the convective operator $\rho u\ \partial/\partial x$, are multiplied by this factor. Terms that act on a surface area whose normal is in the y direction, or that act on a unit volume, should be multiplied by $1 - \pi D^2/4\ell^2$. However, the distributed roughness source or sink terms are already based on the total volume, rather than available flow volume, and need no such factor. If the entire equation is divided by $B(y) = 1 - \pi D^2/4\ell^2$, a relatively simple change occurs. For example, the mean momentum equation becomes

$$f(y)\ \rho\ U \frac{\partial U}{\partial x} + \rho\ V \frac{\partial U}{\partial y} = - f(y) \frac{\partial p}{\partial x} + \frac{1}{B} \frac{\partial}{\partial y} B\mu \frac{\partial U}{\partial y}$$

$$- \frac{\partial}{\partial y} (\rho \overline{u'\ v'}) - \frac{1}{2}\ \rho\ U^2\ C_D \frac{D}{\ell^2} B^{-1} \tag{3}$$

where

$$f(y) = (1 - D/\ell)/(1 - \pi D^2/4\ell^2) \tag{4}$$

The function f(y) contains the main effect of blockage and may be handled by merely redefining the standard stream function which is used to eliminate the normal velocity.

$$\frac{\partial\psi}{\partial y} = f(y)\rho\ U, \quad \frac{\partial\psi}{\partial x} = -\ \rho V \tag{5}$$

Note that if the elements are packed so tightly that they are touching over some range of y, then $D = \ell$ and $f(y) = 0$. The stream function would be forced to be independent of y over that range [from Eq. (5)]; the velocity would remain zero from the bottom of the elements up to the height where $D < \ell$ and flow is unblocked. Of course, common sense would dictate redefining $y = 0$ as the lowest point where the flow is unblocked. However, the model does yield the limiting result that $U = 0$ if there is no space between the elements.

A major advantage of the present Reynolds stress model is that solutions are obtained for both velocity and thermal variables. Heat transfer is obtained directly, without invoking a Reynolds analogy. Finite-difference solutions are obtained using the obvious boundary conditions: fluctuating quantities are zero at the solid wall and in the freestream. It is important to note that the boundary conditions at the wall are not treated as empirical functions of roughness as has been done in other approaches.[8] For numerical solutions, the equations are first transformed to the stream function coordinate, guaranteeing mass conservation and eliminating the normal velocity V. The transverse coordinate is normalized by the edge value of the stream function, so that additional mesh points need not be carried in the free stream to allow for boundary-layer growth. For proper resolution of the region near the wall, a linear mesh in the logarithm of the stream function is used. The finite difference equations are solved with a block tridiagonal Newton-Raphson technique.

Comparison with Experiments on Roughness Character

The first measurements to be examined are those of Schlichting,[2] which were obtained in a 4 cm x 17 cm water channel. Various arrangements of roughness elements, shown in Fig. 1, were used on the wall. Velocity profiles were measured, and the skin friction or equivalent sand grain roughness k_s was derived from the logarithmic portion of the profile. The simple shapes and regular spacing of the roughness elements can be simulated quite well by our roughness model, with the exception of the short angles, which were not investigated. It is unfortunate that there have apparently been no other basic experiments in a well characterized low-speed

Roughness Elements	Dimension	D cm	d cm	k cm	k_s cm
Spheres	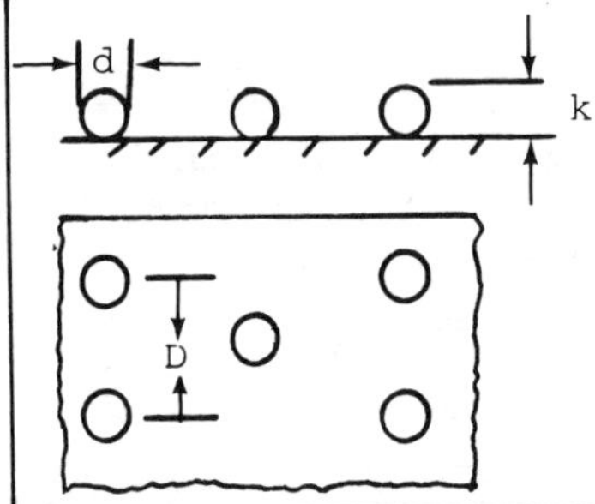	4.0	0.41	0.41	0.093
		2.0	0.41	0.41	0.344
		1.0	0.41	0.41	1.26
		0.6	0.41	0.41	1.56
		1.0	0.21	0.21	0.172
		0.5	0.21	0.21	0.759
Spherical segments	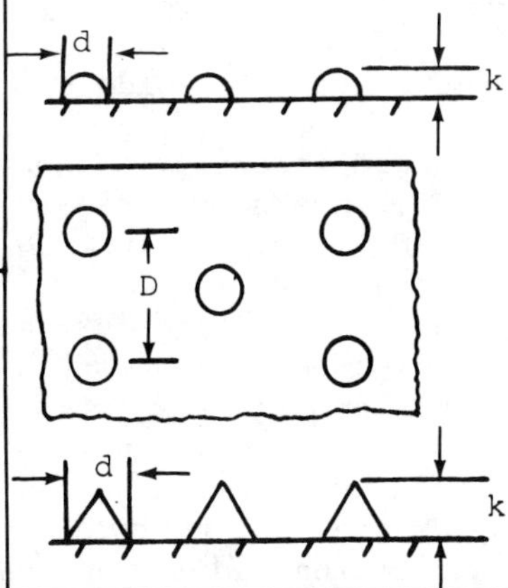	4.0	0.8	0.26	0.031
		3.0	0.8	0.26	0.049
		2.0	0.8	0.26	0.149
Cones		4.0	0.8	0.375	0.059
		3.0	0.8	0.375	0.164
		2.0	0.8	0.375	0.374
Short angles	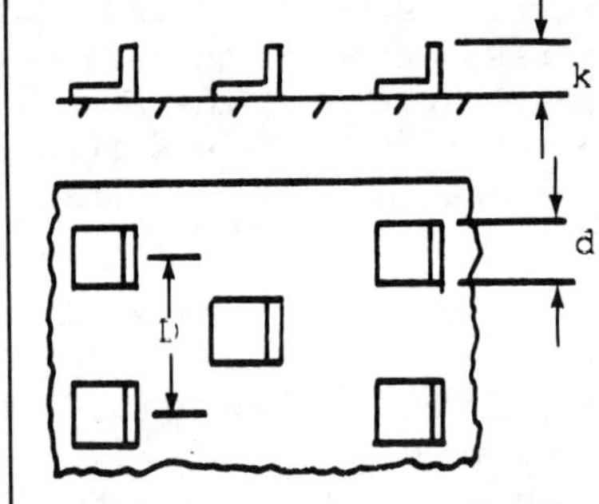	4.0	0.8	0.30	0.291
		3.0	0.8	0.30	0.618
		2.0	0.8	0.30	1.47

Fig. 1 Schlichting roughness experiments.[2]

flow where roughness shape and spacing were varied over a wide range. Several studies (e.g. Refs. 9, 10) have been reported where spacing was varied over a limited range for a fixed shape, and it is hoped to analyze such data with the present method at a later date.

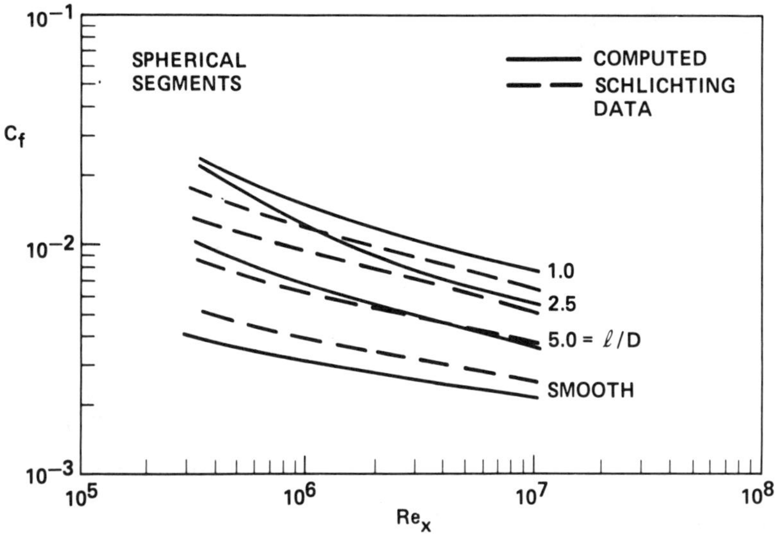

Fig. 2 Computed and measured skin friction coefficients vs distance Reynolds number, as a function of roughness spacing for spherical segments (≅ hemispheres).

Figure 2 compares the skin friction computed with our model against Schlichting's data for the spherical segments. As indicated in Fig. 1, the segments are nearly hemispherical (height = 0.26 cm, radius = 0.40 cm). Further, the dashed lines were not actually presented by Schlichting.[2] He determined the equivalent sand grain roughness, based on the observed increase in C_f over the smooth wall value at selected downstream stations. We took the liberty of using the k_s values to obtain the augmentation of C_f as a function of distance. In so doing, the appropriate upstream initial conditions are ignored. The computations were started with a fully turbulent smooth wall boundary layer at $x/\ell = 0.03$. Given the arbitrary treatment of initial conditions, the mild disagreement at upstream locations is not significant. Otherwise, the model reproduces fairly well the observed increase in skin friction with increasing roughness density.

A detailed comparison of the mean velocity profiles computed from the present model with Schlichting's data for spherical segments is shown in Fig. 3, in semilogarithmic coordinates ($U_\tau = \sqrt{\tau_w/\rho_w}$). Agreement is again seen to be quite good, with a 10%-15% error for the most dense packing. The data and curves of Fig. 3 agree well with the velocity pro-

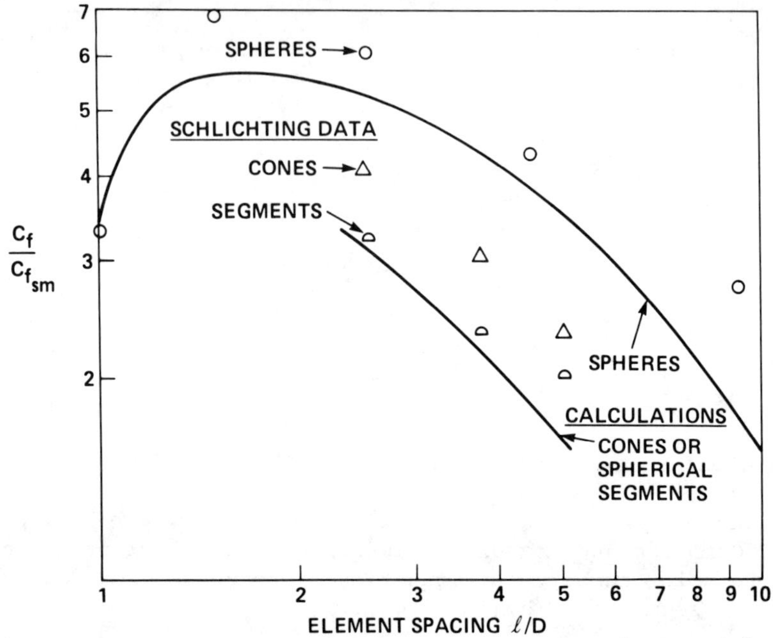

Fig. 3 Computed mean velocity profiles compared to Schlichting[2] measurements for spherical segments at three relative spacings.

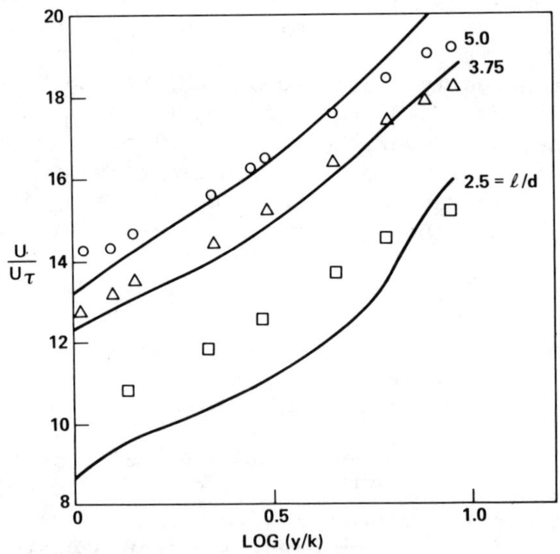

Fig. 4 Comparison of calculated skin friction augmentation vs Schlichting experimental results,[2] as a function of roughness spacing for three types of elements.

file of classic rough wall pipe flow when replotted against y/k_s.

Figure 4 shows the computed skin friction coefficients and Schlichting's measurements for three different element shapes as a function of spacing. The calculations were evaluated at $Re_x \cong 10^7$, but the increase in skin friction is essentially independent of distance. The decrease in C_f for spheres as $\ell/D \to 1$ is easy to explain qualitatively. At large ℓ/D the flow "sees" the entire sphere. With substantial element spacing, the drag increases with decreasing ℓ/D because there are more elements per unit area. But as $\ell/D \to 1$ the flow becomes negligible below the centers of the spheres and the drag is due only to the upper half of the elements. The cones and segments were not investigated at such close spacings. The calculations for the cones and segments are almost identical, although the cones are slightly taller, 0.375 cm vs 0.26 cm. However, the data for the cones fall above that for the segments. This implies that the effective drag coefficient for the flow about a conical roughness element is somewhat larger than that for flow about spherical elements. The same value was used in all of our calculations. If one were to allow such a higher value of C_D for cones, the influence of roughness element shape would be well understood.

The second set of interesting tests was carried out by Acurex Corporation in AEDC Tunnel F,[11] using 45 deg conical models with a variety of surface roughnesses. This facility was an arc-driven hot shot tunnel, in which the test section pressure decreased during the run (total time ≅ 200 ms). The most useful tests were performed on sharp 45 deg cones at $M_\infty \cong 7$ at a freestream Reynolds number of 45 x 10^6/ft. The resulting boundary-layer edge Mach number was 1.7. N_2 was the

Table 1
Roughness characteristics for Acurex AEDC tests at Mach 7

Designation	Mean k, mil	Mean spacing ℓ, mil	$\rho_e u_e$, kg/m^{2}s
Smooth	-	-	2038
Grit blasted	1.63	7.70	2235
Bonded grit	2.00	4.00	2292
4 mil wide	3.00	23.0	2157
4 mil close	2.50	13.0	2357
10 mil wide	10.32	56.0	1998
10 mil close	9.50	31.0	2315

test gas. The first 0.75" of wetted length was roughened to 4-5 mil to ensure rapid transition.

Seven surface finishes were used on the remainder of the cone: essentially smooth, grit blasted to almost 2 mil, 2 mil bonded grit, and four chemically etched roughness patterns (wide and close spacing at nominal heights of 4 and 10 mil). The etching process resulted in roughness elements that are best approximated as truncated cones (top radius ≅ 1/4 base radius) whereas the grit roughness elements are simulated with hemispheres. The roughness characteristics of these chemically milled surfaces varied by as much as ± 30% in mean roughness height over the surface of the cones. Table 1 lists the average element height and spacing for the various surfaces. Note that the "4 mil" etched roughness is actually considerably less rough. Also, the "4 mil" roughness have larger relative roughness spacing than the "10 mil" surfaces. In fact, the "4 mil close" and "10 mil wide" roughness have essentially the same ℓ/k.

The primary measurements for this test series are the heat-transfer rates, determined by thin wall calorimetry with thermocouples on the back wall. Some coaxial heat-transfer gages, as well as skin friction gages, were also used, although there may be uncertainties regarding how faithfully the roughness is reproduced on the surface of the gages. The skin friction measurements are limited and show such scatter that few conclusions are possible (see Ref. 12 for more discussion).

Figure 5 compares the Stanton numbers, $q/\rho_e u_e C_p(T_r - T_w)$, calculated by the present model with the Acurex data. [The data points have actually been derived by dividing the reported heat-transfer coefficients by our computed values of $\rho_e u_e$ and the specific heat of N_2. Table 1 includes the values used for $\rho_e u_e$ for the various cases. Note that $\rho_e u_e$ varies by as much as 18% from one case to another, and examination of the heat-transfer rate (q) rather than the Stanton number could possibly lead to inaccurate conclusions on the effect of roughness.] The agreement is seen to be good, with the theory generally well within the scatter of the data. Several trends are evident from either data or calculations. The bonded grit and grit-based surfaces cause a similar heating augmentation, although the bonded grit is slightly taller and considerably more densely packed; a blockage effect must be counteracting the more obvious effect of element spacing. However, for the chemically etched surfaces, spacing appears to be more important than height. The 10 mil wide spacing yields

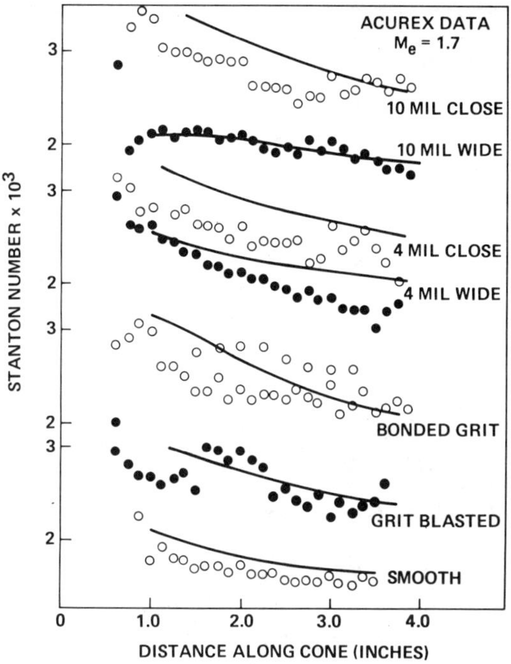

Fig. 5 Heat-transfer coefficient comparisons for various surface roughnesses — 45 deg cones at $M_\infty = 7$.

a greater heating rate than the 4 mil (actually 3 mil) wide case only at larger distances, and the 4 mil and 10 mil close spacing results are also quite similar. As already noted, the relative roughness spacing is not constant between the 4 mil and 10 mil heights, and a more detailed discussion of the dependence of height, shape, and spacing is presented below.

Another interesting experiment on roughness effects has been conducted in the hypersonic shock tunnel at Calspan.[13] Only one roughness was studied — a bonded grit similar to that employed by Acurex in the series discussed above. The mean roughness height was 3.8 mil, with a spacing of 10-15 mil (we specified 12.5 mil for the spacing). As with the Acurex series, the models were 45 deg cones. The freestream Mach number was 11-13, although this larger value has little effect on the edge Mach number, due to the large cone angle. T_w/T_e is considerably less in the Calspan conditions. Heat transfer was measured with thin film gages covered with the surface roughness. Figure 6 shows our comparison with Holden's results, for smooth and roughened surfaces, on a sharp 45 deg cone at the highest Reynolds number tested. The smooth wall boundary layer was naturally turbulent near the nose. It is

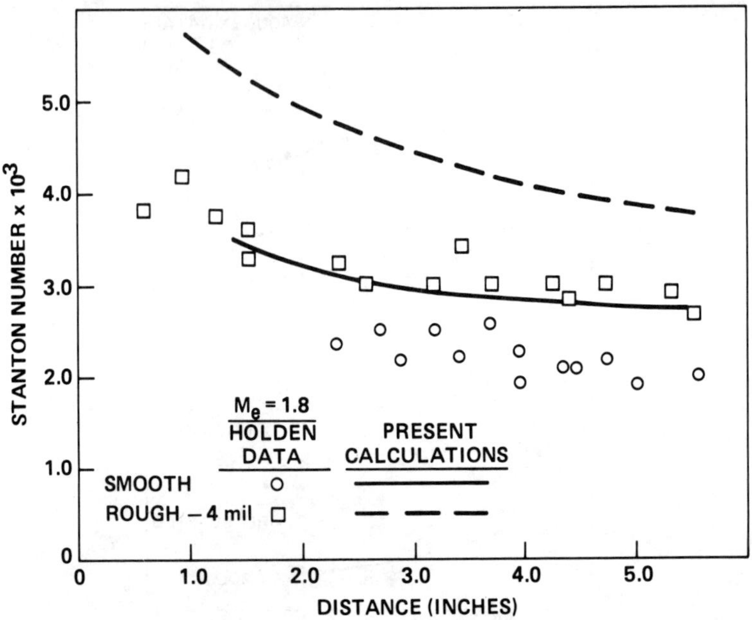

Fig. 6 Comparison of present theory with Holden's data for smooth and rough (4 mil bonded grit).

evident that the model yields heating rates about 15% higher than measured, even for the smooth wall. The measured smooth wall Stanton number is consistent with values predicted from accepted engineering methods such as that of Van Driest[14] ($St \cong 2.56 \times 10^{-3}$). The reason for this discrepancy is not clear, although it may be associated with numerical inaccuracies that arise with large density differences across the boundary layer. In any case, the present model predicts a degree of roughness heating augmentation comparable to that observed.

The first high Mach number tests with distributed surface roughness were conducted by Keel[15] on 5 deg sharp cones in tunnel No. 2 at the Naval Surface Weapons Center (NSWC) at $M_\infty = 5$ or $M_e = 4.77$. Sand grains were uniformly applied to the model with epoxy, yielding roughness heights of either 23 mil or 43 mil. Element spacing measurements were not reported, and we assumed a value ($\ell/D = 2.5$) which is typical of the bonded grit surfaces constructed by Acurex[11] and Calspan.[13] Skin friction and heat transfer were measured with floating element balances and slug calorimeter gages, respectively.

One noteworthy aspect of Keel's experiment is that the measurements were obtained at a fixed station on the cone, x = 2 ft. The Reynolds number was varied by decreasing the

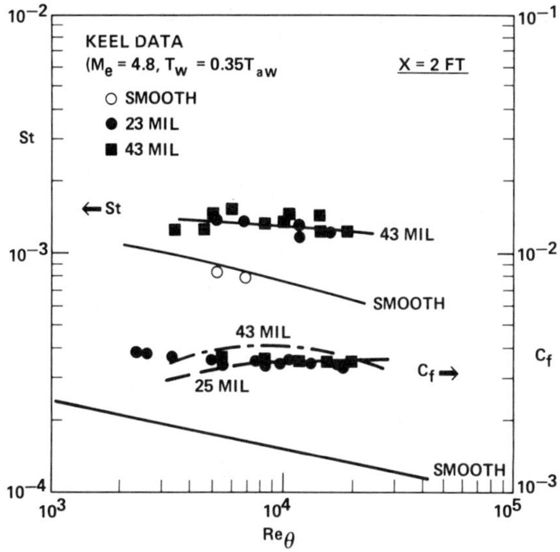

Fig. 7 Comparison of present theory with Keel's heat-transfer and skin friction data – M_e = 4.8, k = 23 and 43 mil. The computed Stanton number for k - 25 mil is essentially identical to that shown for k - 43 mil.

tunnel pressure, and the results plotted as C_f or St vs. Re_θ (see Fig. 7). The manner by which the data were collected must be recognized to properly interpret these results. With a smooth wall, C_f is a function only of Re_θ (assuming fixed values of M_e and T_w/T_e) and one may equally well traverse the C_f vs Re_θ curve by varying pressure (i.e., density) or distance (i.e., θ). However, a second independent parameter enters with rough walls. For simplicity, let us use

$$k^+ = \rho_w u_\tau k/\mu_w$$

as the fundamental parameter to describe the roughness augmentation effect. Furthermore, u_τ/u_e is a very weak function of Re_θ and may be considered constant for the purposes of this argument. Thus, if we consider increasing distance along a flat plate or cone, Re_θ increases because $\theta \sim x^{0.8}$, while k^+ remains constant. Conversely, when Keel raises the pressure at a fixed station, both Re_θ and k^+ increase together, linearly with the density. This offers an explanation for the almost complete absence of slope for the data in Fig. 7; the increasing roughness augmentation (with increasing k^+) tends to cancel the natural tendency of C_f or St to decrease with increasing Re_θ. Agreement with the data is very good - k^+ is apparently sufficiently large so that the roughness augmentation is nearly saturated and there is little difference between 23 and 43 mil. It should be noted that the computed Mach number at the tops of the elements is barely supersonic in this case.

For hypersonic edge conditions, Fig. 8 shows a comparison with the heat-transfer data obtained by Hill[16] at NSWC on a 7 deg sharp cone at $M_e = 8$ and $Re_\infty = 7 \times 10^6$/ft. The roughness heights are based on the nominal sizes of the grits that were applied to the surface. We simulated the roughness with hemispheres of height equal to the nominal values, with a spacing of 2.9k, as indicated by surface profilometer measurements. It is evident from Fig. 8 that the computer model yields approximately the observed heat-transfer augmentation as a function of roughness height, although the data show less dependence on distance. The reasons for this discrepancy are not clear.

An even more intriguing set of hypersonic data have been obtained by Holden[17] in the Calspan shock tunnel. In this case the edge Mach number is 9.4, the cone half-angle is 6 deg, and the freestream Reynolds number for the cases of interest is 1.1×10^7/ft. Again, we simulated the roughness with hemispheres with height equal to the nominal grit sizes. In this case, however, the elements are rather tightly packed, $\ell/k \cong 2.25$. Figure 9 shows that the theory yields a trend of increasing heat transfer with increasing roughness, while the data show significant decreases. There are several possible explanations for the observed reductions in heat transfer. One is associated with the tight packing of the elements. However, our calculations show that there might be a modest reduction in the heating augmentation as $\ell/D \to 1$ (cf. Fig. 4), but not below smooth wall values. Furthermore, Holden's data on 45 deg cones at $M_e = 1.8$, with a similar 4 mil bonded grit roughness, were presented above in Fig. 6 and showed no reduction below smooth wall values. A second possibility has to do with upstream behavior; the rough wall boundary layer may simply be much thicker at a given station. Again, the present calculations do not indicate any such reversal effect. A third possibility has to do with the high Mach number, which could result in locally supersonic flow, shock waves around the elements, and perhaps an alteration of the drag or heating laws. In fact, Holden's[17] shadowgraph observations indicate a significant wave structure emanating from the rough surface. But, our computed local Mach numbers are only barely supersonic (1.14) at the tops of the elements, no more supersonic than in the Keel case discussed above. A related Mach number effect, discussed in more detail below, is that roughness causes increased boundary-layer temperatures at high edge Mach numbers;

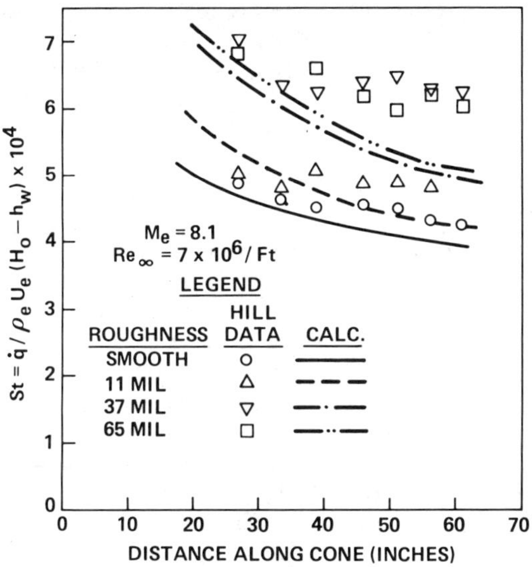

Fig. 8 Comparison of present theory with Hill's data for smooth and rough surfaces at an edge Mach number of 8.

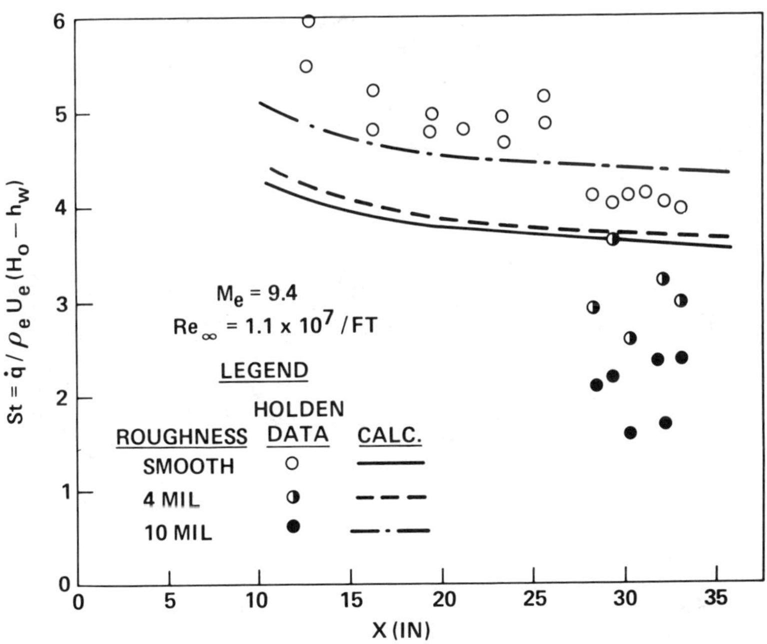

Fig. 9 Comparison of present theory with Holden's data for smooth and rough surfaces at an edge Mach number of 9.4.

perhaps more careful calculations would show a reduction below smooth wall values. In any case, the matter is not resolved, particularly in light of the similar conditions of the experiments of Hill and Holden. Clearly, more measurements and calculations are needed.

Roughness Scaling Behavior

A major advantage of the present type of numerical study is that the solutions can be scrutinized to determine the nature of roughness effects. Of course, the data comparisons shown in the previous section left several issues unresolved, but, nevertheless, a detailed examination of the numerical results is justified. The results are quite interesting.

The most striking aspect of our solutions is that the mean velocity is computed to be quite uniform over much of the range $y < k$. A typical example is shown in Fig. 10. Of course, very near the wall (i.e., at the bottom of the elements) the velocity must be zero, and near the tops of the elements the velocity tends to increase. This constant velocity, which was unexpected, is evident in almost all of the cases considered, the sole exceptions being cases with very small roughness (say $k^+ \lesssim 10$) or very large roughness ($k/\theta >$ 1). Within the range of constant velocity, production of turbulent energy is negligible; the turbulence simply diffuses toward the wall and dissipates. In the mean velocity equation

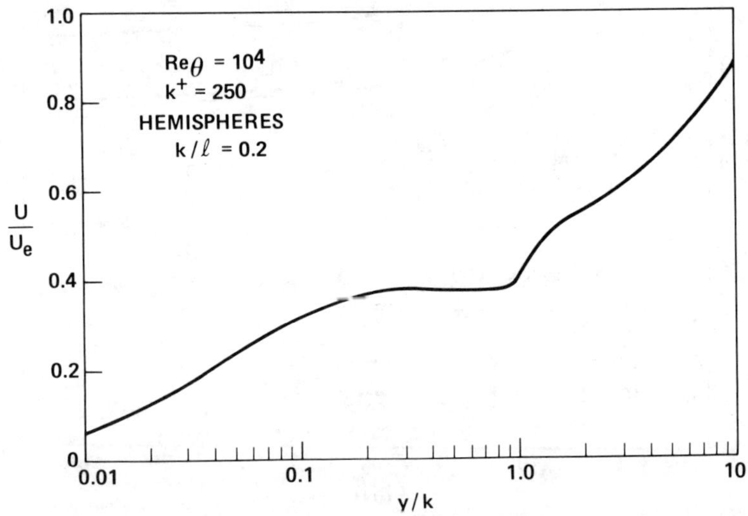

Fig. 10 Calculated mean velocity profile for a typical rough wall case.

[Eq. (3)] only the final two terms, representing turbulent shear and form drag on the elements, are important. The total drag on the elements, which should be close to the actual rough wall skin friction, is easily related to the element shape and spacing

$$C_f \sim f(y) \int_0^k \frac{\rho U^2}{\rho_e U_e^2} C_D \frac{D(y)\,dy}{\ell^2} \cong f(y) \frac{\rho_R U_R^2}{\rho_e U_e^2} C_D \int_0^k \frac{D(y)\,dy}{\ell^2} \tag{6}$$

Here we use $(\ )_R$ to denote conditions in the region of constant velocity.

The integral in Eq. (6) is the frontal area of the elements per unit superficial (smooth) surface area. This quantity is easily related to the shape and density of roughness elements. However, to compute the skin friction the roughness velocity U_R is also needed (determination of ρ_R, as well as the relation between heat transfer and skin friction, will be addressed below). We have been unable to predict U_R by any simple theoretical arguments. Accordingly, the computer model was systematically run over a range of parameters, including element shape, spacing, height, T_W/T_e, and M_e. To be specific, U_R was defined to be the mean velocity at $y = k/2$.

Figure 11 shows the computed roughness velocity for hemispheres in incompressible conditions. The roughness spacing

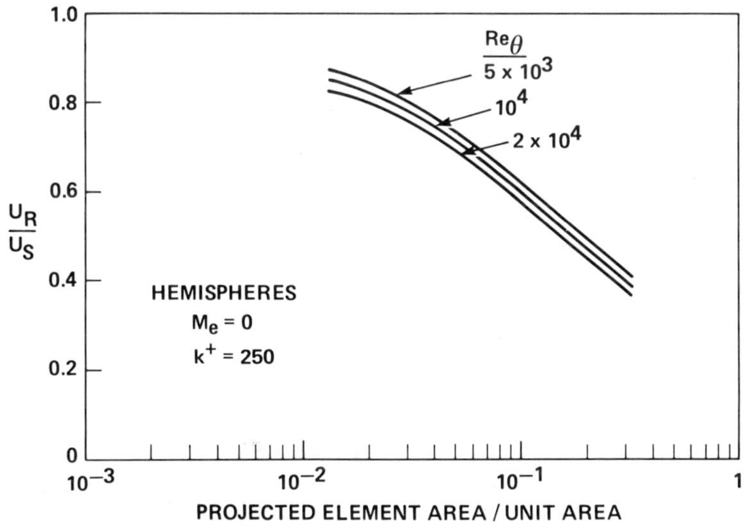

Fig. 11 Ratio of computed roughness velocity ($y = k/2$) to smooth wall velocity at the same height, as a function of projected roughness element area and Re_θ.

varies along the abscissa; U_R is normalized by the smooth wall velocity at y = k/2 (easily computed from the classic law-of-the-wall). Only a slight dependence on Re_θ is apparent. It should be emphasized that only roughness spacing and downstream distance (Re_θ) were varied to obtain these curves. However, the really significant finding is that the same curves apply to variations in element shape, height, T_W/T_e, and M_e. In addition to hemispheres, we also considered 30 deg conical elements (base diameter = k), 45 deg cones (base diameter = 2k), truncated 30 deg cones (base diameter = k, top diameter = k/2), and cylinders (base diameter = k). For each of these shapes the results were indistinguishable from those of Fig. 11. There is a dependence on roughness height (k^+), as illustrated in Fig. 12, but the trend is slight unless k^+ is less than about 20 (as $k^+ \rightarrow 0$, $U_R/U_S \rightarrow 1$, but the smooth-wall shear would dominate anyway). Varying the wall temperature over the range T_W/T_e = 0.2 - 1.0 has little effect, as does increasing the edge Mach number up to at least 8.

The substantial effect of roughness on the velocity profile is not observed with the enthalpy profile. This undoubtedly results from the fact that there is no heat-transfer mechanism analogous to form drag. In Fig. 13 we compare normalized total enthalpy values at y = k/2 with the roughness velocity. At M_e = 0, the rough and smooth wall enthalpies at y = k/2 are identical within the accuracy of our computations,

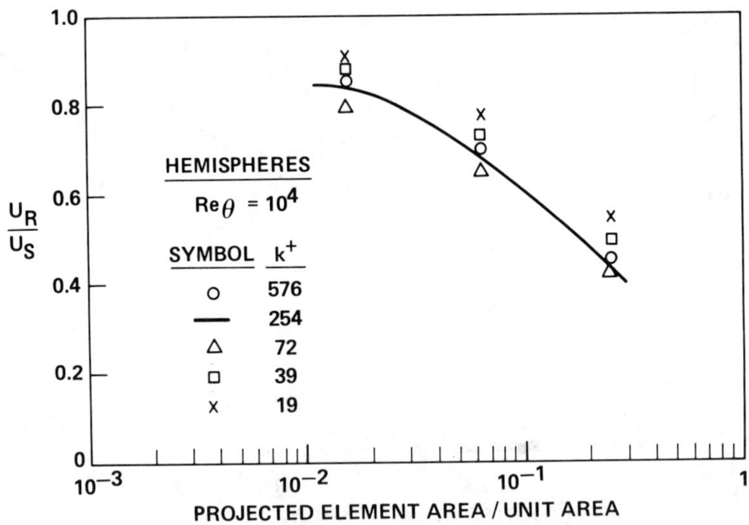

Fig. 12 Ratio of computed roughness velocity (y = k/2) to smooth wall velocity at the same height, as a function of projected roughness element area and k^+.

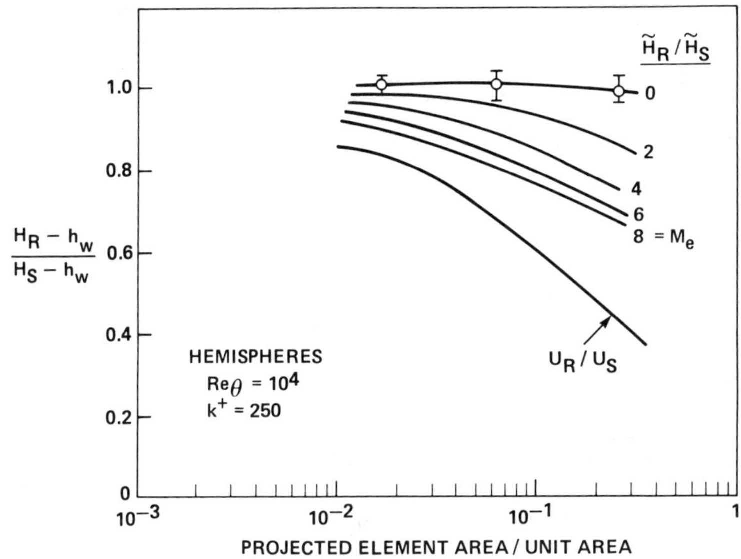

Fig. 13 Ratio of computed total enthalpy (y = k/2) to smooth wall enthalpy at the same height, as a function of projected element area and edge Mach number.

even for large temperature differences across the boundary layer ($T_W = T_e/5$). At supersonic velocities, the total enthalpy is obviously reduced by smaller margins than is the velocity. There appears to be a hypersonic limit, with $\tilde{H}_R/\tilde{H}_S \sim (U_R/U_S)^{1/2}$, although more computations would be required to confirm this.

An important consequence of Fig. 13 is that roughness will tend to increase the static enthalpy for $y < k$ at supersonic velocities. The resulting decrease in density will reduce the form drag on the elements, and the increased temperature will reduce the underlying smooth wall skin friction. Thus our calculations suggest a specific Mach number dependence for roughness friction or heating augmentation. The compensating effect of increased temperatures reduces roughness augmentation at high Mach numbers; if this is somewhat greater than we have estimated, it could offer an explanation for the decreases in heat transfer observed by Holden.

Finally, we might note two additional trends that are well established by our numerical results. First, the underlying smooth wall skin friction is, to a good approximation, unchanged by roughness (fixed value of Re_θ), except for the high Mach number effect just mentioned. Secondly, the increase in heat transfer is nicely described by the square root of the skin friction augmentation. The reason was given in our previous study[7]: roughness increases fluctuating velocities but not fluctuating temperatures, and the turbulent stress goes as $-\rho \overline{u'v'}$ while the turbulent heat flux goes as $-\rho u'T'$. Hence,

$$St/St_S = (C_f/C_{f_S})^{1/2} \tag{7}$$

Roughness Scaling Formulae

It is useful to summarize the relations needed to predict roughness effects on skin friction and heat transfer, particularly since the present results are in a very different form from existing methods. First, the skin friction and heat-transfer augmentations are given by

$$St/St_S = (C_f/C_{f_S})^{1/2} \tag{7}$$

$$C_f = C_{f_S} + \frac{\rho_R U_R^2}{\rho_e U_e^2} C_D \int_0^k \frac{D(y)\,dy}{\ell^2} f(k/2) \tag{8}$$

The smooth wall values may be obtained from a variety of well-known sources, such as Ref. 14. The integral in Eq. (8) requires the roughness element shape and spacing, and $f(k/2)$ follows from Eq. (4). We recommend $C_D = 0.6$. The ratio U_R/U_S may be obtained from Figs. 11 or 12, and the smooth wall velocity at $y = k/2$ from the standard law-of-the-wall:

$$U_S/U_{\tau_S} = 2.5 \ln (U_\tau y/\nu_w) + 5 \; , \; U_{\tau_S}/U_e = \sqrt{(\rho_e (C_{f_S}/\rho_w)/2)} \tag{9}$$

The roughness density appearing in Eq. (8) requires slightly more algebra. Assuming a perfect gas,

$$(\rho_R/\rho_S)^{-1} = h_R/h_S = (1 + \frac{\gamma-1}{2} M_S^2)\; H_R/H_S - (\gamma-1)/2 M_S^2 \; U_R^2/U_S^2 \tag{10}$$

$$M_S = M_e \; U_S/U_e \; (h_S/h_e)^{-1/2} \tag{11}$$

It is adequate to use a Reynolds analogy for the smooth wall total enthalpy, $(H-h_w)/(H_0-h_w) = U/U_e$, to estimate the smooth wall static enthalpy:

$$\frac{h_S}{h_e} = \frac{h_w}{h_2} + \left(1 + \frac{\gamma-1}{2} M_e^2 - \frac{h_w}{h_e}\right) \frac{U_S}{U_e} - \frac{\gamma-1}{2} M_e^2 \frac{U_S^2}{U_e^2} \qquad (12)$$

And, the total enthalpy ratio in Eq. (10) is directly related to the normalized total enthalpy difference

$$\frac{H_R}{H_S} = \frac{h_w/h_S + \left(1 + \frac{\gamma-1}{2} M_S^2 - h_w/h_S\right) \tilde{H}_R/\tilde{H}_S}{1 + \frac{\gamma-1}{2} M_S^2} \qquad (13)$$

Taking $\tilde{H}_R/\tilde{H}_S$ from Fig. 13, Eqs. (10-13) are sufficient to provide ρ_R/ρ_e.

Finally, at high Mach numbers, the temperature increase due to roughness can reduce the smooth wall friction coefficient or Stanton number. To estimate this effect, we suggest modifying a simple reference enthalpy method. The smooth wall coefficients are inversely proportional to the reference enthalpy, which in turn depends on the peak boundary-layer enthalpy

$$h_{ref}/h_e \cong 1 + 0.7\ (h_{peak}/h_e - 1) \qquad (14)$$

For a smooth wall, h_{peak} may be estimated by taking the maximum of Eq. (12) with respect to U_S:

$$\frac{h_{peak,S}}{h_e} = \frac{h_w}{h_e} + \frac{\left[1 + (\gamma-1)/2\ M_e^2 - (h_w/h_e)\right]^2}{4(\gamma-1)/2\ M_e^2} \qquad (15)$$

However, if the rough wall enthalpy resulting from Eq. (10) is greater than the value obtained from Eq. (15), then the smooth wall friction coefficient and Stanton number should be reduced by the factor

$$[1 + 0.7\ (h_{peak,S}/h_e - 1)]/[1 + 0.7\ (h_R/h_e - 1)] \qquad (16)$$

Discussion

The present rough wall Reynolds stress model yields results that compare acceptably with a variety of available measurements. Of the several discrepancies noted, the most significant is that with Holden's high Mach number tests. The matter may not be resolved until additional high Mach number data are obtained. The observed decrease in heating rates could be due to the extreme close packing of the roughness elements, although our model fails to confirm this. The model does suggest that the increased temperatures generated by roughness at high edge Mach numbers tend to reduce skin friction and heat transfer. However, these findings are only suggestive, and one cannot at this time rule out an experimental artifact or a significant change in the nature of flow about roughness elements in supersonic conditions.

The roughness scaling behavior derived from our systematic variation of the several relevant parameters suggests scaling laws that are a major departure from previous approaches. The influence of roughness element shape and spacing is almost totally defined by the projected frontal area per unit surface area. The velocity that gives rise to form drag is largely constant over the range $y \leq k$. When normalized by the smooth wall velocity at $y = k/2$, this roughness velocity is primarly a function of this projected element area/unit area.

A simple method is thus provided for predicting and analyzing roughness character effects. However, it must be admitted that the relations given here need additional evaluation and refinement before they can be used with great confidence. For example, the minor dependence of U_R/U_S on k^+ and Re_θ should be better defined by additional calculations, and much more hypersonic data ought to be studied to substantiate the Mach number dependence.

Appendix: Reynolds Stress Model Equations

The governing partial differential equations for the variour turbulent quantities are listed here. For more details, the reader is referred to our previous report.[7] In practice, it is convenient to replace $\overline{u'^2}$, $\overline{v'^2}$, $\overline{w'^2}$ by the kinetic energy $q^2 = (\overline{u'^2} + \overline{v'^2} + \overline{w'^2})/2$ and two measures of the degree of anisotropy $S_{11} = \overline{u'^2} - 2/3\, q^2$, $S_{22} = \overline{v'^2} - 2/3\, q^2$. In all of the following equations, the diffusion terms (second partial derivatives with respect to y) should be understood to be multiplied and divided by B(y) as in Eq. (3).

For steady flow, the governing equations include continuity:

$$\frac{\partial}{\partial x_i} (\rho U_i) = 0 \qquad \text{(A1)}$$

the mean momentum equation:

$$\rho U_k \frac{\partial U}{\partial x_k} = - \frac{\partial \bar{p}}{\partial x} - \frac{\partial}{\partial y} (\rho \overline{uv}) + \frac{\partial}{\partial y} (\mu \frac{\partial U}{\partial y}) + R_u \quad \text{(A2)}$$

and, for the five second-order quantities:

$$\rho U_k \frac{\partial q^2}{\partial x_k} = - \rho \overline{uv} \frac{\partial U}{\partial y} - \rho \Phi + 0.2 \frac{\partial}{\partial y} [\rho \frac{q^2 v^2}{\Phi} \frac{\partial}{\partial y} (q^2 + v^2)]$$

$$+ \frac{\partial}{\partial y} \mu \frac{\partial q^2}{\partial y} + (S_{22} - S_{11}) \rho \frac{\partial U}{\partial x} + R_q \quad \text{(A3)}$$

$$\rho U_k \frac{\partial S_{11}}{\partial x_k} = - \frac{14}{33} \rho \overline{uv} \frac{\partial U}{\partial y} - C_E \rho \frac{\Phi}{q^2} S_{11} + 0.2 \frac{\partial}{\partial y} [\rho \frac{q^2 v^2}{\Phi} \frac{\partial}{\partial y} (S_{11}$$

$$- \frac{2}{3} v^2)] + \frac{\partial}{\partial y} \mu \frac{\partial S_{11}}{\partial y} - \rho (\frac{8}{15} q^2 + \frac{2}{33} S_{11} + \frac{1}{33} S_{22}) \frac{\partial U}{\partial x} \quad \text{(A4)}$$

$$\rho U_k \frac{\partial S_{22}}{\partial x_k} = \frac{13}{33} \rho \overline{uv} \frac{\partial U}{\partial y} - C_E \rho \frac{\Phi}{q^2} S_{22} + 0.2 \frac{\partial}{\partial y} [\rho \frac{q^2 v^2}{\Phi} \frac{\partial}{\partial y} (S_{22}$$

$$- \frac{4}{3} v^2)] + \frac{\partial}{\partial y} \mu \frac{\partial S_{22}}{\partial y} + \rho (\frac{8}{15} q^2 + \frac{1}{33} S_{11} + \frac{2}{33} S_{33}) \frac{\partial U}{\partial x} \quad \text{(A5)}$$

$$\rho U_k \frac{\partial \overline{uv}}{\partial x_k} = - \rho (\frac{4}{15} q^2 - \frac{2}{11} S_{11} + \frac{5}{22} S_{22}) \frac{\partial U}{\partial y} - C_E \rho \frac{\Phi}{q^2} \overline{uv}$$

$$+ 0.4 \frac{\partial}{\partial y} (\rho \frac{q^2 v^2}{\Phi} \frac{\partial \overline{uv}}{\partial y}) + \frac{\partial}{\partial y} \mu \frac{\partial uv}{\partial y} \quad \text{(A6)}$$

$$\rho U_k \frac{\partial \Phi}{\partial x_k} = - 1.25 \rho \frac{\overline{uv}}{q^2} \frac{\partial U}{\partial y} \Phi - C_\Phi \rho \frac{\Phi^2}{q^2} + 177.6 \rho \frac{\nu^2 q^2}{y^4}$$

$$+ 0.322 \frac{\partial}{\partial y} (\rho \frac{q^2 v^2}{\Phi} \frac{\partial \Phi}{\partial y}) + \frac{\partial}{\partial y} \mu \frac{\partial \Phi}{\partial y} - 1.25 \rho \frac{u^2}{q^2} \frac{\partial U}{\partial x} \Phi + R_\Phi \quad \text{(A7)}$$

where

$$C_E = (1.2 + 12.5 \, \pi/Re_\Lambda)/(1 + 12.5 \, \pi/Re_\Lambda)$$

$$C_\Phi = (0.288 + 6.6\ \pi/Re_\Lambda + 35\ \pi^2/Re_\Lambda{}^2)/(0.4 + 5\ \pi/Re_\Lambda)^2$$

and Re_Λ is the turbulent Reynolds number $q\Lambda/\nu$, with Λ being related to the dissipation rate by

$$\Phi = 0.4\ (q^3/\Lambda) + 5\ \pi\nu/(q^2/\Lambda^2) =$$

$$0.4\ (q^3/\Lambda)\ (1 + 12.5\ \pi/Re_\Lambda) \tag{A8}$$

the corresponding equations for enthalpy-related quantities are:

$$\rho \frac{D\bar{h}}{Dt} = \bar{U}_i \frac{\partial \bar{p}}{\partial x_i} - \frac{\partial}{\partial y} (\rho \overline{v'h'}) + \frac{1}{Pr} \frac{\partial}{\partial y} (\mu \frac{\partial \bar{h}}{\partial y}) + \mu (\frac{\partial U}{\partial y})^2 + \rho\Phi \tag{A9}$$

$$\rho \frac{D\overline{h'^2}}{Dt} = - 2\ \rho\overline{v'h'} \frac{\partial \bar{h}}{\partial y} - C_{T_1}\ \rho \frac{\Phi}{q^2} \overline{h'^2}$$

$$+ 0.40 \frac{\partial}{\partial y} (\rho \frac{q^2 v^2}{\Phi} \frac{\partial \overline{h'^2}}{\partial y}) + \frac{1}{Pr} \frac{\partial}{\partial y} (\mu \frac{\partial \overline{h'^2}}{\partial y}) \tag{A10}$$

$$\rho \frac{D\overline{v'h'}}{Dt} = - \rho v^2 \frac{\partial \bar{h}}{\partial y} - 0.09835\ \rho\ \overline{u'h'} \frac{\partial U}{\partial y} - C_{T_2}\ \rho \frac{\Phi}{q^2} \overline{v'h'}$$

$$+ 0.80 \frac{\partial}{\partial y} (\rho \frac{q^2 v^2}{\Phi} \frac{\partial}{\partial y} \overline{v'h'}) + \frac{1}{Pr} \frac{\partial}{\partial y} (\mu \frac{\partial \overline{v'h'}}{\partial y}) \tag{A11}$$

$$\rho \frac{D\overline{u'h'}}{Dt} = - 0.3989\ \rho\overline{v'h'} \frac{\partial U}{\partial y} - \rho\overline{uv} \frac{\partial \bar{h}}{\partial y} - C_{T_2}\ \rho \frac{\Phi}{q^2} \overline{u'h'}$$

$$+ 0.40 \frac{\partial}{\partial y} (\rho \frac{q^2 v^2}{\Phi} \frac{\partial \overline{u'h'}}{\partial y}) + \frac{1}{Pr} \frac{\partial}{\partial y} (\mu \frac{\partial \overline{u'h'}}{\partial y}) \tag{A12}$$

where

$$C_{T_1} = (0.8 + 7.5\ \pi/Re_\Lambda)/(1 + 12.5\ \pi/Re_\Lambda)$$

$$C_{T_2} = (1.165 + 12.5\ \pi/Re_\Lambda)/(1 + 12.5\ \pi/Re_\Lambda)$$

The terms R_u, R_h, R_q, and R_Φ contain the effect of roughness on the boundary layer. For the mean velocity and enthalpy equations, R_u and R_h were presented in Eqs. (1) and (2). The other two terms are source terms for kinetic energy and dissipation, describing the fluctuations introduced in the wakes of elements. For the fully turbulent boundary layers considered in this study, these terms are generally small compared to the natural turbulence production terms. As described in Ref. 7, the terms used are:

$$R_q = 0.04\ \rho U^3\ D/\ell^2 \qquad \text{(A13)}$$

$$R_\Phi = 0.04\ \rho U^3\ \nu/D\ell^2 \qquad \text{(A14)}$$

Acknowledgment

Research sponsored by the Air Force Office of Scientific Research (AFSC), United States Air Force, under Contract F49620-80-C-0015.

References

1. Nikuradse, J., "Strömungsgesetze in rauhen Rohren," VDI Forschungsheft, SerB, Vol. 4, No. 361, 1933; English translation, NACA TM1292, 1950.

2. Schlichting, H., "Experimental Investigation of the Problem of Surface Roughness," NACA TM823 (1937); also *Boundary Layer Theory*, McGraw-Hill, New York, 1968.

3. Betterman, D., "Contribution a l'Etude de la Connection Forces Turbulente le Long de Plaques Rugueuses," *International Journal of Heat & Mass Transfer*, Vol. 9, 1966, pp. 153-164.

4. Dvorak, F. A., "Calculation of Turbulent Boundary Layers on Rough Surfaces in Pressure Gradient," *AIAA Journal*, Vol. 7, No. 9, Sept. 1969, pp. 1752-1759; also *AIAA Journal*, Vol. 4, No. 11, Nov. 1972, pp. 1447-1451.

5. Simpson, R. L., "A Generalized Correlation of Roughness Density Effects on the Turbulent Boundary Layer," *AIAA Journal*, Vol. 11, No. 2, Feb. 1973, pp. 242-244.

6. Dirling, R. B., Jr., "A Method for Computing Rough Wall Heat Transfer Rates to Re-entry Vehicle Nosetips," AIAA Paper 73-763, Palm Springs, Ca., July 1973.

7. Finson, M. L. and Wu, P. K. S., "Analysis of Rough Wall Turbulent Heating with Application to Blunted Flight Vehicles," AIAA Paper 79-008, January 1979, New Orleans, La.; also PSI TR-158, AFOSR-TR-79-0199.

8. Saffman, P. G. and Wilcox, D. C., "Turbulence-Model Predictions for turbulent Boundary Layers," AIAA Journal, Vol. 12, No. 4, April 1974, pp. 541-546.

9. Chan, C. K. and Roberson, J. A., "Turbulence in Wakes of Roughness Elements," Proceedings of the ASCE, Vol. 100, No. HY1, January 1974, pp. 53-67.

10. Raupach, M. R., Thom, A. S., and Edwards, I., "A Wind-Tunner Study of Turbulent Flow Close to Regularly Arrayed Rough Surfaces," Boundary-Layer Meteorology, Vol. 18, 1980, pp. 373-397.

11. Foster, T., Read, D., and Murray, A., "Reduced Data Report: Surface Roughness Heating Augmentation Tests in AEDC Tunnel F., Vol. II," Acurex Report TR-79-183, Mountain View, Ca., 1979.

12. Finson, M. L., Clarke, A. S., and Wu, P. K. S., "Effect of Surface Roughness Character on Turbulent Boundary Layer Heating," PSI TR-204, Woburn, Mass., 1979.

13. Holden, M. S., "Studies of the Effects of Transitional and Turbulent Boundary Layers on the Aerodynamic Performance of Hypersonic Re-entry Vehicles in High Reynolds Number Flows," Calspan Report AB-5834-A-2, AFOSR-TR-79-0125, 1978.

14. Hopkins, E. J. and Inouye, M., "An Evaluation of Theories for Predicting Turbulent Skin Friction and Heat Transfer on Flat Plates at Supersonic and Hypersonic Mach Numbers," AIAA Journal, Vol. 9, No. 6, June 1971, pp. 993-1003.

15. Keel, A. G., Jr., "Influence of Surface Roughness on Skin Friction and Heat Transfer for Compressible Turbulent Boundary Layers," AIAA Paper 77-178, Los Angeles, Ca., Jan. 1977.

16. Hill, J. A. F., "Measurements of Surface Roughness Effects in the Heat Transfer to Slender Cones at Mach 10," AIAA Paper 80-0345, Pasadena, Ca., Jan. 1980.

17. Holden, M. S., "Accurate Vehicle Experimental Dynamics Program. Final Report. Studies of Aero-Thermodynamic Phenomena Influencing the Performance of Hypersonic Re-entry Vehicles," Calspan Corporation, Buffalo, N.Y., SAMSO-TR-79-47 1979.

THREE-DIMENSIONAL PROTUBERANCE INTERFERENCE HEATING IN HIGH-SPEED FLOW

F.T. Hung*
The Aerospace Corporation, El Segundo, Calif.

and

J.M. Clauss†
Rockwell International Corporation, Downey, Calif.

Abstract

Extensive experimental study was conducted on interference heating caused by three-dimensional cylindrical protuberances protruding from a flat plate. The parametric study includes variation of cylinder height and diameter, unit Reynolds number, flow length, and, consequently, boundary-layer thickness. The protuberance can be either submerged in the boundary layer or be infinitely long with the flow on the flat plate either laminar or turbulent. The heating data on the flat plate were correlated in such a way that generalized heating prediction methods were developed for thermal design purposes. One of the important findings from this study is the effect of the protuberance on the flat-plate boundary-layer state which is closely related to the heating levels. The data indicate that flow on the flat plate right in front of the protuberance is always laminar, even though the oncoming flow is turbulent. This is in contrast to the flow in the downstream region, where the laminar flow can be tripped to turbulent resulting in higher heating.

Presented as Paper 80-0289 at the AIAA 18th Aerospace Sciences Meeting, Pasadena, California, Jan. 14-16, 1980.

*Engineering Specialist, Aerothermodynamics Office, Vehicle Engineering Division.

†Member of Technical Staff, Aerosciences Department, Space Division.

Nomenclature

D = cylinder diameter
H = heat-transfer coefficient
K = cylinder height
M = Mach number
P = pressure
Re = Reynolds number
S = primary separation distance
T = temperature
U = velocity
X = streamwise distance measured from flat-plate leading edge
Z = vertical distance measured from flat-plate surface
δ = boundary-layer thickness
δ^* = displacement thickness
θ = momentum thickness

Subscripts

e = boundary-layer edge
i = interference
o = stagnation
REF = reference (Fay-Riddell value)
SL = cylinder stagnation line
u = undisturbed
∞ = freestream

Introduction

In practice, protuberances exist on almost all aerodynamic vehicle surfaces. Protuberances can be two-dimensional or three dimensional with different sizes. The flowfield around these protuberances is of importance, especially when the speed of the vehicle is supersonic or hypersonic. The local heating rate on and in the vicinity of the protuberance increases appreciably besides aerodynamic load increase. For three-dimensional protuberances, the complex inviscid-viscid interaction flow makes a theoretical approach very difficult. The problem is further complicated by the unsteady nature of the separated flow in front of the protuberance.[1] Also, flow transition caused by the protuberance further complicates the problem.[2] The paucity of analytical heating prediction methods is not surprising, even though the flowfield has been described qualitatively by different authors such as Kaufman et al.,[1] Sydney,[3,4] Halprin,[5] Westkaemper,[6,7] Waltrup et al.,[8] Whitehead,[9] Hiers and Loubsky,[10] and Couch.[11]

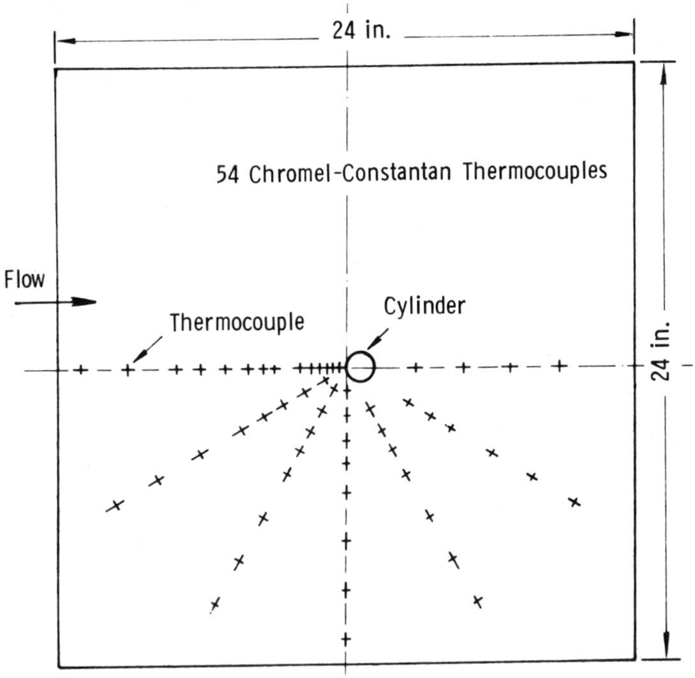

Fig. 1 Test model configurations.

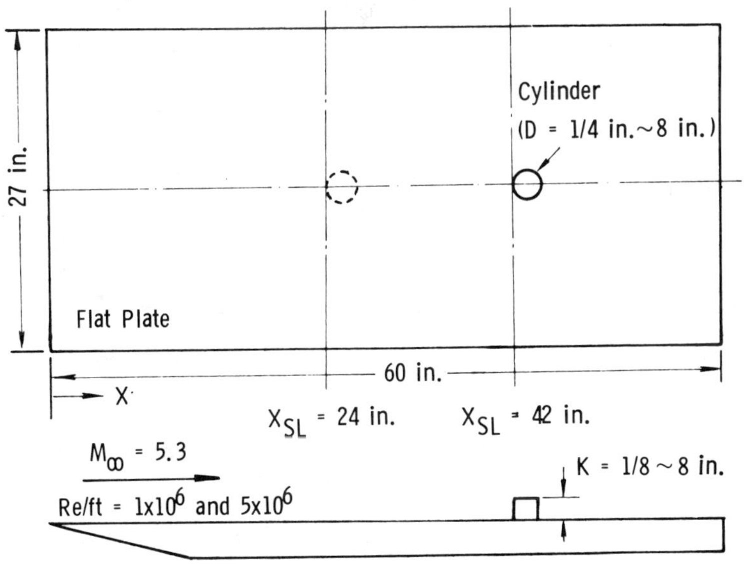

Fig. 2 Thermocouple locations on flat-plate insert.

Experimental data on three-dimensional protuberance heating also are limited. Burbank et al.[12] and Surber and Luning[13] presented long cylinder data, and there are almost no short cylinder data in existence. (Criteria used to define long and short cylinders are discussed below.) The limited data and the number of parameters involved in the problem also make the dimensional analysis very difficult. It is generally expected that interference heating varies with cylinder height, diameter, boundary-layer thickness (or displacement or momentum thickness), boundary-layer state (laminar or turbulent), Reynolds number, and Mach number. Heating distribution in front of the protuberance also depends on the number and the position of the vortices, which is in turn a function of all the parameters just mentioned. This explains why there are no generalized heating prediction methods in existence.

In this study, an experimental investigation was conducted with cylindrical protuberances protruding from a flat-plate surface. The parametric study covered wide ranges of cylinder heights, diameters, and boundary-layer thicknesses with the oncoming laminar or turbulent flow. The cylinders were either submerged in the boundary layer or protruding out of the boundary layer with "infinite" length. Heating measurements were taken on both the cylinder and on the flat plate in the vicinity of the cylinder. Only the flat-plate centerline heating data analysis will be discussed in this paper. The experimental results first were analyzed qualitatively and then were correlated to establish generalized heating prediction methods.

Experimental Program

The test was conducted at the 3.5-ft Hypersonic Wind Tunnel located at NASA Ames Research Center with freestream Mach number of 5.3. Cylinders with different heights (K = 1/8 to 8 in.), and diameters (D = 1/4 to 8 in.) were placed on the flat plate at two different locations (X_{SL} = 24 or 42 in.) and were tested at two different unit Reynolds numbers (Re/ft = 1.0×10^6 and 5.0×10^6) as shown in Fig. 1. This provides a wide range of boundary layer thicknesses with both laminar and turbulent flow conditions. Thermocouples were placed both on the cylinders and the flat plate, Fig. 2. All the instrumented areas on the models were 0.03 in. stainless steel thin skins. During the test the models were injected into the flow, and heat-transfer coefficient, H, was derived from the thin skin temperature response. Thermal insulators were applied between the flat plate and the cylinder to minimize data reduction errors due

Table 1 Test matrix

K (in.)	D (in.)	Re/ft = 1.0×10^6		Re/ft = 5.0×10^6	
		X_{SL} = 24 in. (Laminar)	X_{SL} = 42 in. (Transitional)	X_{SL} = 24 in. (Turbulent)	X_{SL} = 42 in. (Turbulent)
1/8	1/4	X			X
1/8	1/2	X	X		X
1/8	1	X			X
1/8	2	X	X	X	X
1/4	1/4	X	X	X	X
1/4	1/2	X	X	X	X
1/4	1	X	X	X	X
1/4	2	X	X	X	X
1/4	4	X			X
1/4	8	X			X
1/2	1/4	X			X
1/2	1/2	X	X		X
1/2	1	X			X
1/2	2	X	X	X	X
1	1/4	X	X	X	X
1	1/2	X	X	X	X
1	1	X	X	X	X
1	2	X	X	X	X
1	4	X			X
1	8	X			X
2	1/4	X	X	X	X
2	1	X			X
2	2	X	X	X	X
3	1/2	X	X	X	X
5	1	X	X	X	X
8	2	X	X	X	X
8	4	X			X
Smooth Plate		X	X	X	X

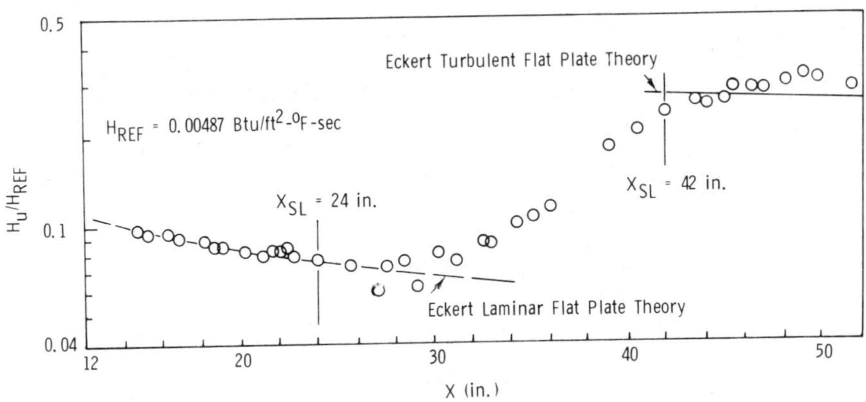

Fig. 3 Undisturbed flat-plate heating data (Re/ft = 1×10^6).

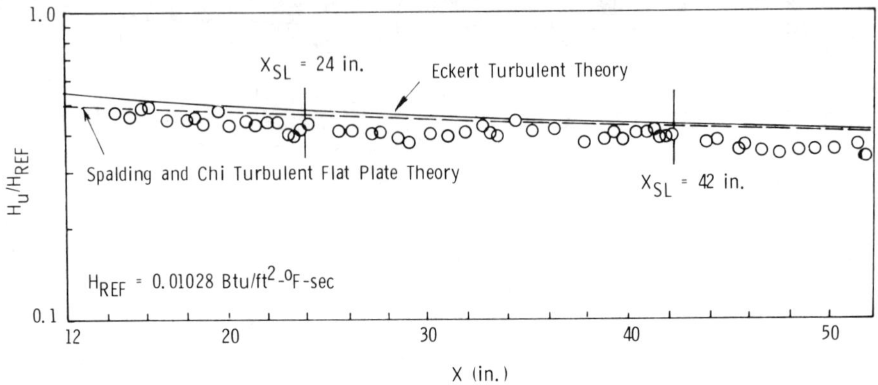

Fig. 4 Undisturbed flat-plate heating data (Re/ft = 5 x 10^6).

to heat conduction. Nominal flow conditions are:

Re/ft	M_∞	P_o (psia)	To (^{o}R)
1.0 x 10^6	5.3	82	1300
5.0 x 10^6	5.3	405	1300

The test matrix is shown in Table 1.

Test Results and Data Analysis

The undisturbed flat-plate heating data are presented in Figs. 3 and 4 where X is the flow length measured from the flat-plate leading edge. The measured undisturbed heat-transfer coefficients, H_u, are compared with Eckert[14] and Spalding-Chi[15] theories with good agreement. H_{REF} in these figures is the stagnation point heating value of a 1-ft radius sphere computed with the Fay-Riddell equation.[16] H_{REF} values for the two flow conditions are given in Figs. 3 and 4. Test results indicate that at low-unit Reynolds number (Re/ft = 1.0 x 10^6, Fig. 3) the flow at the forward position (X_{SL} = 24 in.) is laminar with flow transition occurring downstream; whereas at high-unit Reynolds number (Re/ft = 5.0 x 10^6) turbulent flow covers the entire test model as indicated in Fig. 4.

Typical laminar and turbulent interference data, H_i/H_{REF}, on the flat-plate centerline are presented in Figs. 5 through 8 and Figs. 9 through 12, respectively. The terms laminar and turbulent refer to the condition of the undisturbed oncoming flow upstream of the interference region. The faired H_u/H_{REF} data also are presented for comparison purposes. It should be noted that the laminar and turbulent recovery factors chosen to derive H_i values from thermocouple measurements are based on the boundary-layer state of the undisturbed flow, since flow transition in the interference region cannot be easily determined.

Qualitative Behavior

A background as to the known literature dealing with protuberance flowfield phenomena is in order before any significant data correlation attempts can be made. The interference flow phenomena described in detail in Refs. 1 through 11 have been examined, and it can be concluded that three common features exist regardless of the speed of the flow, the shape and size of the protuberance, or whether the flow is laminar or turbulent: 1) The vortex system is formed in the upstream separated flow. 2) These vortices are stretched around the protuberance in a horseshoe fashion and persist as streamwise vortices far downstream. 3) Twin tornado-like vortices rise up from the plate surface in the near wake as discussed in Ref. 4. The number of vortices in the upstream region can be two, four, or six, as shown in Fig. 13, depending mainly on Reynolds number. In general, the number of vortices decreases with increasing Reynolds number.[4] Note that the sketches in Fig. 13 are not to scale. For the two-vortex pattern, the streamwise extent of the large vortex is about 25 times that of the small one.[4]

The typical data in Figs. 5 through 12 first can be analyzed qualitatively. Heating increases with K and D are obvious for both upstream and downstream regions. Also, as expected, the laminar interference area in front of the cylinder is much larger than the turbulent case. For laminar flow, part of the upstream interference data can be lower than the undisturbed values, whereas for turbulent flow almost all H_i/H_{REF} data are higher than the H_u/H_{REF}. In the downstream region, the laminar heating increases owing to the presence of the protuberance, Figs. 5 through 8, are much larger than the turbulent cases as shown in Figs. 9 through 12. This indicates that the laminar flow downstream of the protuberance has been tripped to turbulent or transitional.

The heating distribution in the upstream region is of particular interest and can be related with the number of vortices. Assuming that heating peaks occur at attachment

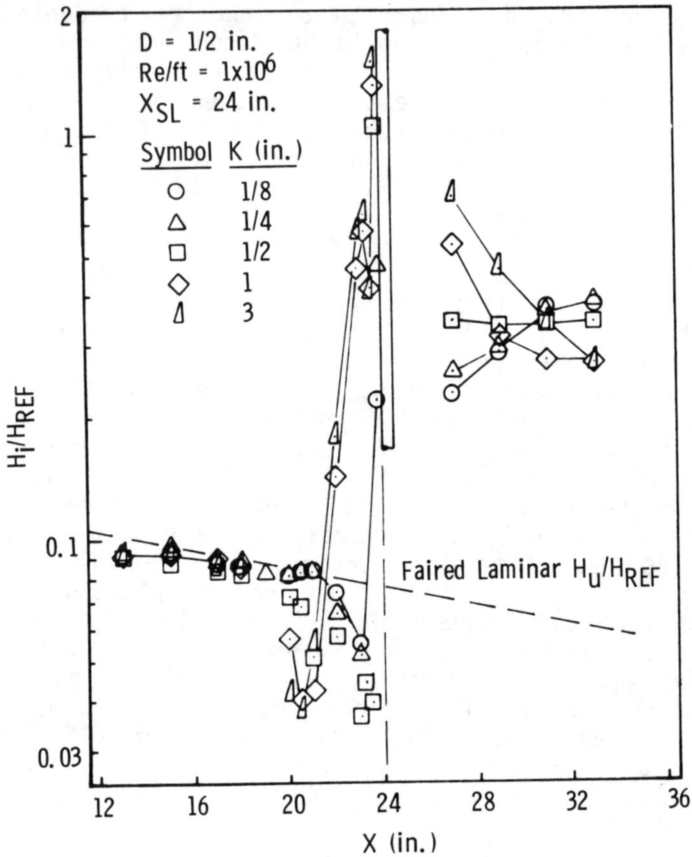

Fig. 5 Heating on flat-plate centerline (laminar undisturbed flow, D = 1/2).

points (A, A_1, and A_2 in Fig. 13) as indicated in Winkelman's experiments[17] and low heating spots correspond to the separation regions (S, S_1, and S_2 in Fig. 13), the heating distribution curves can be used to determine the number of vortices. Results (Figs. 5 throgh 12) indicate that the number of vortices in the upstream region varies only with cylinder height. There are two vortices for K = 1/8 and 1/4 in. and four for K = 1/2 to 8 in. for both laminar and turbulent cases.

Peak Heating Correlations

For thermal design purposes, the peak heating in front of the protuberance is of particular interest. In general,

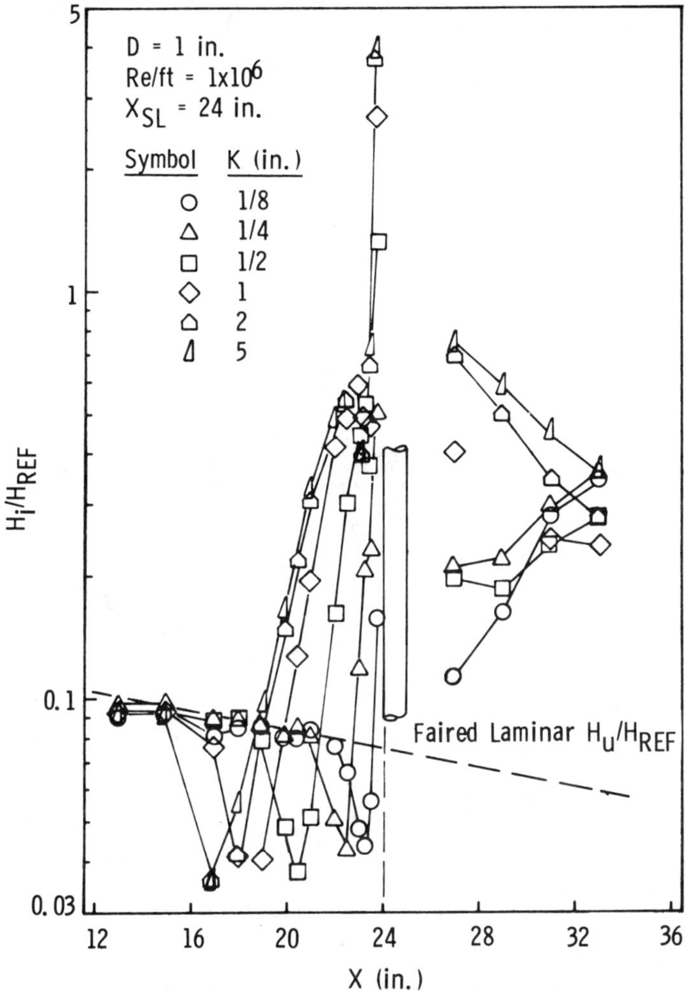

Fig. 6 Heating on flat-plate centerline (laminar undisturbed flow, D = 1).

the peak interference heating factor, max H_i/H_u, can be assumed as a function of D, K, boundary-layer thickness, δ (or displacement thickness, δ^*, or momentum thickness, θ), and freestream velocity U_∞ (or Mach number, M_∞). A typical dimensional analysis will yield

$$\max H_i/H_u = f(D/\delta,\ K/\delta,\ M_\infty) \qquad (1)$$

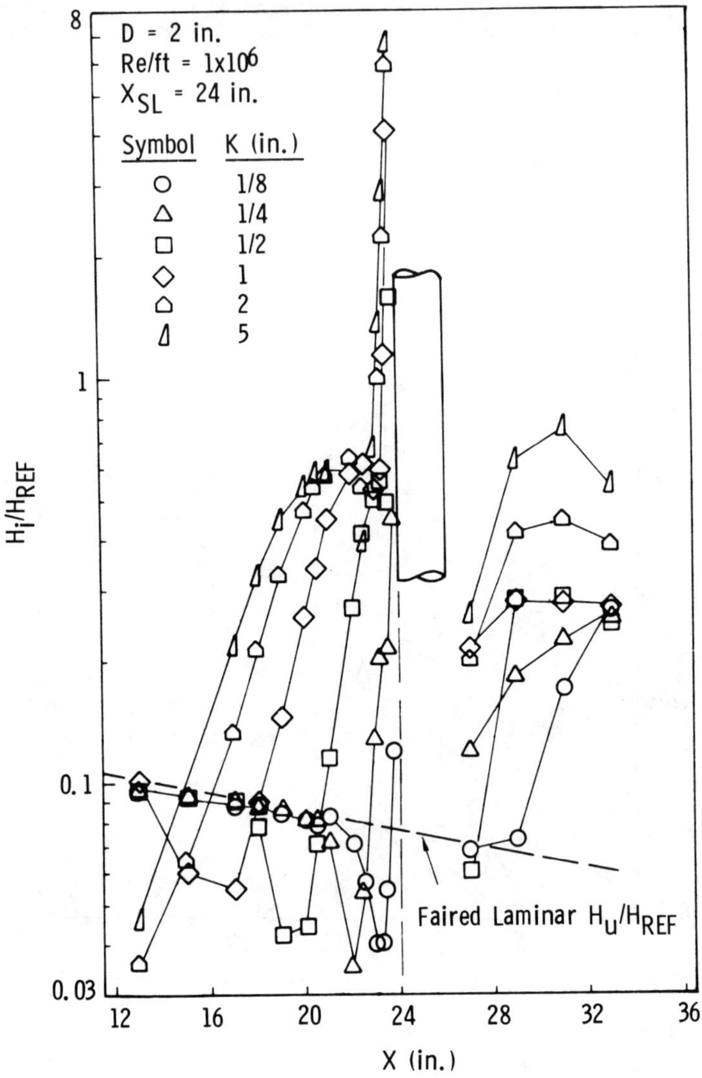

Fig. 7 Heating on flat-plate centerline (laminar undisturbed flow, D = 2).

where δ can be replaced by δ^* or θ. The computed δ, δ^*, and θ values are presented in Fig. 14 and typical laminar and turbulent heating correlations are presented in Figs. 15 and 16, respectively. For the case of small K/δ, diameter effect on heating is not significant. However, heating increase with K is obvious until K reaches a certain value where further increase in K causes no significant heating

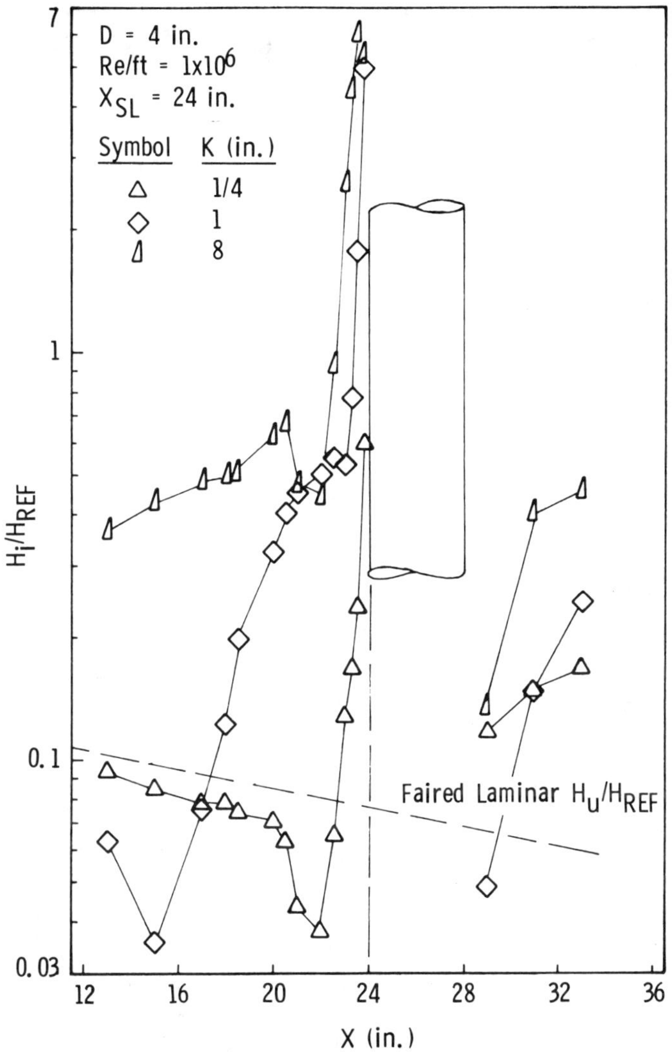

Fig. 8 Heating on flat-plate centerline (laminar undisturbed flow, D = 4).

increase. This can be used as criteria for defining short and long protuberances as shown in Fig. 17 where data are taken from Figs. 15 and 16 when the heating curves start to level off. A protuberance is short when $K < D$ and $K > 2D$ for a long protuberance.

The long protuberance data (the asymptote values from Figs. 15 and 16) are presented in Fig. 18, which indicates

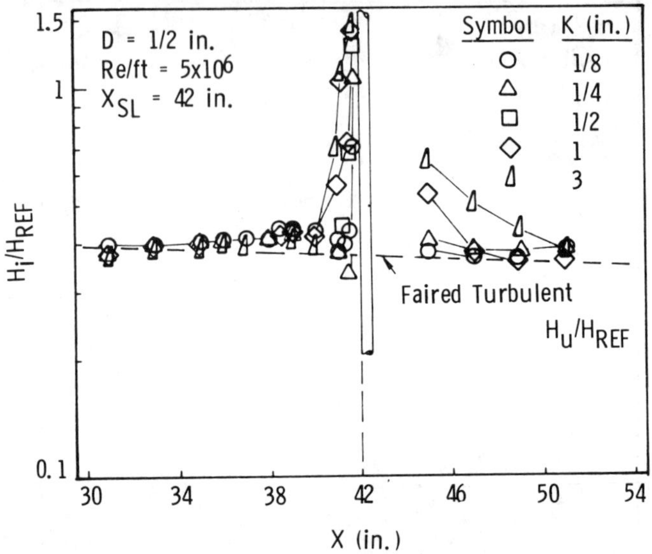

Fig. 9 Heating on flat-plate centerline (turbulent undisturbed flow, D = 1/2).

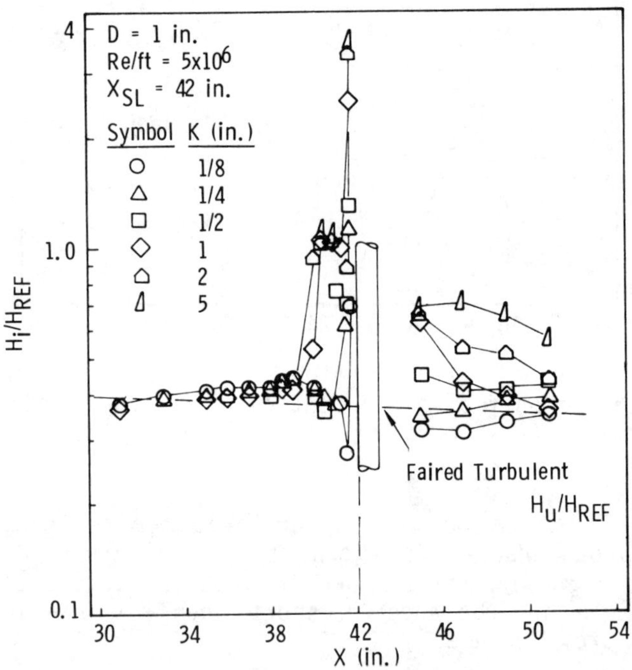

Fig. 10 Heating on flat-plate centerline (turbulent undisturbed flow, D = 1).

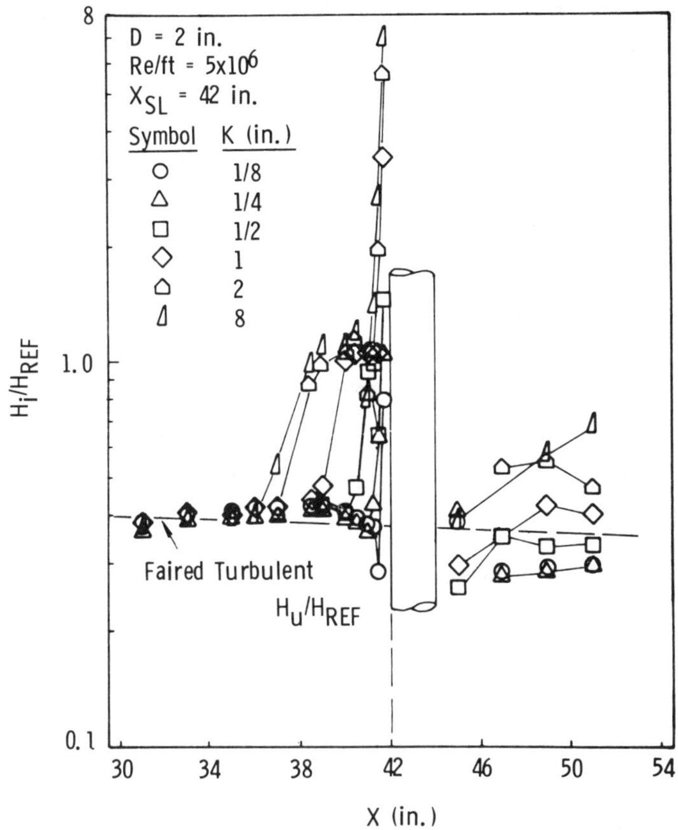

Fig. 11 Heating on flat-plate centerline (turbulent undisturbed flow, D = 2).

that peak heating on the flat plate increases with long protuberance diameter, D, and levels off when D reaches a certain value. This can be used to define large and small protuberances as shown in Fig. 18. The heating even starts to decrease when D reaches 4 in. This is expected since heating has to decrease to zero for a long cylinder ($K = \infty$) with D approaching to infinity (a two-dimensional protuberance with $K = \infty$). The above discussion can be summarized as below:

The criterion for categorizing long and short protuberances is (Fig. 17)

$$K = (1 \sim 2)\ D \tag{2}$$

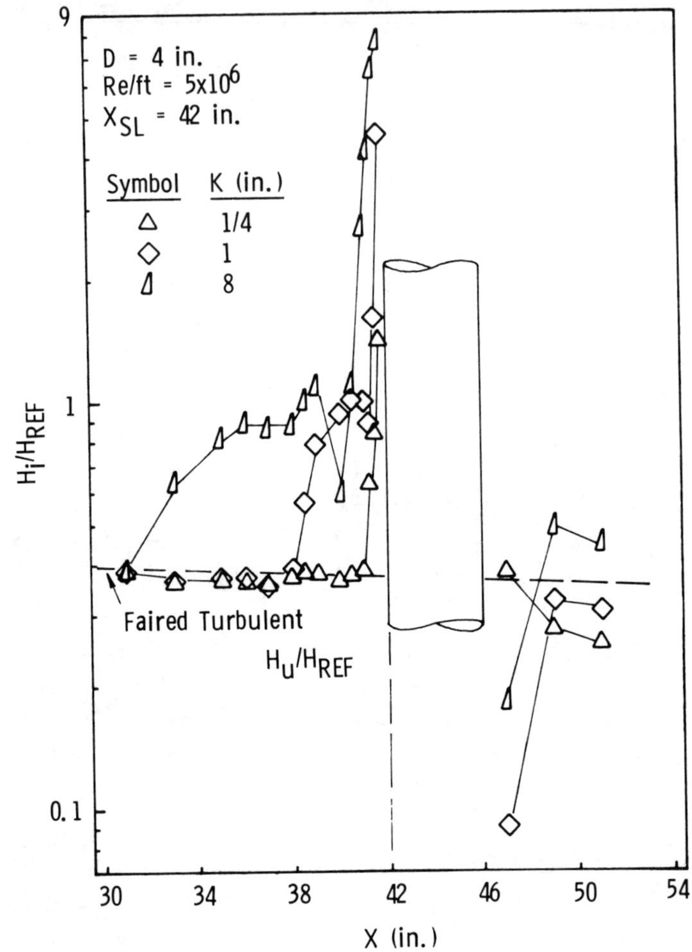

Fig. 12 Heating on flat-plate centerline (turbulent undisturbed flow, D = 4).

For short protuberances with laminar flow (Fig. 15)

$$\max H_i/H_u = (2.5 \sim 4.25)\ (K/\delta)^{1.5} \tag{3}$$

$$\max H_i/H_u = (1.4 \sim 2.5)\ (K/\delta^*)^{1.5} \tag{4}$$

For short protuberances with turbulent flow (Fig. 16)

$$\max H_i/H_u = (5.2 \sim 7.3)\ (K/\delta)^{0.863} \tag{5}$$

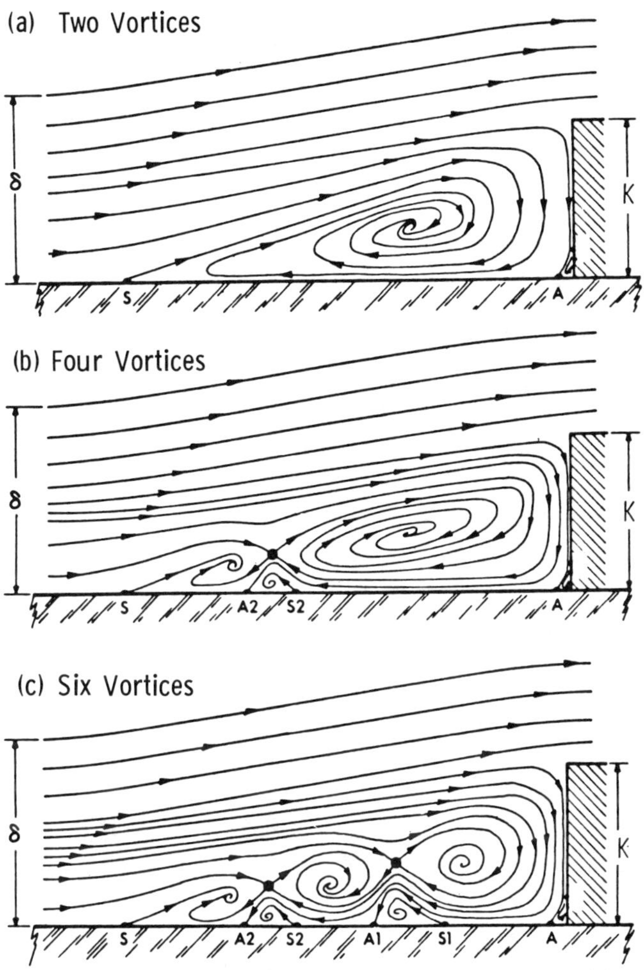

Fig. 13 Vortex patterns in front of protuberance (ref. 4).

$$\max H_i/H_u = (2.4 \sim 3.4)\ (K/\delta^*)^{0.863} \tag{6}$$

For long protuberances (Fig. 18)

$$\max H_i/H_u = 4.2\ (D/\delta)^{1.38} \tag{7}$$

The most interesting result of this analysis is the comparison between the laminar and the turbulent peak heating data shown in Fig. 19. Except for the very short cylinders (K = 1/8 and 1/4 in.), the laminar and turbulent H_i/H_{REF} values for a given cylinder are almost identical

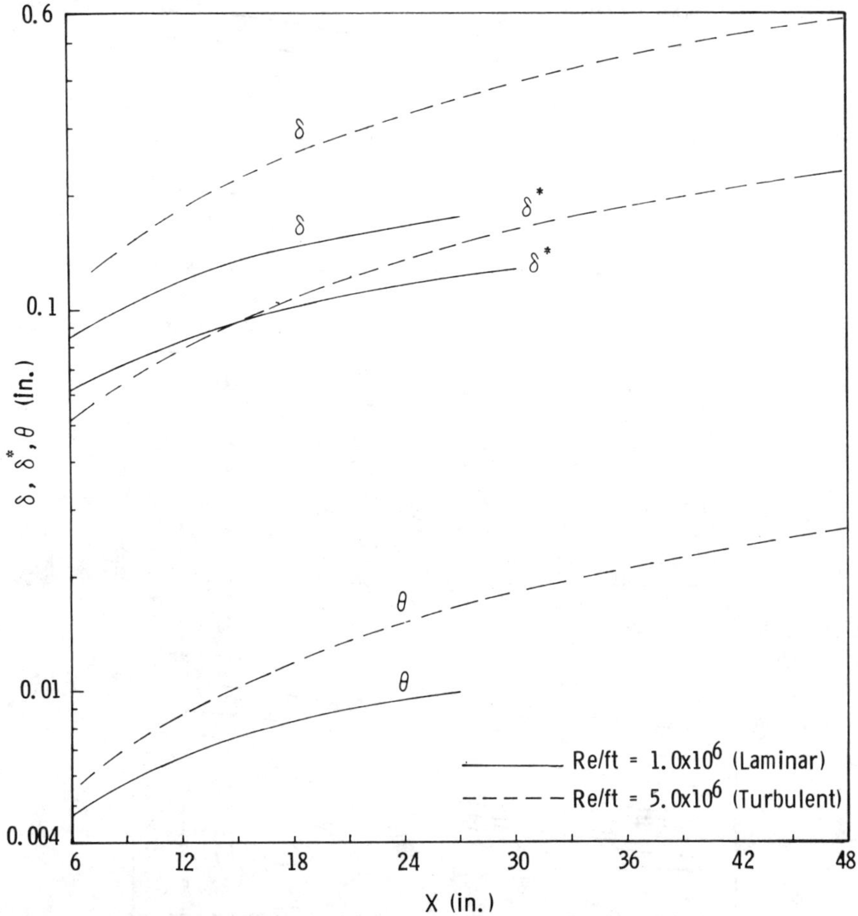

Fig. 14 Boundary layer, boundary-layer displacement and momentum thickness values on flat plate.

regardless of boundary-layer state of the oncoming flow, unit Reynolds number, and flow length. This suggests that the flow on the plate right in front of the cylinder is stagnated with stagnation laminar heating. This is in contrast to the flow downstream of the cylinder where the laminar flow can be tripped by the protuberance to turbulent. The above discussion indicates that the laminar correlations shown in Eqs. (3) and (4) should be independent of Reynolds number as both H_i and H_u are laminar values. However, the turbulent correlations, Eqs. (5) and (6), may only hold for the Reynolds number tested as the laminar H_i and turbulent H_u vary with the Reynolds number in different manners. This undesirable feature can be eliminated by correlating the nondimensionalized max H_i/H_{REF} values, as both H_i and H_{REF} are laminar values. Consequently, for laminar flow, the heating correlation can be either in the form of H_i/H_u as shown in Eqs. (3) and (4) or in the form of H_i/H_{REF} as shown below (Fig. 15)

$$\max H_i/H_{REF} = (0.9 \sim 0.323)\ (K/\delta)^{1.5} \tag{8}$$

$$\max H_i/H_{REF} = (0.11 \sim 0.19)\ (K/\delta^*)^{1.5} \tag{9}$$

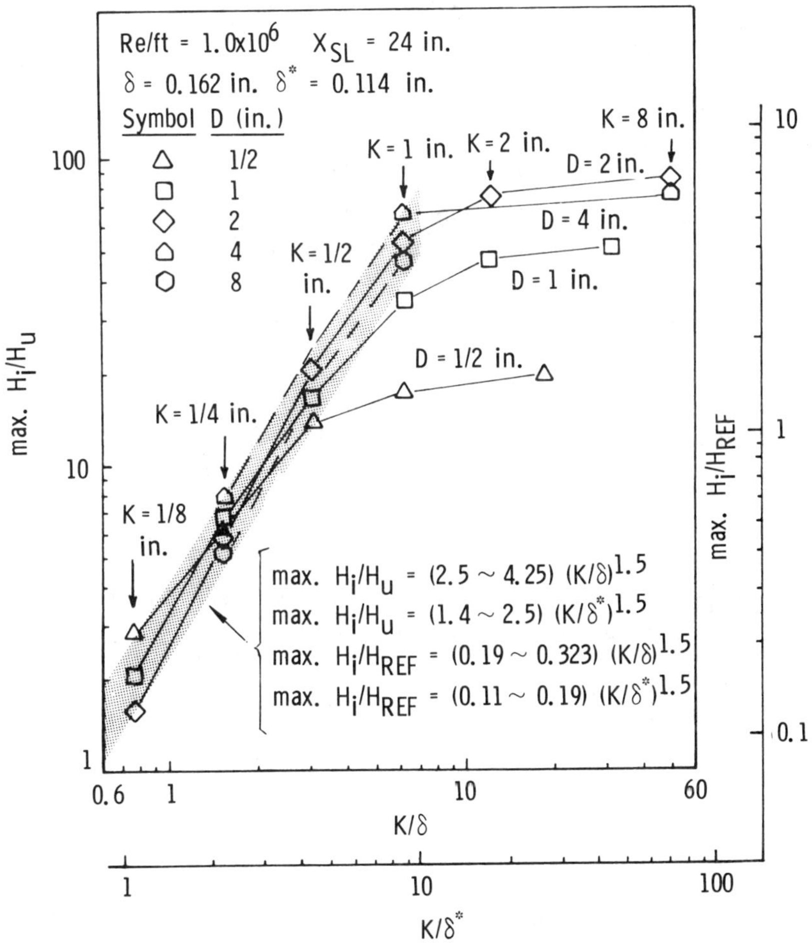

Fig. 15 Laminar heating correlation.

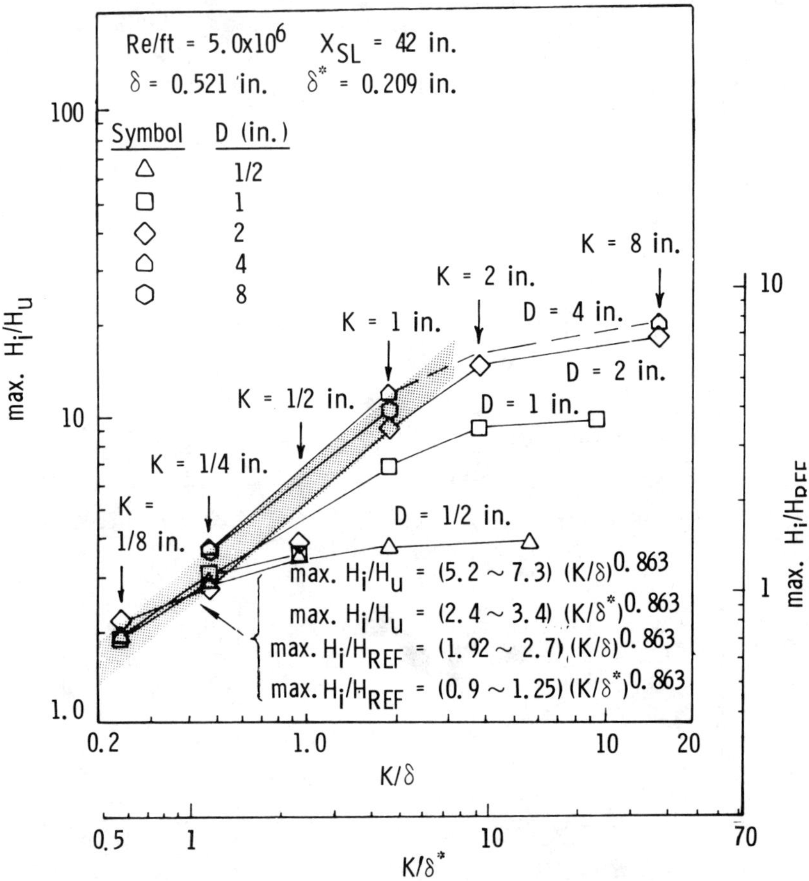

Fig. 16 Turbulent heating correlation.

However, for turbulent flow, Eqs. (5) and (6) should be replaced by the following expressions (Fig. 16):

$$\max H_i/H_{REF} = (1.92 \sim 2.7)\ (K/\delta)^{0.863} \tag{10}$$

$$\max H_i/H_{REF} = (0.9 \sim 1.25)\ (K/\delta^*)^{0.863} \tag{11}$$

In comparison with turbulent data, the laminar H_i/H_{REF} values in Fig. 19 drop sharply for the very short cylinders as mentioned previously. This may be caused by the difference between the laminar and turbulent boundary-layer velocity profiles as shown in Fig. 20. It is known

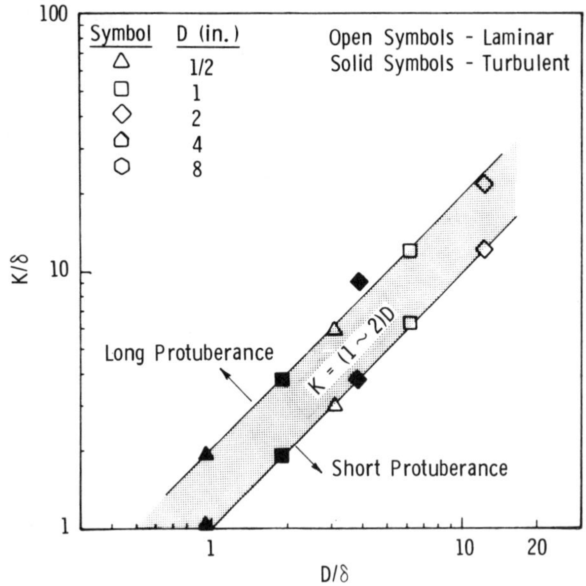

Fig. 17 Criteria for defining short and long protuberances.

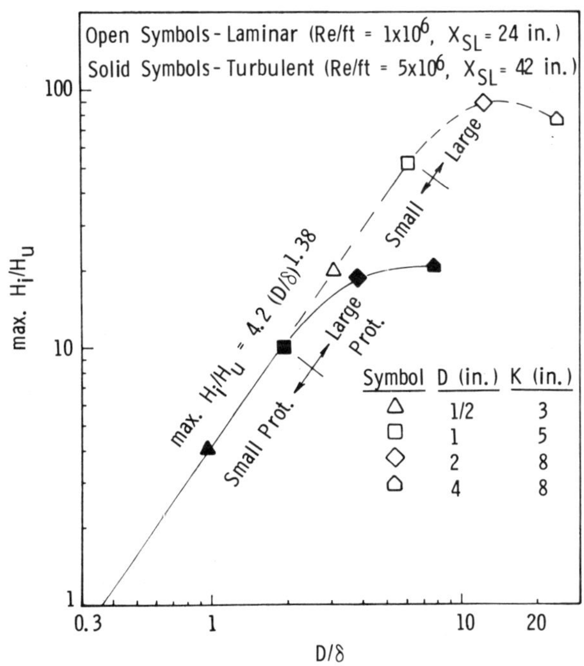

Fig. 18 Heating correlation with long cylinder ($K = \infty$) and criteria for defining small and large protuberances.

that the local high pressure, and thus the local heating, in front of the protuberance is caused by the blockage (or stagnation) of the oncoming flow. Consequently, it can be speculated that in comparison with the "full" turbulent velocity profile, the "lean" laminar velocity profile in front of the short protuberance should result in lower local pressure and thus lower heating.

Typical heating data with all four flow conditions are presented in Fig. 21. Again, the lower laminar and transitional heating values for very short cylinders (k = 1/8 and 1/4 in.) are obvious. For the rest of the cylinders, the heating data do not show significant change with flow condition.

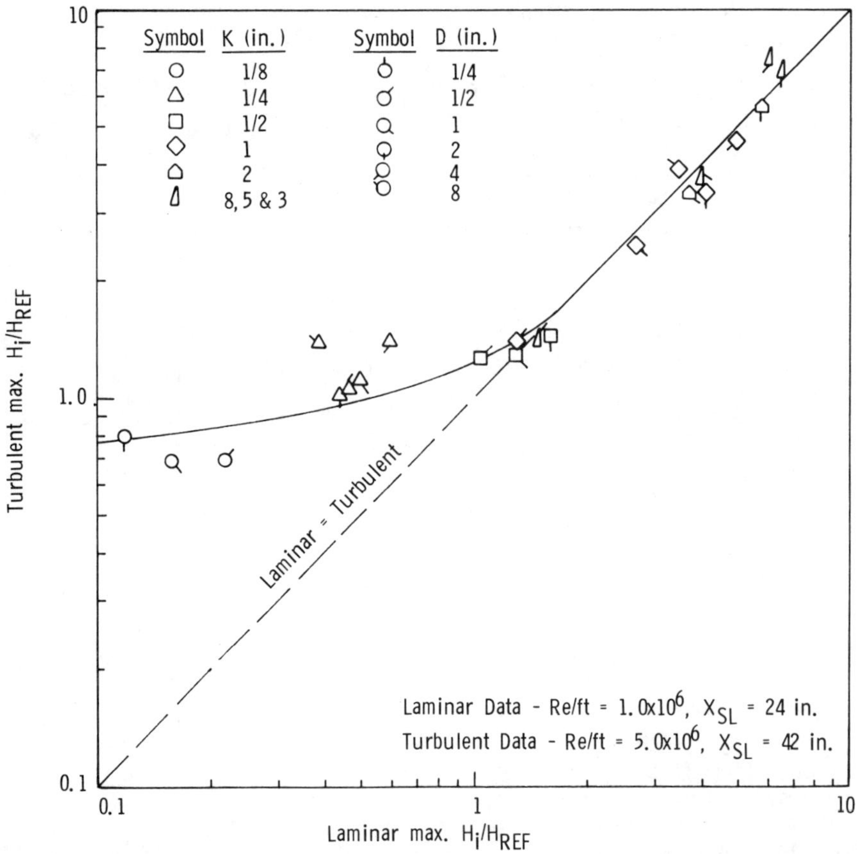

Fig. 19 Laminar and turbulent heating-data comparison.

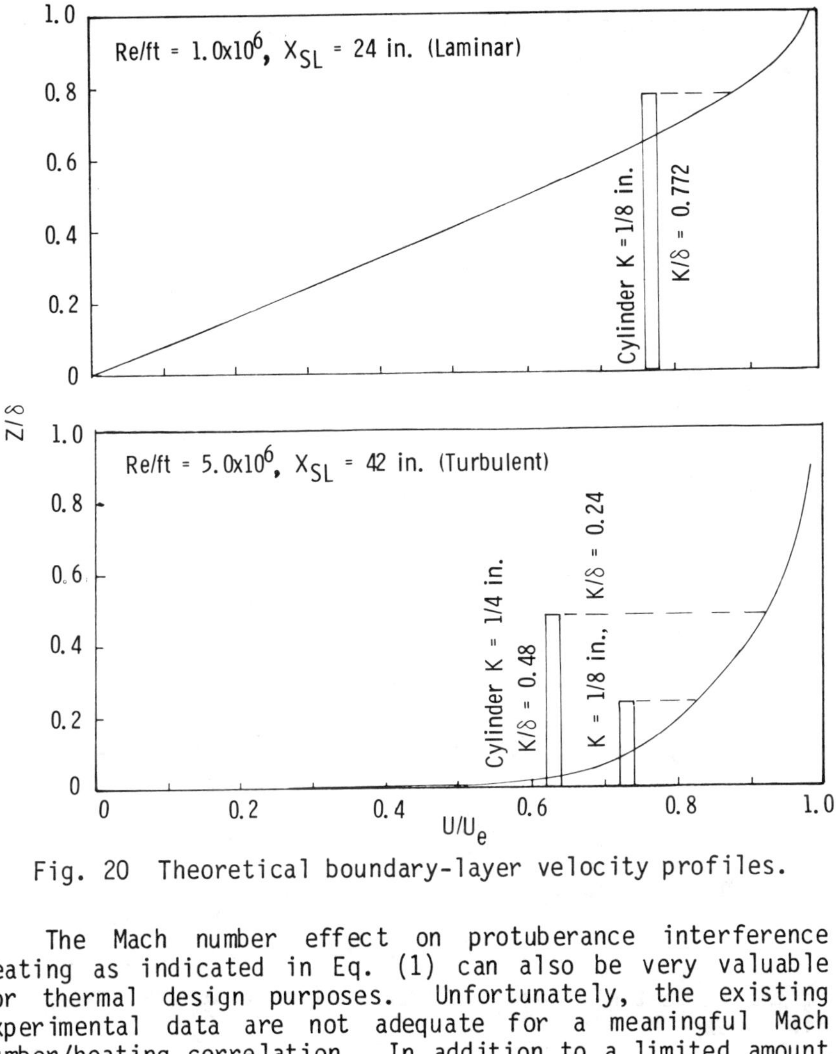

Fig. 20 Theoretical boundary-layer velocity profiles.

The Mach number effect on protuberance interference heating as indicated in Eq. (1) can also be very valuable for thermal design purposes. Unfortunately, the existing experimental data are not adequate for a meaningful Mach number/heating correlation. In addition to a limited amount of data, the problem is further complicated by the fact that different test models have different thermocouple locations, which directly affect the maximum heating measurements. Attempts to combine the current Mach 5.3 long cylinder data with test results from Burbank et al.[12] (M_∞ = 2.65 - 4.44) and Surber and Luning[13] (M_∞ = 12 - 21) have been unsuccessful.

Extent of Upstream Interference

Finally, for practical design purposes it is of interest to predict the size of the upstream interference region

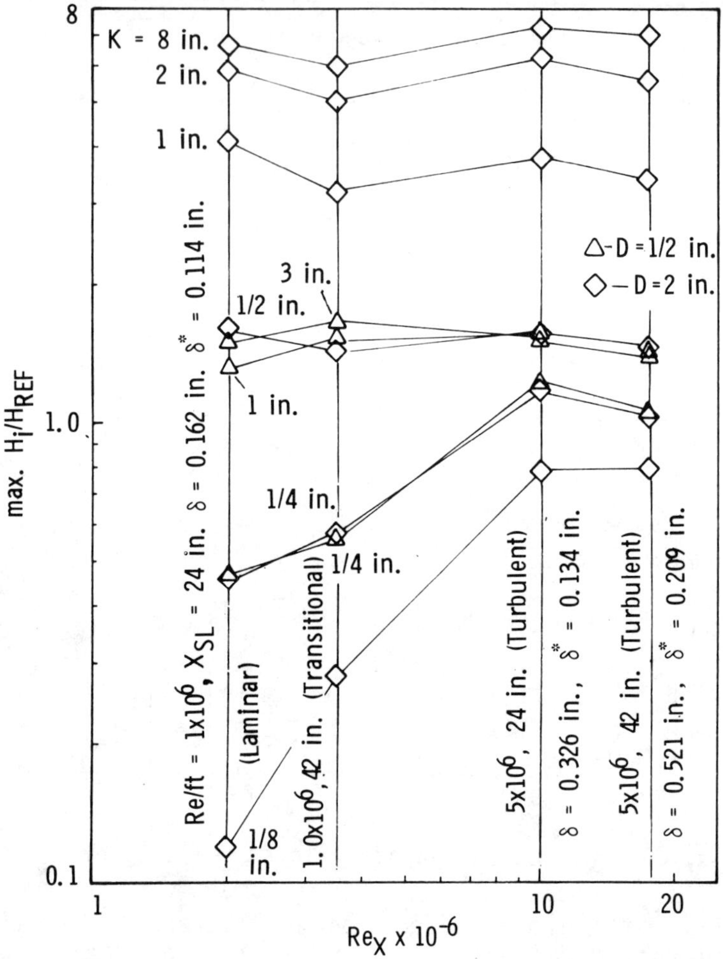

Fig. 21 Effect of flow condition on flat-plate maximum interference heating.

on the plate. The primary separation distance, S, as described in Ref. 4 is defined as the distance from the cylinder stagnation line to the most upstream separation location on the plate. However, different methods of measuring the separation distance have been used by different authors, Refs. 4, 5, 7, 8, 12, 13, and 18 through 21. Methods include surface oil film, side-view shadowgraph or Schlieren photograph, and heating and pressure measurement. The separation boundary on the plate can be defined as the location where values of H_i/H_u or P_i/P_u start to deviate from 1.0 as used in this study. As expected, the laminar S values are much higher than the turbulent data, Fig. 22. It is also obvious that S approaches to an asymptote when K reaches a certain value. A typical correlation of S should yield

$$S/D = f(K/\delta,\ D/\delta) \tag{12}$$

and the final results are given below.

For laminar flow

$$S/D = 6.0(K/\delta)^{0.7}\ (D/\delta)^{-0.435}$$

$$\text{- Short protuberance} \tag{13}$$

$$S/D = 9 \text{ to } 12 \text{ - Long protuberance} \tag{14}$$

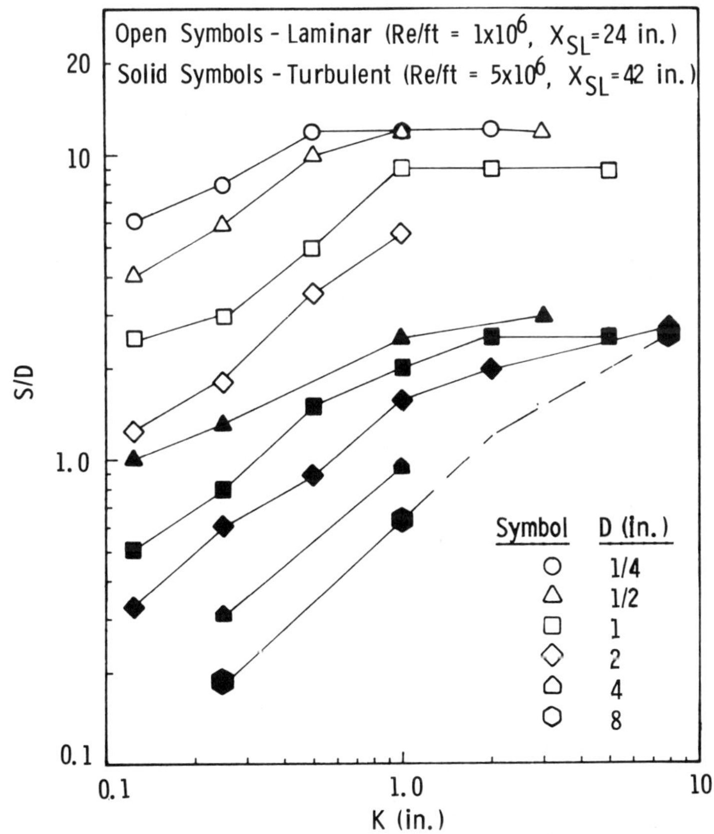

Fig. 22 Primary separation distance data.

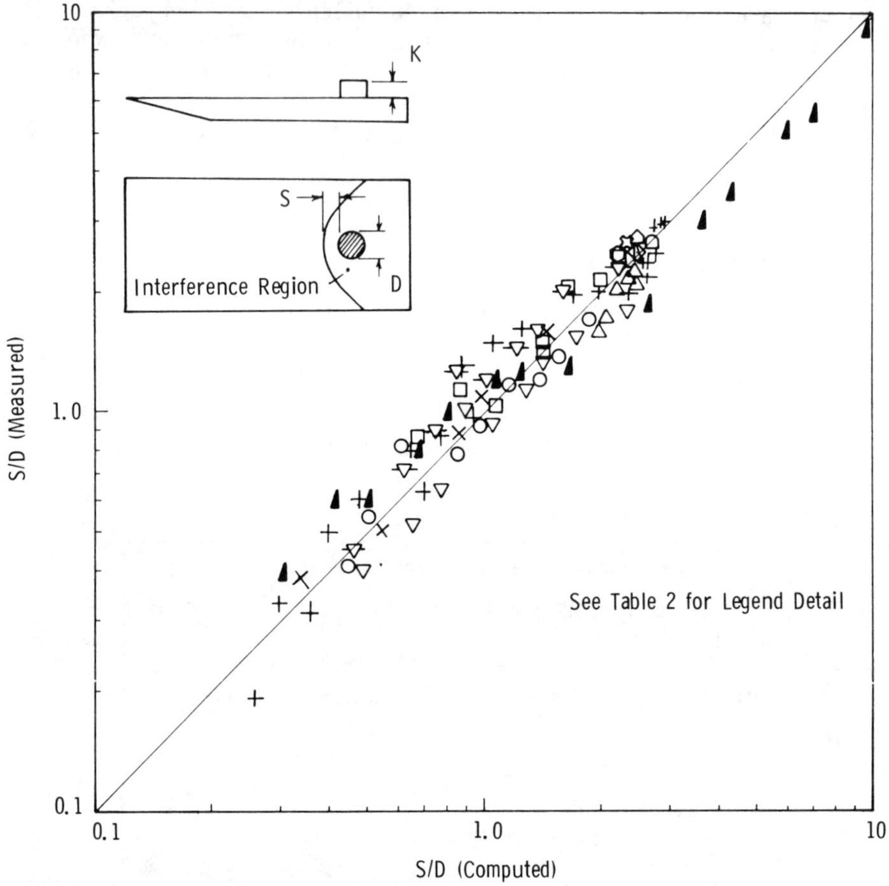

Fig. 23 Comparison between measured and computed primary separation distance values.

For turbulent flow

$$S/D = 1.45(K/\delta)^{0.7}\ (D/\delta)^{-0.435}$$

$$\text{- Short protuberance} \qquad (15)$$

$$S/D = 2.5 \text{ to } 3.0 \text{ - Long protuberance} \qquad (16)$$

Comparison between the measured S data with computed values [Eqs. (13) through (16)] is given in Fig. 23 and Table 2. The agreement is quite fair considering the errors and differences involved in different methods of measuring S values.

Table 2 Separation distance test summary (Fig. 2)

Legend symbol	Reference	M_∞	Re/ft	δ (in.)	D (in.)	K (in.)	Measurement
○	Halpin	2.71	------	0.19	0.5, 0.75, 1.0	0.1-0.5	Schlieren
□	Lucerc	2.71	$2.4(10^6)^*$	0.17	0.6, 1.0	0.05-0.75	Pressure
△	Waltrup, et al.	2.2	$3.0(10^6)$	0.173	0.6	0.8-2.1	Pressure
✣	Westkaemper	4.9	$11.5(10^6)$	0.25	0.25-1.5	0.03-1.5	Oil flow
◇	Burbark, et al.	2.65-4.44	$2.8(10^6)$	1.5, 6	2.8	12.5	Pressure
△	Hung and Scottoline	5.3	$3.5(10^6)$	0.59	2	8	Oil flow
○	Kaufman, et al.	2.5, 3, 4	$9.6(10^6)$	0.1	3/8, 3/4, 1	2.5	Schlieren, pressure, oil
∩	Surber and Luning	11.6-21.4	$0.05\text{-}5(10^6)$	3.9-6.3	1.5	7	Heat transfer
⌂	Price and Stallings	2.3-4.4	$0.5\text{-}4.5(10^6)$	6.07	0.58-3.5	1-10	Oil flow pressure
▽	Sedney	2.5, 3.5	$2.99(10^6)$	0.86-1.08	0.75-3.0	0.4-4	Oil, schlieren, shadowgraph
+	Present study	5.3	$5(10^6)^{**}$	0.521	1/4-8	1/8-8	Heat transfer
×	Present study	5.3	$5(10^6)^+$	0.326	1/4-8	1/8-8	Heat
▲‡‡	Present study	5.3	$1(10^6)^+$	0.162	1/4-8	1/8-8	Heat

* Re_x

** X_{SL} = 42 in.

$^+$ X_{SL} = 24 in.

‡‡ laminar flow, others turbulent flow

Conclusions

The three-dimensional protuberance interference heating test data were analyzed and correlated in such a way that generalized heating prediction methods were developed for thermal design purposes. The cylinder can be categorized as long or short, based on its nondimensionalized height. Large and small cylinders also were defined based on the nondimensionalized diameter. For a short cylinder, heating on the flat plate increases sharply with cylinder height until a long cylinder is reached. For a long cylinder, heating on the flat plate increases with cylinder diameter until a large cylinder is reached. One of the important findings from this study is the effect of the protuberance on the flat-plate boundary-layer state which is closely related to the heating levels. The data indicate that flow on the flat plate right in front of the protuberance is always laminar, even though the oncoming flow is turbulent. This is in contrast to the flow in the downstream region, where the laminar flow can be tripped to turbulent, resulting in higher heating.

V. References

[1] Kaufman, L.G., Korkegi, R.H., and Morton, L.C., "Shock Impingement Caused by Boundary Layer Separation Ahead of Blunt Fins," AIAA Journal, Vol. 10, 1973, pp. 1363-1364.

[2] Korkegi, R.H., "Effect of Transition on Three-Dimensional Shock-Wave/Boundary Layer Interaction," AIAA Journal, Vol. 10, 1972, pp. 361-363.

[3] Sedney, R., "A Survey of the Effect of Small Protuberances on Boundary-Layer Flows," AIAA Journal, Vol. 11, 1973, pp. 782-792.

[4] Sedney, R., "The Structure of Three-Dimensional Separated Flows in Obstacle, Boundary-Layer Interactions," AGARD Conference Proceedings No. 168, "Flow Separation," 1975, pp. 37.1-37.15.

[5] Halprin, R.W., "Step Induced Boundary-Layer Separation Phenomena," AIAA Journal, Vol. 3, 1965, pp. 357-359.

[6] Westkaemper, J.C., "Turbulent Boundary-Layer Separation Ahead of Cylinders," AIAA Journal, Vol. 6, 1968, pp. 1352 1355.

[7] Westkaemper, J.C., "Step Induced Boundary-Layer Separation Phenomena," AIAA Journal, Vol. 4, 1966, pp. 1147-1148.

[8] Waltrup, P.J., Hall, D.G., and Schetz, J.A., "Flowfield in the Vicinity of Cylindrical Protuberances on a Flat Plate in Supersonic Flow," Journal of Spacecraft, Vol. 5, 1968, pp. 127-128.

[9] Whitehead, A.H., "Flow-Field and Drag Characteristics of Several Boundary-Layer Tripping Elements in Hypersonic Flow," NASA TN D-5454, 1969.

[10] Hiers, R.S., and Loubsky, W.J., "Effects of Shock-Wave Impingement on the Heat Transfer on a Cylindrical Leading Edge," NASA TN D-3859, 1967.

[11] Couch, L.M., "Flow-Field Measurements Downstream of Two Protuberances on a Flat Plate Submerged in a Turbulent Boundary Layer at Mach 2.49 and 4.44," NASA TN D-5297, 1969.

[12] Burbank, P.B., Newlander, R.A., and Collins, I.K., "Heat-Transfer and Pressure Measurements on Attached Protuberances in a Supersonic Turbulent Boundary Layer at Mach Numbers of 2.65, 3.51, and 4.44," NASA TN D-1372, 1962.

[13] Surber, T.E., and Luning, R.B., "Hypersonic Heat Flux and Pressure Measurements near a Surface Protuberance," North American Aviation Report NA-64-1194-1, 1965.

[14] Eckert, E.R.G., "Survey of Boundary Layer Heat Transfer at High Velocities and High Temperatures," WADC Technical Report 59-624, 1960.

[15] Spalding, D.B., and Chi, S.W., "The Drag of a Compressible Turbulent Boundary Layer on a Smooth Flat Plate with and Without Heat Transfer," Journal of Fluid Mechanics, Vol. 18, Part 1, 1964, pp. 117-143.

[16] Fay, J.A., and Riddell, F.R., "Theory of Stagnation Point Heat Transfer in Dissociated Air," Journal of Aeronautical Sciences, Vol. 25, 1958, pp. 73-85, 121.

[17] Winkelman, A.E., "Flow Visualization Studies of a Fin Protuberance Partially Immersed in a Turbulent Boundary Layer at Mach 5," Naval Ordnance Laboratory Report NOLTR 70-93, 1970.

[18] Lucero, E.F., "Turbulent Boundary Layer Separation Induced by Three-Dimensional Protuberances on a Flat Plate," M.S. Thesis, University of Maryland, 1966.

[19] Hung, F.T., and Scottoline, C.A., "Aerodynamic Heating in Gaps in the Presence of Protuberances and Shock Wave Impingement," AIAA Paper 75-670, 1975.

[20] Kaufman, L.G., Korkegi, R.H., and Morton, L.C., "Shock Impingement Caused by Boundary Layer Separation Ahead of Blunt Fins," AIAA Paper 73-236, 1973.

[21] Price, E.A., and Stallings, R.L., "Investigation of Turbulent Separated Flows in the Vicinity of Fin-Type Protuberances at Supersonic Mach Numbers," NASA TN D-3804, 1967.

HYPERSONIC FLOW OVER SMALL SPAN FLAPS IN A THICK TURBULENT BOUNDARY LAYER

L. A. Cassel*
TRW Defense & Space Systems Group, San Bernardino, Calif.
and
T. W. Jarrett+
Science Applications, Inc., Irvine, Calif.

Abstract

Pressure and aeroheating distributions on finite span trailing flaps in hypersonic flow were measured. The flaps were trailing from a flat plate with a sharp leading edge and cold wall under a turbulent boundary layer with a Mach number of 6.1 at the edge. The flap chord length was the order of the approaching boundary-layer thickness. Three flap spans were tested at compression deflection angles of 5, 10, and 15 deg. Both chordwise and spanwise aeroheating and pressure distributions are reported. Methods of correlating the chordwise pressure and aeroheating distributions on the flap with the flow properties profiles in the approaching boundary layer are described. The influence of the proximity of finite span edges on both chordwise and spanwise aeroheating and pressure distributions is analyzed. The span edge influence on pressure distributions is shown to be characteristically different than the edge influence on aeroheating distributions.

Nomenclature

C_{P1} = pressure coefficient $(P - P_1)(\gamma M_{1_e}{}^2 P_1/2)$
C_{Pe} = $(P_{2_e} - P_{1_e})/(\gamma M_{1_e}{}^2 P_1/2)$
L_f = flap chord length
M = Mach number
P = pressure
$\bar{S}t_1$ = Stanton number, $\dot{q}/\rho_1 u_1 c_p (T_w - T_r)$
St_1 = centerline value on widest span flap

Presented as Paper 80-1475 at the AIAA 15th Thermophysics Conference, Snowmass, Colo., July 14-16, 1980.

*Section Head, Aeromechanics Projects.
+Staff Scientist.

Re = Reynolds number, $\rho ux/\mu$
T = temperature
b = flap span
c_p = specific heat at constant pressure
Δx = chordwise distance along flap (Fig. 12)
Δy = spanwise distance from flap edge (Fig. 12)
$\dot{q}$ = rate of heat transfer to wall
r = recovery factor ($r = 0.89$)
x = chordwise distance from leading edge
x_t = end of transition (maximum heating)
x_{t_1} = beginning of transition
y = spanwise distance from centerline
u = velocity
u_τ = friction velocity $(\tau_w/\rho_w)^{1/2}$
δ = boundary-layer thickness
θ = boundary-layer momentum thickness
λ = flap deflection angle
μ = dynamic viscosity; or Mach angle, $\sin^{-1}(1/M)$
ν = kinematic viscosity
ρ = density
τ = shear stress

Subscripts

1 = condition upstream of flap shock
2 = condition downstream of flap shock
o = flow stagnation condition
∞ = condition in wind-tunnel stream/ambient flow
℄ = condition or measurement at centerline
c = evaluated at compression corner location
e = condition at boundary-layer edge
f = referred to flap
r = recovery (adiabatic wall) value
w = evaluated at wall or wall temperature
D = leading edge diameter or thickness

Introduction

Trailing edge flaps are favored control concepts for a wide range of advanced hypersonic missiles, aircraft, and reentry vehicles. The design of these controls and evaluation of their effectiveness is a continuing contemporary problem in hypersonic flight vehicle development. Fundamental to the design process is the prediction of aeroheating and pressure distributions over the flap surface. There are three design parameters of interest. The integral of the pressure distribution provides the control forces. The moment of that distribution is required to design actuator systems and predict dynamics. Finally, knowledge of the variation in overheating over the surface is required to design thermal protection systems.

In many applications of interest, the critical design environments occur under conditions when the boundary layer approaching the flap is turbulent, since these conditions coincide with high dynamic pressures. For hypersonic flight vehicles, this condition may be impossible to simulate on geometrically scaled full vehicle models in wind tunnels. Reynolds number limitations often preclude successful attainment of turbulent boundary layers upstream of controls on such models. This is particularly the case for the "thick" turbulent boundary-layer condition when the flap chord is the same order of magnitude as the approaching boundary-layer thickness, which is a common design condition. Attainment of turbulent boundary layers through employment of boundary-layer trips which effect the boundary-layer development may introduce significant error into the simulation of flow over the flap. Consequently, a source of data defining the behavior of pressure and aeroheating distributions is a valuable aid to the design aerodynamicist.

A number of efforts involving either extensive experiments or complex analytical model development have addressed the problem of defining flow in compression corners. Those which have dealt with turbulent boundary layers have, for the most part, addressed flows which are two-dimensional in nature. They have modeled the effectively infinite span flap. The most extensive experiments have been principally directed toward the separated flow problem. A frequent objective of those investigations has been the characterization of incipient separation conditions. They have produced analytical models and experimental data pertinent to the attached boundary layer with high flap deflection angle.

In the effort reported here, the investigation has been directed to compression flap deflection angles much smaller than those resulting in turbulent boundary-layer separation. An experiment was conducted to define the pressure and heat-transfer distributions over flaps, under flow conditions approaching two-dimensional at the centerline. The chordwise and spanwise variations in pressures and aeroheating rates were measured as functions of flap deflection angle. Finally, the

Fig. 1 Finite Span Flap Flat Plate Model.

influence of reduction in flap span to the point where opposite edge influence was clearly present at the centerline was investigated. Details concerning the experiment conducted, and data beyond that presented here can be found in the paper and report by Cassel, et al.[1,2] The experiments were conducted in the Continuous Flow Hypersonic Tunnel (CFHT) at the National Aeronautics and Space Agency (NASA) Langley Research Center (LaRC).

Facility and Apparatus

Continuous Flow Hypersonic Tunnel

The LaRC CFHT is a fixed nozzle, return duct facility which is frequently operated in an intermittent or blowdown mode, as it was for this experiment. The fixed nozzle transitions from a circular throat to a square test section, 78.74 cm on a side, operating at a nominal freestream Mach number of $M_\infty = 10$. The working fluid is air which is heated after compression by electrical resistance tube bundles. For the experiment reported here, the tunnel was operated at a nominal Mach number of $M_\infty = 10.2$ with stagnation temperature, T_{o_∞}, of 978°K and reservoir pressure, P_{t_∞}, of 11.38×10^6 N/m^2. This provided a freestream Reynolds number of $Re_\infty = 7.2 \times 10^6$/m.

Finite Span Flap Model

The model employed for the finite span flap experiment is illustrated in Fig. 1. The geometry of the model is a flat plate surface over a rectangular profile with a 20 deg wedge angle leading edge. Three leading edges were tested, representing sharp and cylindrically blunted wedges. Only data for the effectively sharp leading edge is reported here. The diameter of that leading edge was 0.94 mm. The trailing edge flaps were mounted with 0, 5, 10, and 15 deg compression angles relative to the plate surface. The flaps are mounted with 15 deg boat-tail expansion bevels on the plate trailing edge adjacent to their span extremes. The flat plate length was 81.28 cm from leading edge to flap compression corner. Each flap had the same chord length, 10.16 cm. The three flap spans illustrated, b = 5.08, 10.18, and 12.7 cm were interchangeable. The flaps were constructed as the surfaces of closed boxes which provided thermal protection to instrumentation. Two sets of model surface plates and two sets of flaps were interchangeable. One set was instrumented with backface thermocouples and the other with pressure orifices.

The flap model was mounted to the tunnel sidewall on a pylon at an angle of attack of 12 deg compression on the plate surface. The pylon attachment is designed to keep the leading edge clear of the 18.3 cm thick tunnel wall boundary layer, and the trailing edge of the flap within the reflected bow shock of the sharp leading edge. The width of the plate is limited to maintain the span extremes inside the tunnel sidewall boundary layer.

Tunnel Sidewall Flap Model

In addition to the data taken on the flat plate model, a limited set of pressure and heat-transfer measurements were made on a plate mounted at a 15 deg compression angle on the tunnel sidewall. The sidewall flap model is illustrated in Fig. 2. The chord of the sidewall flap model is 60.46 cm. The span is 30.48 cm.

Instrumentation

Heat-Transfer Measurements

Chromel-alumel thermocouples were welded to the backface of nominally 0.081 cm thick counter-bored sections of the flat plate. The surface of each flap was nominally 0.081 cm thick, with thermocouples welded to the backface at measurement locations. Thermocouple leads were welded with approximately 0.08 cm separation perpendicular to the mean flow direction. Data were taken with 2-5 s residence times in the flow. Thermocouples were located no closer than 1.27 cm from the conduction heat sinks and no closer than 0.635 cm to each other, except at a few control points on the flaps. At these points a few thermocouples were mounted within 0.635 cm of the flap edges to detect the signature of the edge effects on the temperature rise. The nonlinearity of the temperature rise at these points, where conduction losses were known present, was small, but noticeable. These data were ignored in the analyses. At all other points, temperature rises were linear within measurement accuracy. No corrections were made for conduction or radiation.

One row of 11 thermocouples was located along the plate centerline, upstream of the flaps. A transverse row of thermocouples was located 7.62 cm forward of the compression corner to detect spanwise heating gradients approaching the flap. Measurements were made 5.08 cm from the plate edges to validate symmetry of the lateral heating rate distribution approaching the flap. There was a slight aft facing step discontinuity in the flat plate surface at a joint between the forward and aft plate sections, 17.78 cm forward of the flap

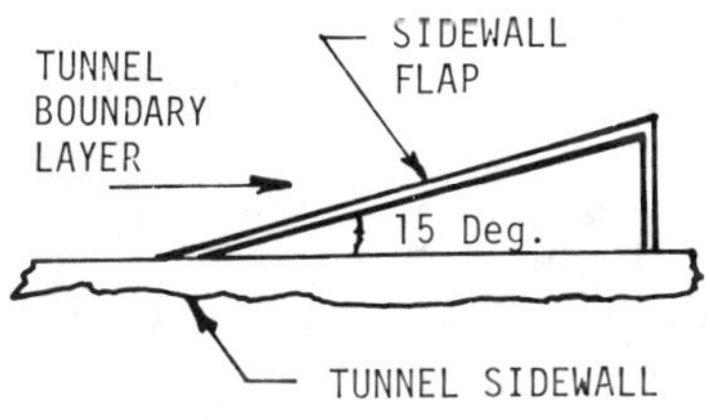

Fig. 2 Sidewall Ramp Model.

compression corner. Rows and columns of measurement points on the flap were separated by 0.635 cm from one to the next.

The sidewall ramp model employed the same type of instrumentation as the flat plate model. Pressure taps were located at 5.08 cm intervals along the chord, 1.27 cm from the centerline. Thermocouples were attached at image locations on the other side of the centerline.

Surface Pressure Measurements

Surface pressure measurement locations on the flat plate model are separated by intervals which are multiples of 0.635 cm. The surface pressure measurements were made with a multiple range, self-ranging, capacitance type transducer system. The total range of the system is $0 - 69.8 \times 10^3$ N/m^2. The minimum range increment is $0 - 68.9$ N/m^2. The transducers are located in the model injection system, outside the tunnel sidewall which the model is attached to. The pressure on the model surface is sensed through 0.178 cm i.d. tubing approximately 2 m long. Because of limitations in recording channels, fewer pressure measurements were made than heat-transfer measurements. In most cases, two tunnel runs with different pressure orifice hookups were employed to provide a complete set of pressure measurements for each model configuration condition. The pressure measurement locations duplicate some heat-transfer measurement locations. On the sidewall flap model, the same pressure instrumentation was employed as on the finite span flap model.

Boundary-Layer Pitot Pressure Surveys

The pitot pressure distribution perpendicular to the plate surface was measured through a rake of tubes. The rake was constructed with flattened tube leading edges 0.079 cm tall (O.D.) and 0.198 cm wide. Minimal centerline separation of tubes was 0.198 cm. The rake was mounted in place of a flap, with a flat surface under the bottom tube. The leading edge of the rake was at the flap compression corner location.

Approaching Flow Conditions

In the experiment reported here, surface pressure and heat-transfer rates forward of the flap were measured in an attempt to verify boundary layer transition to turbulent flow upstream of the interaction. The objective of the experiment design was to achieve the point of maximum heating rate at a location at least ten times the boundary-layer thickness upstream of the compression corner. As described in Ref. 1, this condition was actually achieved, and the testing reported in this paper was carried out, with the plate at 12 deg angle of attack and a surface Mach number of $M_{1_e} = 6.1$. Under these conditions, the length Reynolds number at the boundary layer edge, at the compression corner was $Re_{x_e} = 8.8 \times 10^6$.

Aeroheating Data Quality

The centerline distribution of Stanton number along the flat plate model surface upstream of the flap is shown in Fig. 3. The data from two different tunnel runs are shown. In one of these runs, the aft facing step between the forward and aft plate sections was 0.76 mm instead of the nominally acceptable less than 0.25 mm. As is evident, the data is repeatable and the higher step (at 63.5 cm station) has little influence on the downstream heating data. As will be described below, the boundary-layer thickness at the compression corner was found to be 0.79 cm. Therefore, as indicated in Fig. 3, the point where the surface heating rate was a maximum occurred at a distance larger than ten times the boundary-layer thickness upstream of the compression corner.

Boundary-Layer Profile

The pitot pressure distribution measured in the boundary layer on the finite span flap model, is illustrated in Fig. 4a. It was measured on the centerline, at the location of the compression corner, with no compression present (zero flap deflection). The static pressure measured at this location was 2.3 kN/m^2. This agrees very closely with the value predicted on the basis of inviscid flow compression from freestream conditions to the sharp flat plate value at 12 deg angle of attack. Assuming this to be constant through the boundary

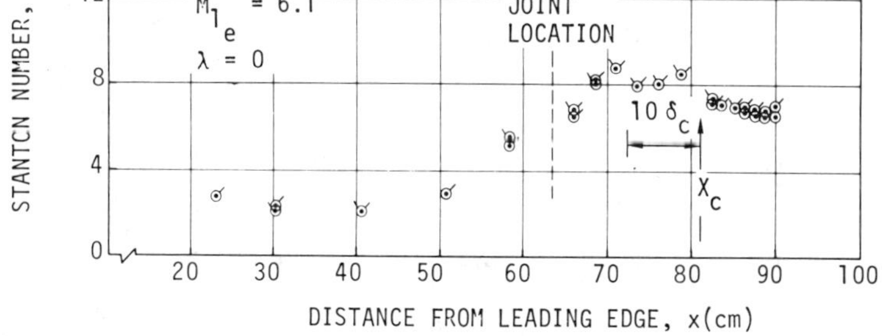

Fig. 3 Aeroheating Data Consistency.

layer, the Mach number distribution can be determined from isentropic flow equations.[3] The velocity distribution in the boundary layer can then be calculated with the assumption of the linear relationship between velocity and total temperatures given by the modified Crocco relationship

$$T/T_e = T_w/T_e + (T_r/T_e - T_w/T_e)(u/u_e) + (1 - T_r/T_e)(u/u_e)^2$$

The inferred velocity at survey points above the wall point have been transformed to logarithmic coordinates in Fig. 4b. A fairing through

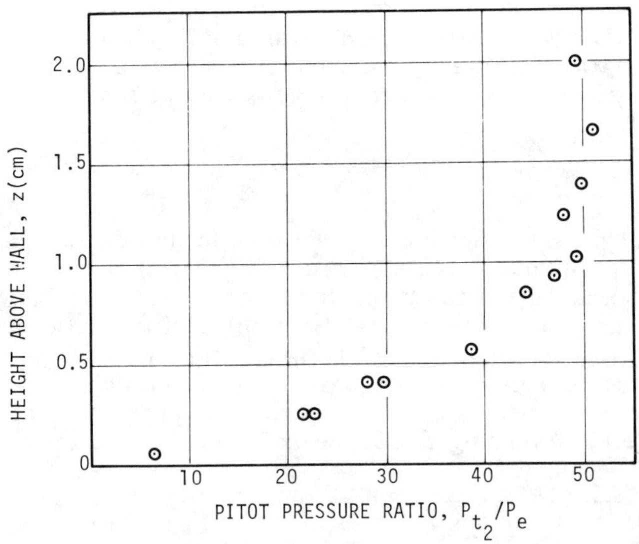

Fig. 4a Pitot Pressure Profiles.

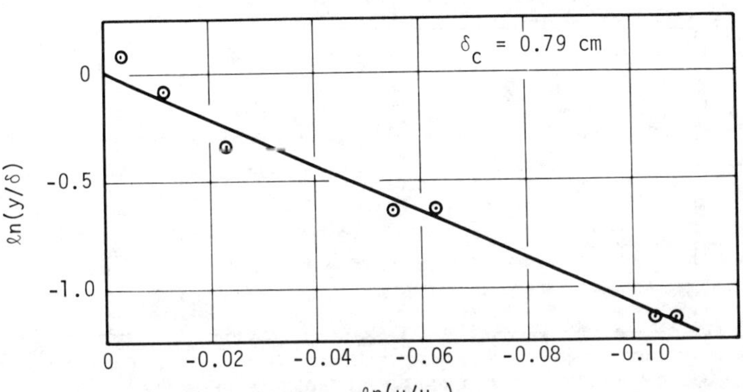

Fig. 4b Velocity Profile.

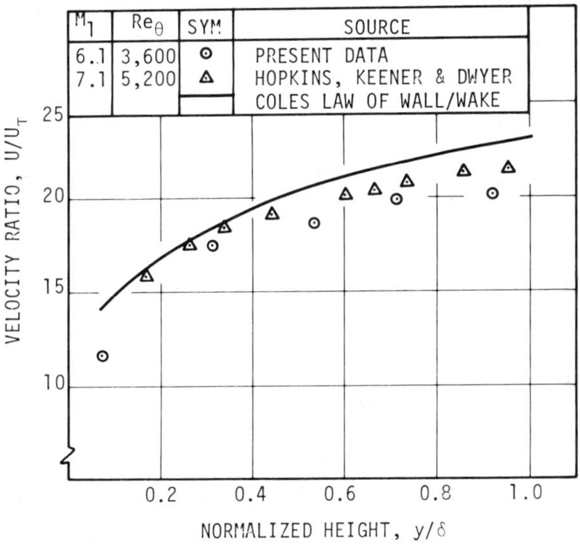

Fig. 4c Transformed Velocity Profiles.

these values yields an approximate power law outer layer in the form

$$u/u_e = (y/\delta)^{1/11.4}$$

where the boundary layer thickness, δ_C is 0.79 cm.

Following the methods employed by Hopkins, Keener, and Dwyer,[4] the velocity distribution in the boundary layer has been transformed to the reduced velocity profile through the second method derived by van Driest.[5] This distribution is compared to the Hopkins, Keener, and Dwyer data transformed the same way in Fig. 4c. Also shown is Coles law of the wall and wake in the form given by Cebeci and Smith[6]

$$u/u_\tau = 1/\kappa \ln (u_\tau y/\nu_w) + C + 1/\kappa \, \Pi W$$

where:

$$\kappa = 0.41, \ C = 5.0$$

$$\Pi = 0.55 \{1 - \exp [- 0.243 \sqrt{Re_\theta/425 - 1} - 0.298 (Re_\theta/425 - 1)]\}$$

$$W = 2 \sin^2 [(\pi/2) (y/\delta)]$$

The data presented in Figs. 3 and 4 indicate that the boundary layer approaching the flap in the present experiment is turbulent. However, the degree of development of a self-similar profile may be suspect. Consequently, it appears necessary to travel more than ten times the boundary-layer thickness downstream of the maximum heating point to achieve a self-similar turbulent boundary-layer profile at the conditions of this experiment.

Lateral Flow Effects

For the flat plate model tested, the spanwise distribution of pressures and aeroheating on the plate, 7.62 cm forward of the compression corner, were reported by Cassel, et al.[1] These appear adequately level over the region of the flap span. While this does not guarantee uniform, parallel flow approaching the flap, it is indicative that the spanwise gradients are small over the span of the flap.

Surface pressure and aeroheating measurements on the NASA LaRC CFHT sidewall, approaching the location where the sidewall flap was mounted for the present experiment, were reported by Throckmorton.[7] Unfortunately, there is significant lateral flow in the tunnel sidewall boundary layer according to Throckmorton's data. Pitot pressure profiles reported by Throckmorton confirm the spanwise variation in the flow. The data are not shown there, but Throckmorton also reports departures from the Crocco integral relating velocity to temperature profiles in the CFHT wall boundary layer. The assumption of a quadratic relationship between velocity and temperature is reported to fit the CFHT boundary-layer data. A similar relationship has been reported to fit other nonadiabatic, nonequilibrium wind tunnel wall boundary layers.[8]

Flap Centerline Flow Properties

Despite the low aspect ratio of the widest flap in this experiment, the distribution of pressure and aeroheating along the centerline appears consistent with distributions measured on nearly two-dimensional flaps by other investigators. The spanwise pressure and aeroheating distributions were constant over a large fraction of the widest span flap, as will be shown below.

Flap Centerline Pressure Distribution

The distribution of pressure coefficients along the flap centerline is illustrated for three deflection angles in Fig. 5. The distribution at zero deflection is shown for comparison.

It has been suggested by Thomke and Roshko[9] and by Elfstrom[10] that the compression interaction is modeled well by an inviscid, rotational flow representing the outer approaching boundary layer. If this is the case, the length scale for the compression interaction is the oncoming boundary-layer thickness. To test this hypothesis, the present data is compared to several sets of compression corner pressure data under various conditions in Fig. 6.

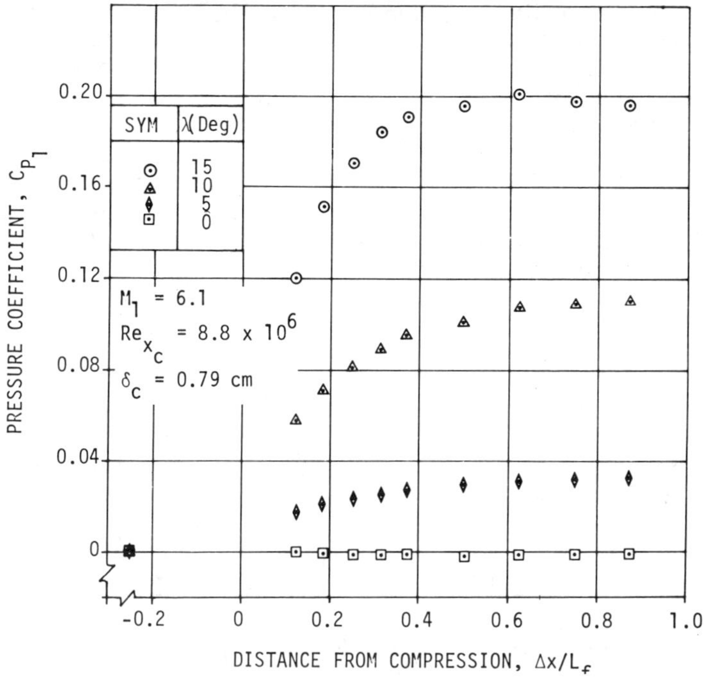

Fig. 5 Flap Centerline Pressure Distributions.

The comparisons in Fig. 6 certainly demonstrate the validity of scaling the pressure distribution with the oncoming boundary-layer thickness. The boundary-layer thickness differences between data represented vary by more than an order of magnitude. The correlation between the cold wall and adiabatic wall data may appear intuitively surprising, however Elfstrom[10] has shown an insensitivity of the attached flow compression pressure distribution to wall temperature for cold walls.

The influence of the nonlinear temperature-velocity relationship typical of nozzle boundary layers in strong pressure gradients is evident in the sidewall data from the present experiment. It can be compared to Elfstrom's data at nearly the same Mach number and wall temperature ratio. The lack of correlation can be traced to the different Mach number profiles illustrated in Fig. 6a. Throckmorton's[7] description of the Mach number profile on the CFHT sidewall should closely approximate the boundary layer in the present sidewall experiment.

It was observed in the present investigation that the pressure coefficient associated with the isentropic flow pressure rise appears to approximately correlate the effect of flap deflection up to high deflection angles, as illustrated in Fig. 7. In that figure, the value C_{P_e} is the observed asymptotic level rather than the calculated inviscid flow pressure rise. They differ at low flap deflection where small deflection angle errors in the experiment are influential. Though these are not known to have been present, some uncertainty must be assumed. An empirical representation of the pressure rise on flaps downstream of turbulent boundary layers in dynamic equilibrium can be derived from the set of data illustrated in Fig. 6b cast in the form of Fig. 7. The empirical equation which fits the data in Fig. 6b reasonably well for $x/\delta_c < 7$ is

$$C_p/C_{p_e} = m_1 + m_2\,(\Delta x/\delta_c) - m_3\,(\Delta x/\delta_c)^2$$

SYM	M_{1_e}	δ_c (cm)	T_w/T_o	EXPERIMENT
X	10.2	18.3	0.33	THROCKMORTON
◇	9.2	0.71	0.28	ELFSTROM
⊙	6.1	0.79	0.33	PRESENT
Δ	3.9	12.1	1	THOMKE/ROSHKO

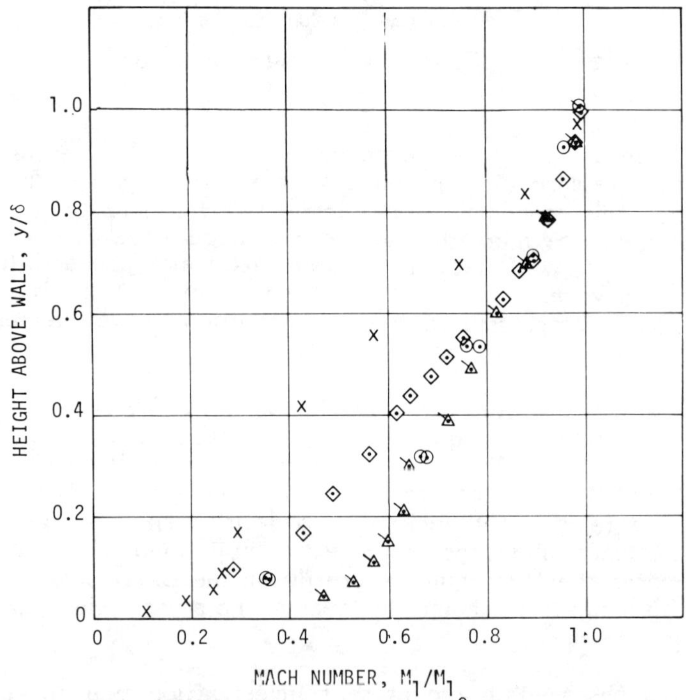

Fig. 6a Approaching Mach Number Profiles.

where

$$m_1 = 0.65 - 0.058\, M_{1_e}$$

$$m_2 = 0.099 + 0.017\, M_{1_e}$$

$$m_3 = 0.0071 + 0.0012\, M_{1_e}$$

Obviously, it is implied that the pressure reaches the inviscid level and remains constant for $x/\delta_c > 7$. Data provided by Elfstrom[10] and Holden[11] show that this assumption fails at high deflection angles. However, this empirical representation may be useful to some deflection angles beyond 15 deg for some estimation purposes. The data for 30 deg deflection in Fig. 7 indicate divergence of observed pressure levels from the correlation at higher deflection angles. Partial explanation for this divergence may be the approach to flow separation which Elfstrom found to occur near 30 deg deflection for the condition of his test. If reasonable estimates of the Mach-number profile in the approaching boundary layer are available, the correlation and transformation methods described below should provide signifi-

SYM	M_{1_e}	δ_c (cm)	Re_δ X 10^5	T_w/T_o	EXPERIMENT
◬	3.9	12.1	55.5	1	THOMKE/ROSHKO
△	4.9	13.9	53	1	THOMKE/ROSHKO
⊙	6.1	0.79	0.9	0.33	PRESENT
◇	9.2	0.71	4.0	0.28	ELFSTROM
X	10.2	18.3	13.9	0.33	PRESENT

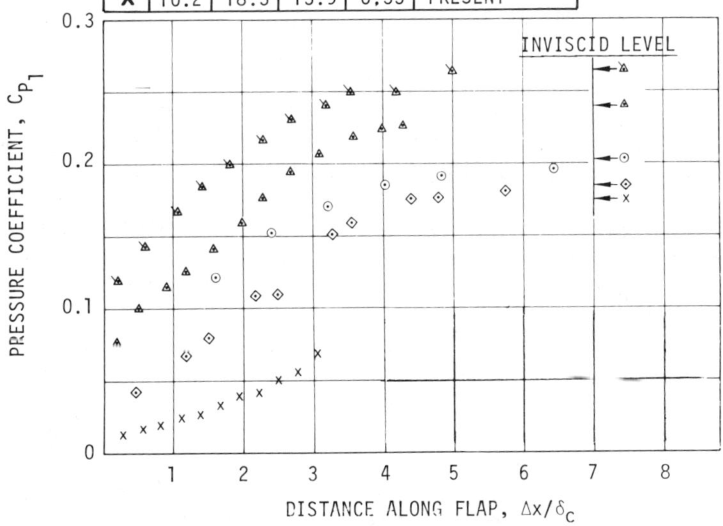

Fig. 6b Flap Pressure Distributions.

cantly better predictions. The methods to be described appear valid for deflection angles approaching those resulting in flow separation.

Flap Centerline Aeroheating Distributions

The distribution of aeroheating rates measured on the widest span trailing flap on the flat plate model is shown in Fig. 8. Stanton number distributions are shown for 0, 5, 10, and 15 deg deflection angles. The aeroheating distribution on the side wall ramp ($\lambda = 15$ deg) is illustrated in Fig. 9. For comparison, the data reported by Coleman and Stollery[12] for a flap trailing from a flat plate at approximately the same edge Mach number are illustrated in Fig. 9. The Coleman and Stollery[12] data were taken in the same approaching flow as the pressure data reported by Elfstrom[10] (Fig. 6). The wide discrepancy between the Mach number profiles (Fig. 6a) make it difficult to draw conclusions from Fig. 9 regarding the adequacy of approaching boundary-layer thickness as a length scale for the aeroheating distribution. However, this discrepancy is analogous to the discrepancy between Elfstrom's data and pressure data from the present sidewall test noted above.

Correlation of Flap Centerline Flow Properties

The behavior of pressure and/or aeroheating distributions along the centerline chords of compression flaps or ramps of effectively infinite span is reported by Thomke and Roshko,[9] Holden,[11,13] Elfstrom,[10,14] and by Coleman and Stollery.[12] The entire bibliography of references in this area (e.g., see Elfstrom[14]) is extensive. These are cited because they provide original data for turbulent boundary-layer interactions and some evidence concerning the establishment of a turbulent boundary layer approaching the interaction in each case. Each of these investigations was directed toward definition of incipient separation and the characterization of parameters variation in separated flow. However, they also provide valuable data on the attached flow case, usually at high deflection angles.

From the earliest of these efforts,[9] it has been suggested that the pressure rise on the flap in attached flow could be calculated by treating the supersonic region of the oncoming boundary layer as an inviscid, rotational flow. This approach had previously been suggested for incident shock boundary layer interactions by Rose, et al.[15] A particularly simple implementation of this concept is discussed by Elfstrom.[10] The geometry employed is illustrated in Fig. 10. At intervals, y, above the wall, in the approaching boundary layer, the Mach number, $M_1(y)$, is determined. The associated shock angle, $\beta(M_1)$, provides the Mach number, $M_2(y)$, downstream of the flap shock. It also provides the static pressure rise across the shock, which is then projected along the Mach line, $\mu(M_2)$, to the flap surface.

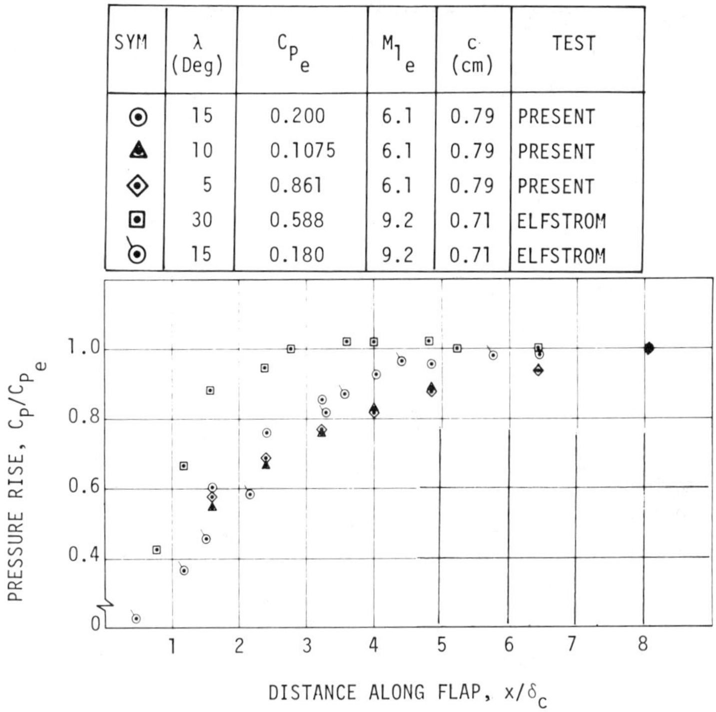

Fig. 7 Deflection Angle Influence on Normalized Pressure.

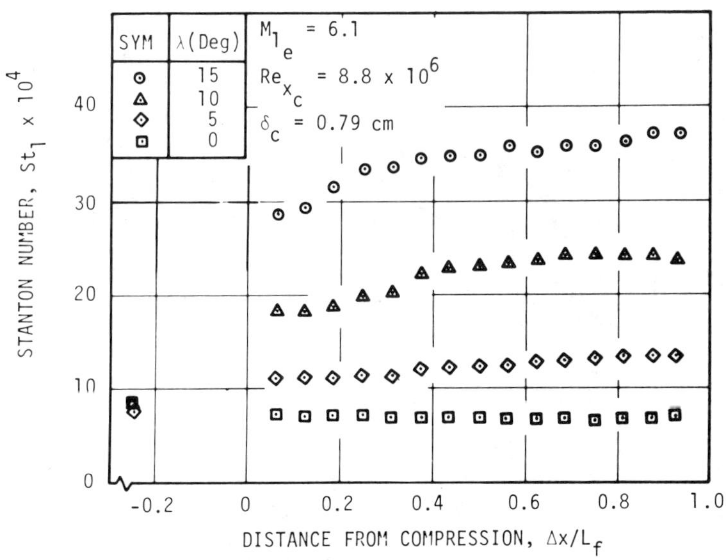

Fig. 8 Flap Centerline Aeroheating Distribution.

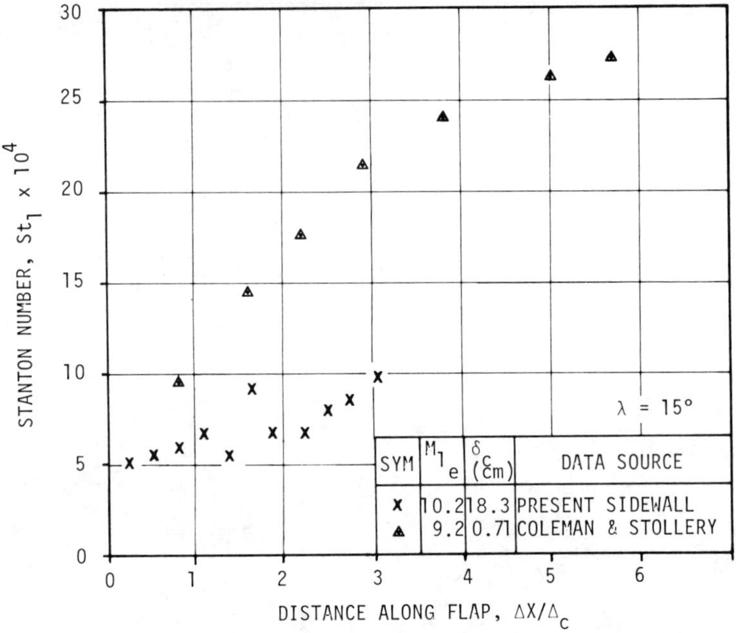

Fig. 9 Sidewall Ramp Aeroheating Distribution.

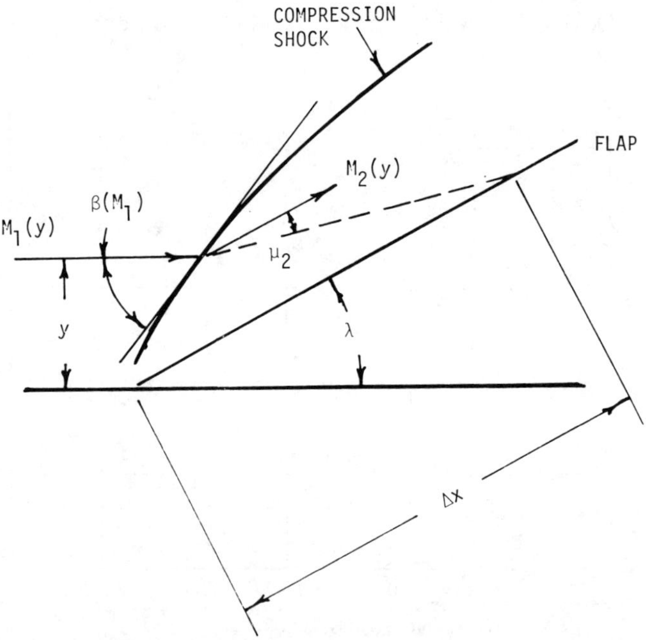

Fig. 10 Elfstrom Transformation Geometry.

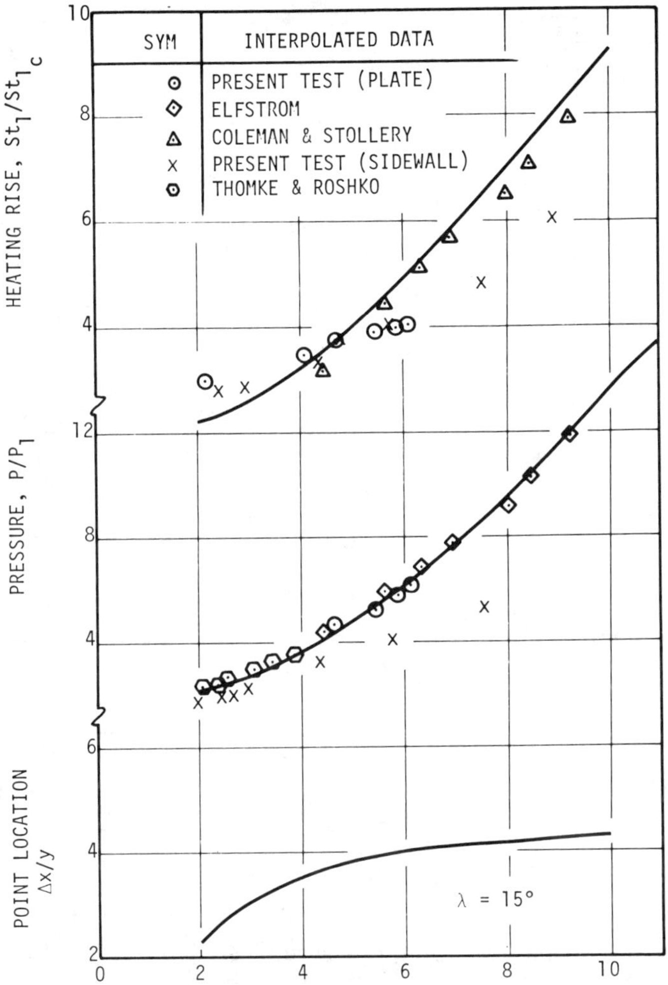

Fig. 11a Aeroheating and Pressure Distribution Correlation, $\lambda = 15$ deg.

Analysis of this geometry provides a relationship between the height of a Mach number, $M_1(y)$, in the approaching flow and the location of the induced pressure rise on the flap which is

$$\Delta x/y = 1/\sin\beta\,[\cos(\beta-\lambda) + \sin(\beta-\lambda)/\tan\mu_2]$$

From this transformation, and oblique shock relations,[3] the pressure and density rise at any point on the flap can be projected from knowledge of the approaching flow properties profile.

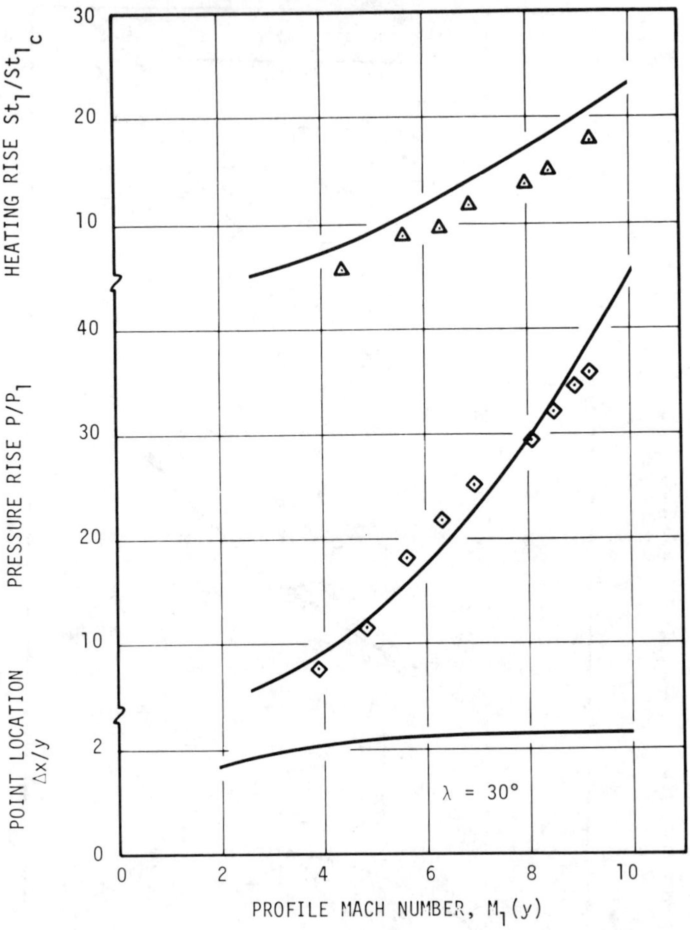

Fig. 11b Aeroheating and Pressure Distribution Correlation, λ = 30 deg.

Coleman and Stollery[12] derived an expression for the rise in heat-transfer rate across the flap shock which they reduce to the low deflection angle approximation

$$\dot{q}/\dot{q}_c = (\rho_2 u_2/\rho_1 u_1)(M_1/M_2)^{1.3}$$

They demonstrate that this approximation, combined with flow properties calculation scheme suggested by Elfstrom,[10] provides good agreement with measurements they made and with Holden's [13] measurements.

The Elfstrom approach to predicting flow properties on the flap has been employed in this work as a transformation from approaching

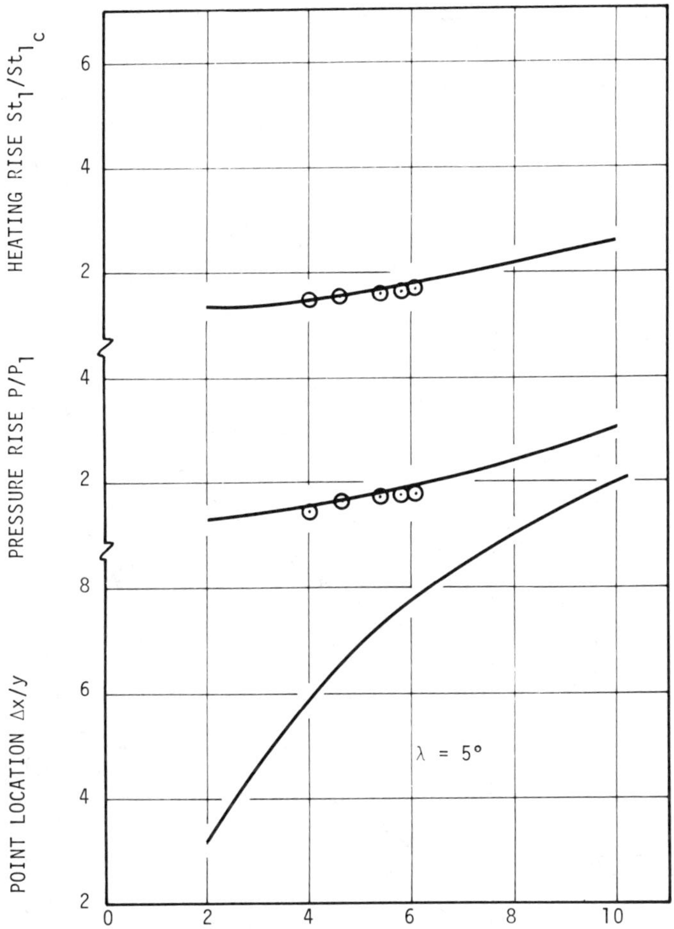

Fig. 11c Aeroheating and Pressure Distribution Correlation, $\lambda = 5$ deg.

flow Mach number profiles to chordwise distribution of properties on the flap. The relationship is illustrated for four flap deflection angles in Fig. 11. The heating rate rise at the transformed flow conditions is calculated by the Coleman and Stollery[12] approximation. Data from a number of experiments, at widely varying approaching flow conditions (see Fig. 6), are compared to the transformed flow conditions in Fig. 11a. Orders of magnitude differences in approaching boundary-layer thickness exist between sets of data shown. Cold wall[10] and adiabatic wall[9] data for the pressure rise are included. Finally, even the data for both the pressure rise and aeroheating rate rise measured on the sidewall ramp in the present experiment are approximately correlated. Consequently, some validity can be expected even when

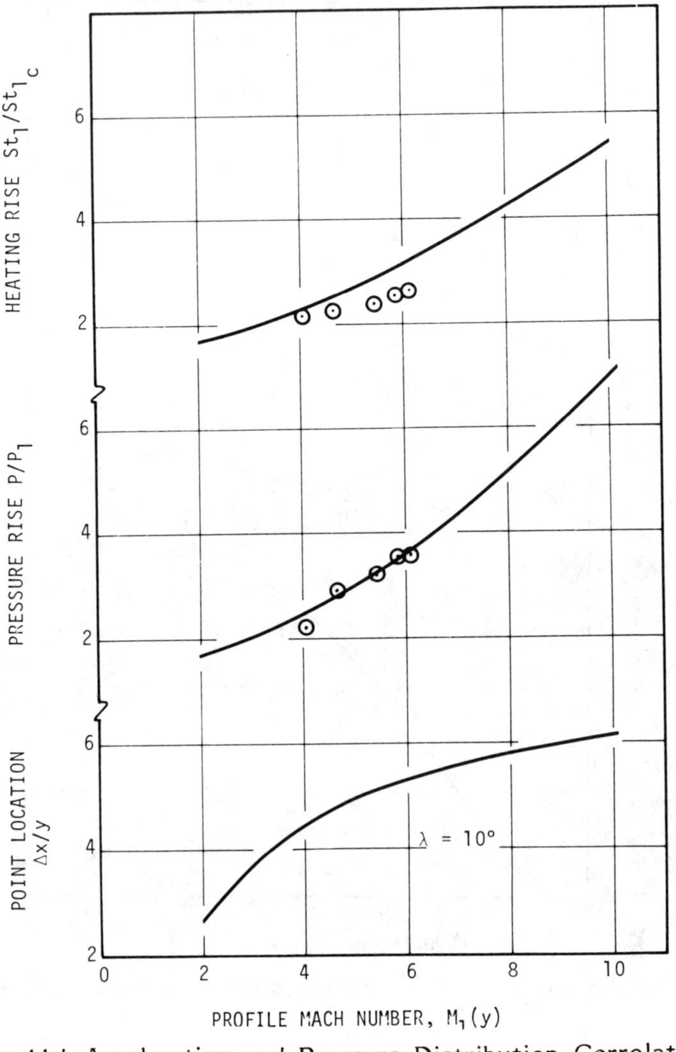

Fig. 11d Aeroheating and Pressure Distribution Correlation, $\lambda = 10$ deg.

the approaching boundary layer lacks dynamic equilibrium or self-similarity.

Variable Span Influence on Flow Properties

The elimination of finite span effects on shock boundary-layer interactions such as in compression corners is a subject of considerable concern in experimental fluid mechanics. Separated flow phenomena have been shown to be particularly sensitive to the proximity of edge

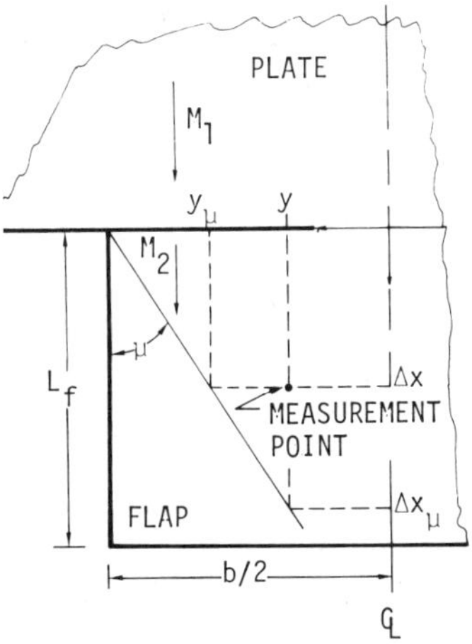

Fig. 12 Nomenclature for Spanwise Pressure and Aeroheating Data Analysis.

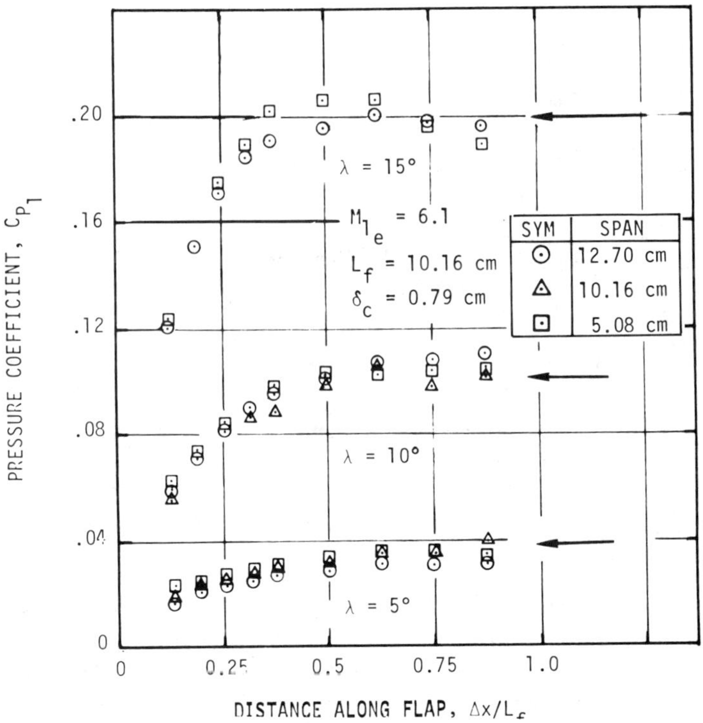

Fig. 13 Influence of Span on Flap Centerline Pressure Distribution.

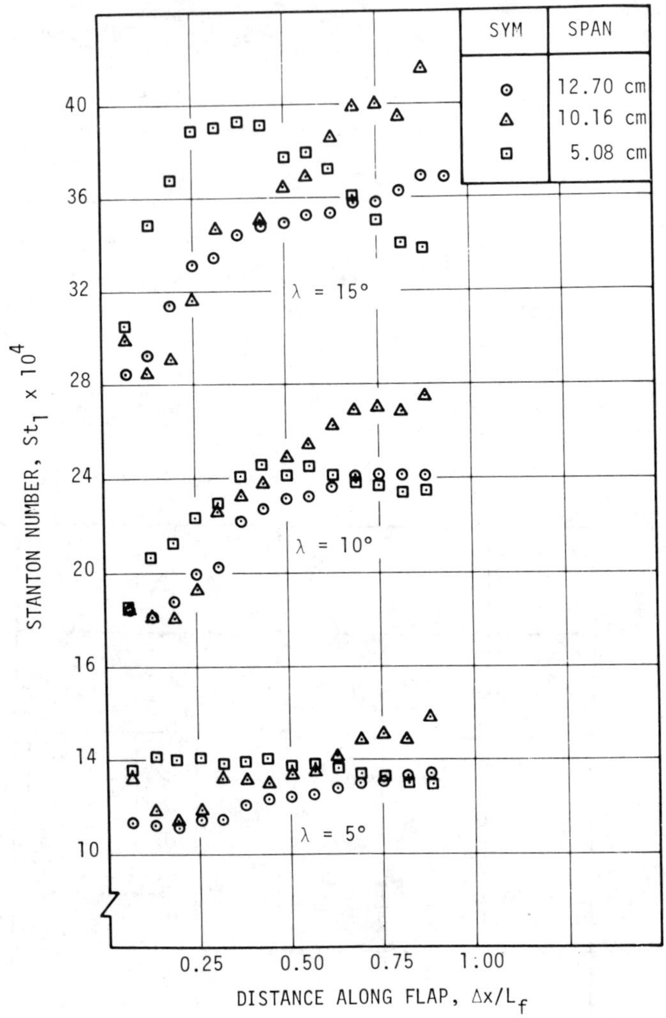

Fig. 14 Influence of Span on Flap Centerline Aeroheating Distribution.

effects. The present investigation was oriented toward flap design application, and therefore did not address the determination of interaction on effectively infinite span flaps. Fortunately, the data for the widest flap tested was in approximate agreement with results from experiments addressing the two-dimensional flow problem, as described above. The principal objective of the present effort was to investigate the influence of changing finite span on the aeroheating and pressure distributions on the flap. Some related effects of finite span geometry on laminar, transitional, and turbulent flow have been reported by Gray and Rhudy.[16]

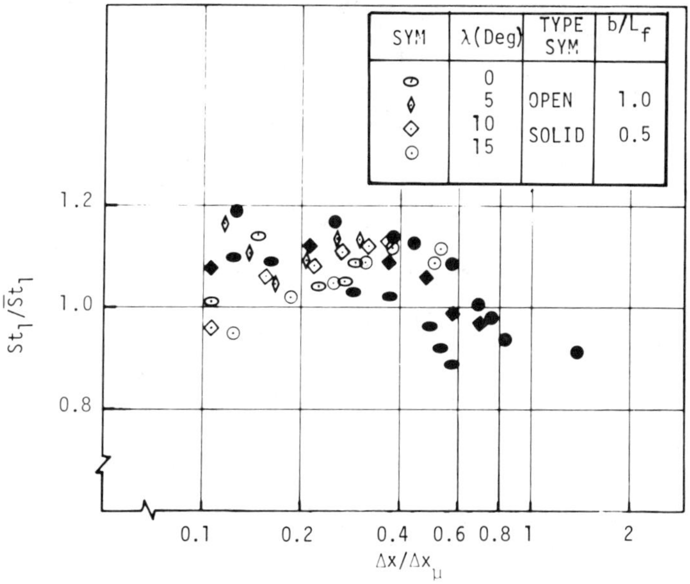

Fig. 15 Span Edge Effect on Centerline Aeroheating.

For purposes of analyses, the effect of variable span on the flow over the flap is viewed from two aspects. First, the influence of variable span on the centerline pressure and heating rate distribution is considered. Then the variation in spanwise distributions caused by varying aspect ratio is considered. The nomenclature for description of spanwise variations in properties is llustrated in Fig. 12. The Mach number downstream of the flap shock in the flow just outside the approaching boundary layer is termed M_{2_e}. The associated Mach line is $\mu_2 = \sin^{-1}(1/M_{2_e})$, which is closely approximated by the arctan or the argument itself in hypersonic flows.

Span Effect on Centerline Flow

The variation in chordwise distribution of pressure on the flap centerline is shown for three different span flaps in Fig. 13. (Data for only two flap spans are shown at λ = 15 deg because of apparently invalid results obtained for that condition with 10.16 cm span flap in the present experiments.) There appears to be very little influence of changes in span on the centerline pressure distribution. It appears that, if they existed, they were of the order of the data accuracy. However, further evidence is provided below in the discussion of spanwise distributions.

In distinct contrast, the influence of span variation on the chordwise distribution of aeroheating is quite significant, as illustrated

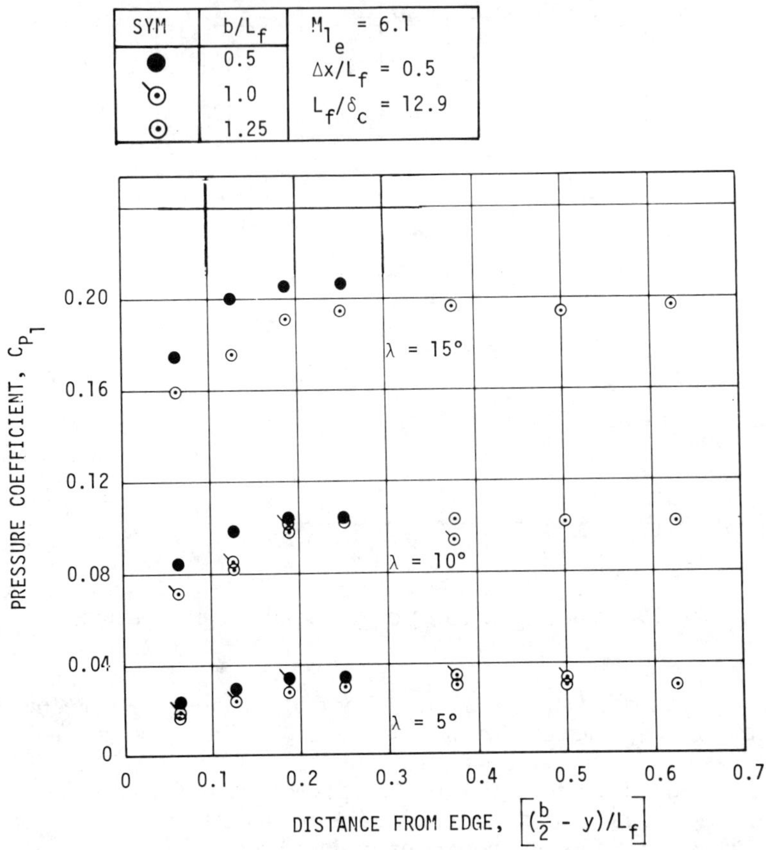

Fig. 16a Spanwise Pressure Distribution at Half Chord .

in Fig. 14. The forward movement of the peak in the chordwise distribution appears to be related to the proximity of edge effects. That is, it moves forward with decreasing span and inviscid flow Mach number. To investigate this edge effect influence, the ratio of measured Stanton number to that recorded on the widest span flap (b = 12.7 cm) is shown as a function of the chordwise approach to the Mach line emanating from the span edge, at the compression corner, in Fig. 15. Within the accuracy of the data, this effect appears related to the variations observed in Fig. 14. It may be the case that the edge effects extend further upstream in the lower Mach number regions of the boundary layer. Therefore, they influence the temperature gradient at the wall (and presumably the shear stress) at an earlier time (further upstream) than the influence is felt on the pressure.

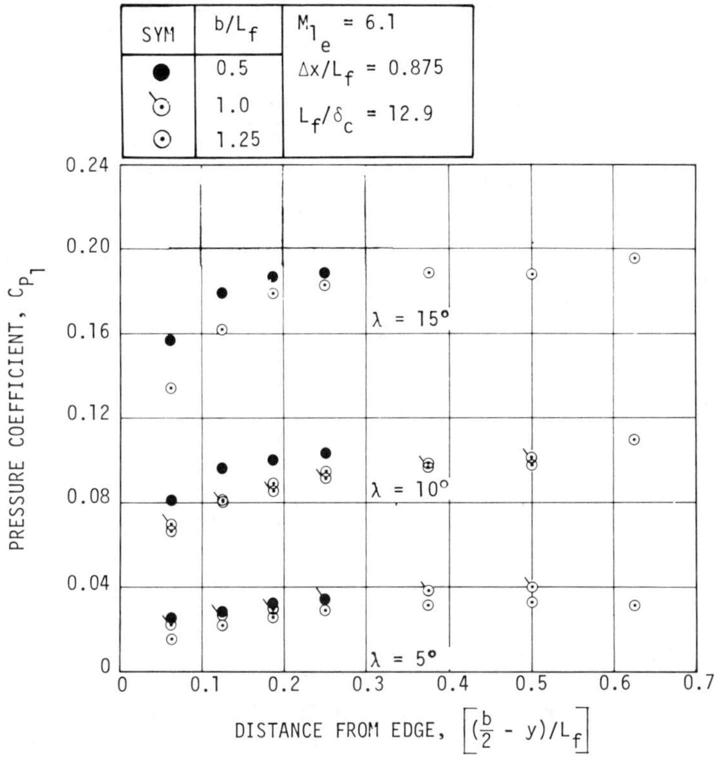

Fig. 16b Spanwise Pressure Distribution at 7/8 Chord.

Span Effect on Spanwise Distributions

The spanwise distribution of pressure coefficients on the flap is shown at two chord locations, for three spans and three deflection angles in Fig. 16. The pressure levels for the smallest span are consistently higher than those for the larger span flaps, indicating the three-dimensional nature of the shock as well as the influence of the flap edge. These indicate that some influence of flap span is probably present in the centerline distributions shown in Fig. 13. However, there it is obscured by the variations in the chordwise distributions. To investigate the influence of finite span edge effects, the ratio of the pressure coefficient to that measured on the flap centerline at the same chordwise station is shown versus approach to the inviscid flow Mach line from the flap leading edge corner in Fig. 17. The qualitative correlation is evident. The accuracy of the data is difficult to judge. However, the correlation shown in Fig. 17 appears sufficiently consistent to assign first-order effects in spanwise pressure variations to span edge effects.

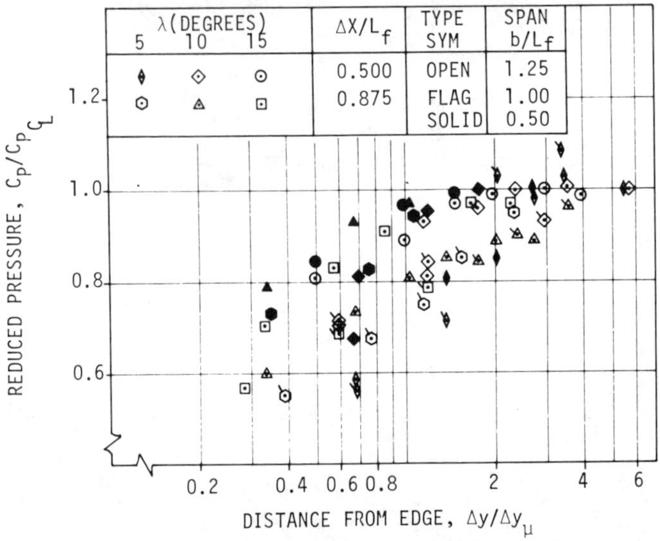

Fig. 17 Span Edge Effect on Pressure Distribution.

The spanwise distribution of aeroheating rates on the flap is illustrated for three flap spans and three deflection angles in Fig. 18. Spanwise distributions are shown for the same chordwise stations that pressure distributions were shown for above. There is one distinction between the pressure and aeroheating data: the closest approach to the edge for the pressure data shown in 0.635 cm; the closest approach for the aeroheating data shown is 1.25 cm. Proximity of the edge aside, the influence of the edge, and of span variations, is less evident in spanwise aeroheating distributions than in the same pressure distributions. This is the opposite contrast from that observed for span influence on chordwise centerline distributions.

The influence of approach to the inviscid flow flap edge effect on the spanwise aeroheating distribution is illustrated in Fig. 19. The influence is weaker than that observed for presure in Fig. 17.

Summary and Conclusions

The principal result of the experiment described has been to provide a consistent set of data for describing spanwise variation of pressure and aeroheating on finite span flaps, in thick turbulent boundary layers, and in uniform hypersonic flow. In addition, data has been provided to describe the influence of flap span, as a parameter, on the pressure and aeroheating distribution. Finally, data has been provided to describe the influence of low-angle flap deflection on the pressure and aeroheating distributions on the flaps.

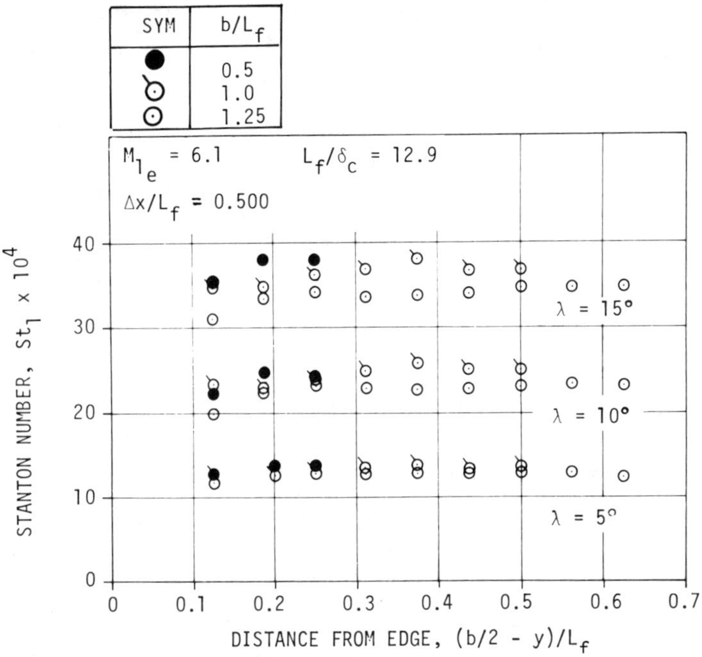

Fig. 18a Aeroheating Distribution at 1/2 Chord.

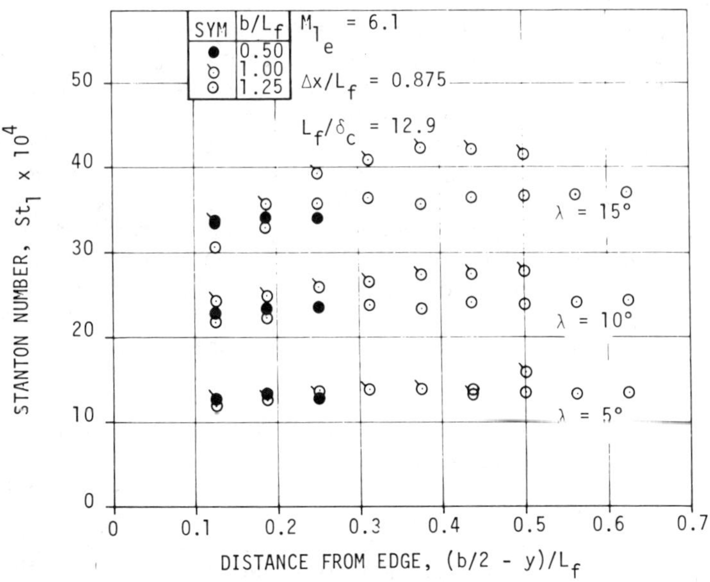

Fig. 18b Aeroheating Distribution at 7/8 Chord.

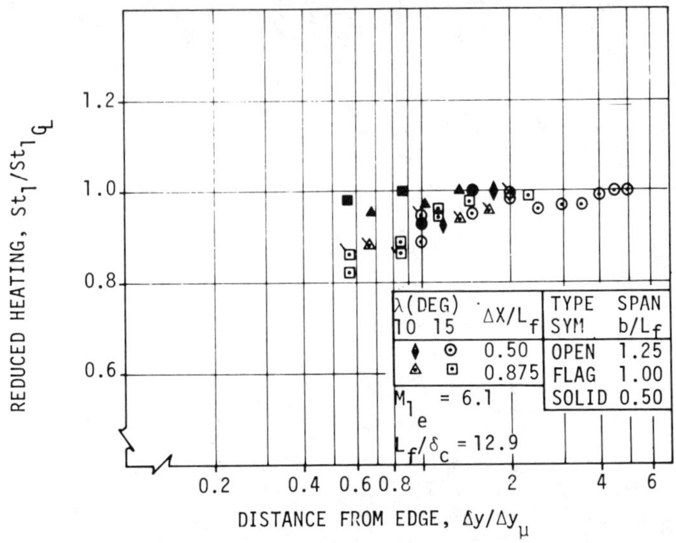

Fig. 19 Span Edge Effect on Aeroheating Distribution.

The comparison of data from the present experiment with other data for centerline pressure and aeroheating distributions has led to, or substantiated, several conclusions concerning turbulent boundary-layer flow through a compression corner. The dominant influence of the approaching flow parameters profile on the downstream pressure and aeroheating is shown to be consistent to low-deflection angles. Comparisons of data reflecting order of magnitude variations in the approaching flow boundary-layer thickness document the significance of that parameter in scaling the centerline pressure or aeroheating rise on the flap. Combining these two conclusions, the influence of nonsimilar approaching turbulent boundary-layer profiles on downstream pressure and aeroheating is illustrated. The adequacy of the relatively simple calculation scheme suggested by Elfstrom for predicting flow properties distributions on the flap downstream of the compression is demonstrated for a wide range of approaching flows. Validity is also demonstrated for the aeroheating rise prediction suggested by Coleman and Stollery to be used with the Elfstrom transformation.

Several conclusions have been suggested by analysis of the data from the present experiment. The chordwise distribution of the pressure rise on the flap centerline appears relatively insensitive to variations in span, or flap aspect ratio. In contrast, the chordwise distribution of aeroheating at the centerline appears to be sensitive to flap span. It appears to be significantly influenced by the proximity of span edge effects. Conversely, the spanwise distributions of pressure appear to be directly influenced, in fact dominated, by the proximity of span edge effects. However, the spanwise distribution of aeroheating appears relatively insensitive to spanwise edge effects. This latter observation is believed to imply a competing mechanism which causes a spanwise increase in aeroheating, such as boundary-layer thinning at the edge or the influence of crossflow.

Acknowledgment

The experiments were sponsored by the Air Force Wright Aeronautical Laboratory under Contract No. F33615-78-C-3001.

References

[1] Cassel, L. A. and Jarrett, T. W., "Hypersonic Flow Over Small Span Flaps in a Thick Turbulent Boundary Layer," AIAA 15th Thermophysics Conference, 1980.

[2] Cassel, L. A., McMillen, L. D., and Taylor, S., "Finite Span Effects on Flap Heating and Effectiveness in a Turbulent Boundary Layer," AFWAL-TR-80-3090, Aug. 1980.

[3] Ames Research Staff, "Equations, Tables, and Charts for Compressible Flow," NACA 1135, 1953.

[4] Hopkins, E. J., Keener, E. R., and Dwyer, H. A., "Turbulent Skin Friction and Boundary-Layer Profiles Measured on Non-Adiabatic Flat Plates at Hypersonic Mach Number," AIAA Paper 71-167, 1971.

[5] van Driest, E. R., "Turbulent Boundary Layers in Compressible Fluids," Journal of the Aeronautical Sciences, Vol. 18, No. 3, 1951, pp. 145-160.

[6] Cebeci, T. and Smith, A. M. O., Analysis of Turbulent Boundary Layers, Academic Press, New York 1974.

[7] Throckmorton, D. A., "Pressure Gradient Effects on Heat Transfer to Reusable Surface Insulation Tile-Array Gaps," NASA TN D-7939, Aug. 1975.

[8] Bertram, Mitchel H. and Neal, Luther, Jr., "Recent Experiments in Hypersonic Turbulent Boundary Layers," NASA TM X 56335, 1965; also AGARD Specialists' Meeting on Recent Developments in Boundary-Layer Research, Naples, Italy, May 10-14, 1965.

[9] Thomke, G. J. and Roshko, A., "Incipient Separation of a Turbulent Boundary Layer at High Reynolds Number in Two-Dimensional Supersonic Flow Over a Compression Corner," DAC 59819, McDonnell Douglas Astronautics Company, Western Division, 1969.

[10]Elfstrom, G. M., "Turbulent Hypersonic Flow at a Wedge-Compression Corner," Journal of Fluid Mechanics, Vol. 53, Part 1, 1972, pp. 113-127.

[11]Holden, M. S., "Shock Wave-Turbulent Boundary-Layer Interaction in Hypersonic Flow," AIAA Paper 77-45, Jan. 1977.

[12]Coleman, G. T. and Stollery, J. L., "Heat Transfer from Hypersonic Turbulent Flow at a Wedge Compression Corner," Journal of Fluid Mechanics, Vol 56, Part 4, 1972, pp. 741-752.

[13]Holden, M. S., "Shock Wave-Turbulent Boundary-Layer Interaction in Hypersonic Flow," AIAA Paper 72-74, Jan. 1972.

[14]Elfstrom, G. M., "Turbulent Separation in Hypersonic Flow," Imperial College, University of London Aero Report 71-16, Sep. 1971.

[15]Rose, W. C., Murphy, J. D., and Watson, E. C., "Interaction of an Oblique Shock Wave with a Turbulent Boundary Layer," AIAA Journal, Vol. 6, No. 9, Sept. 1968, p. 1796.

[16]Gray, D. J. and Rhudy, R. W., "Investigation of Flat-Plate Aspect Ratio Effects on Ramp Induced, Adiabatic, Boundary-Layer Separation at Supersonic and Hypersonic Speeds," Arnold Engineering Development Center Report AEDC-TR-70-235, March 1971.

Chapter II. Thermal Protection

THERMOCHEMICAL ABLATION OF TANTALUM CARBIDE LOADED CARBON-CARBONS

Bernard Laub*

Acurex Corporation/Aerotherm Division,
Mountain View, Calif.

Abstract

A thermochemical ablation model for tantalum carbide loaded carbon-carbons is developed assuming phase equilibrium at the ablating surface with the simplified form of the surface energy balance employing the transfer coefficient approximation for unity Lewis number. The model also utilizes the "fail temperature" concept for all condensed phase species that can form a liquid phase at the surface. The important surface chemical reactions are discussed, and normalized mass loss, B′, curves are presented for several tantalum carbide loaded composites. Steady-state ablation predictions are compared with available arc test data and other existing ablation models.

Nomenclature

B' = normalized total mass loss rate [see Eq. (25)]
B'_{ℓ} = normalized condensed phase mass loss rate [see Eq. (27)]
B'_{v} = normalized gas phase mass loss rate [see Eq. (26)]
C_H = Stanton number for heat transfer
C_M = Stanton number for mass transfer
C_{H_0} = Stanton number for heat transfer with no mass injection
C_{kj} = the number of atoms of element k in a molecule of species j
D = Fick's law diffusion coefficient

Presented as Paper 80-1476 at the AIAA 15th Thermophysics Conference, Snowmass, Colo., July 14-16, 1980.

*Leader, Thermal Sciences Section, Aerospace Technology Department.

g = denotes gas phase species
H = total enthalpy (sensible + chemical + kinetic)
H_{comp} = composite enthalpy of surface species [see Eq. (29)]
H_{∞} = enthalpy of the solid material far from the ablating surface
ΔH_f^o = heat of formation of the solid at the JANNAF 298 K reference state
j_i = diffusional mass flux of species i
$\tilde{j}_k$ = total diffusional mass flux of element k regardless of molecular configuration
K_i = mass fraction of species i
K_k = total mass fraction of element k regardless of molecular configuration
K_{pi} = equilibrium constant
ℓ = denotes melting condensed phase species
$\dot{m}$ = mass flux
P = system pressure
P_i = partial pressure of species i
q_{cond} = heat flux conducted into the solid
q_{conv} = convective heat flux to the surface
q_{rad} = radiative heat flux to the surface
s = denotes solid condensed phase species
T = temperature
T_{fail} = melt (fail) temperature of condensed phase species
u = streamwise velocity
v = velocity normal to the surface
x = streamwise coordinate, mole fraction
y = surface normal coordinate
α_{ki} = mass fraction of element k in species i
ε = emittance
λ = transpiration coefficient
ρ = density
σ = Stefan-Boltzmann constant
ϕ = correlation parameter for transpiration effects [see Eq. (32)]

Subscripts

c = denotes ablating material at the gas/solid interface
e = at the boundary-layer edge
i = gas phase species index
k = element index
ℓ = denotes condensed phase species
w = adjacent to the surface

Introduction

Refractory metals and carbides are often added to composite materials to increase certain performance characteristics in high-temperature and high-pressure environments. While much effort has been devoted to characterizing the relative performance of these materials, the thermochemical ablation performance has been primarily addressed experimentally.

Among the materials considered, tantalum carbide containing carbon-carbon composites have exhibited certain improvements in overall performance relative to standard carbon-carbons with only a small penalty in ablation performance. In this paper, a thermochemical ablation model is presented for carbon-carbon composite materials loaded with uniaxial tantalum carbide (TaC) fibers that can be employed over the entire range of TaC volume fractions.

The model is based on the following simplifying assumptions: 1) All gas phase and condensed phase reaction products are in chemical equilibrium with the ablating surface. 2) Condensed phase species that melt or decompose are assigned a "fail temperature" equal to their respective melt or decomposition temperatures. Such condensed phase species will be removed from the surface (in a condensed phase) if the surface temperature is greater than, or equal to, the fail temperature for this species. For such melting solids, the implication is that a liquid layer is unstable and is removed immediately. 3) The diffusion coefficients for all chemical species are equal and the Prandtl and Lewis numbers are unity. The transfer coefficient approximation for the surface energy equation, as defined by Kendall et al.[1] is employed.

Subject to these assumptions, solution of elemental mass balances at the ablating surface define a series of solutions for the normalized (by the mass transfer coefficient) mass loss rate vs surface temperature and pressure. For any given set of boundary conditions, predictions of mass loss rates require solution of the surface energy equation in which surface enthalpy terms are obtained from the equilibrium mass balance solutions.

Surface Mass Balance

No Condensed Phase Surface Removal

Consider the fluxes of chemical elements, k, entering and leaving a control surface affixed to the ablating surface. The solid material may be visualized as moving into this surface at a rate of $\dot{s}$. If it is assumed that no material is removed in a condensed phase (the normal assumption for homo-

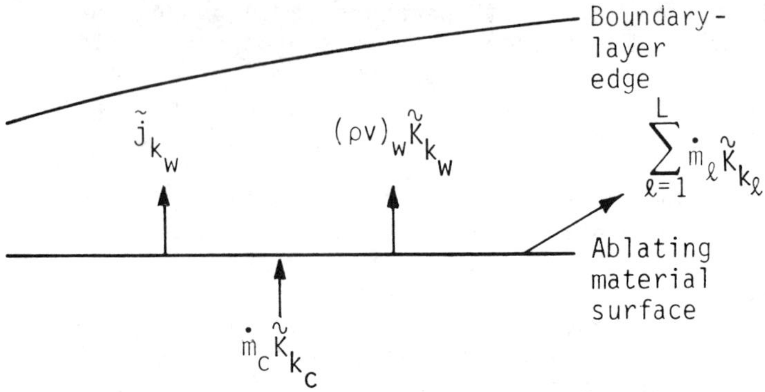

Fig. 1 Surface elemental mass balance.

geneous carbon materials), then Fig. 1 illustrates the surface and the fluxes of the kth chemical element where the condensed phase removal term is neglected. Terms superscripted by a tilde, ~, represent the total mass fraction of element k, independent of molecular configuration. Thus,

$$\tilde{K}_{k_w} = \Sigma \alpha_{k_i} K_{i_w} \tag{1}$$

$$\tilde{j}_{k_w} = \Sigma \alpha_{k_i} j_{i_w} \tag{2}$$

where k pertains to element k, i pertains to species i, and α_{k_i} is the mass fraction of element k in species i. Fluxes of element k away from the surface consist of boundary-layer diffusion and gross motion of the fluid adjacent to the surface due to injection flux $\dot{m}_c$. From Fig. 1, it is seen that conservation of chemical elements requires that

$$\tilde{j}_{k_w} + (\rho v)_w \tilde{K}_{k_w} = \dot{m}_c \tilde{K}_{k_c} \tag{3}$$

Summing Eq. (3) over all elements k, yields the total mass continuity equation for the case with no condensed phase material removal

$$(\rho v)_w = \dot{m}_c \tag{4}$$

An important fundamental of the present mathematical modeling of the ablation process is the expression of the diffusive mass flux in terms of a transfer coefficient formulation. If, as assumed here, the diffusion coefficients for all chemical species are equal, then Lees[2] has shown that an appropriate summation of the boundary-layer species conservation equation yields the following equation for each element k

$$\rho u \frac{\partial \tilde{K}_k}{\partial x} + \rho v \frac{\partial \tilde{K}_k}{\partial y} = \frac{\partial}{\partial y}\left(\rho D \frac{\partial \tilde{K}_k}{\partial y}\right) \tag{5}$$

whose solutions may be correlated in terms of a dimensionless mass transfer coefficient C_M where

$$-\tilde{j}_{k_w} = \rho_e u_e C_M(\tilde{K}_{k_e} - \tilde{K}_{k_w}) \tag{6}$$

Utilizing this transfer coefficient formulation for the diffusion flux in the elemental balance, Eq. (3), yields

$$\rho_e u_e C_M(\tilde{K}_{k_w} - \tilde{K}_{k_e}) + (\rho v)_w \tilde{K}_{k_w} = \dot{m}_c \tilde{K}_{k_c} \tag{7}$$

Defining a dimensionless mass loss rate as

$$B' = (\rho v)_w / \rho_e u_e C_M \tag{8}$$

and utilizing Eq. (4), the total mass fraction of element k at the wall can be solved from Eq. (7) as

$$\tilde{K}_{k_w} = (B' \tilde{K}_{k_c} + \tilde{K}_{k_e})/(1 + B') \tag{9}$$

Given the relative amounts of chemical elements specified by Eq. (9), the chemical and thermodynamic state of the gases adjacent to the ablating surface may be calculated from equilibrium relations. Since the gases are in equilibrium with the ablating surface

$$-\sum C_{k\ell} \ln P_k = \ln K_{p\ell}(T) \tag{10}$$

if ℓ represents the surface species, and

$$-\sum C_{k\ell} \ln P_k < \ln K_{p\ell}(T) \tag{11}$$

for all other candidate condensed phase species. In the present formulation, Eq. (11) may be thought of as being used to identify surface species ℓ for which Eq. (10) applies. The equilibrium relations for gas phase species is of the form

$$\ln P_i - \sum C_{ki} \ln P_k = \ln K_{pi}(T) \tag{12}$$

where the partial pressures must obey the relation

$$\sum P_i = P \tag{13}$$

The temperature and chemical composition of the gases adjacent to the surface may be determined if P and B′ are specified. From the temperature, pressure, and chemical composition, the calculation of other thermodynamic properties is straightfoward. The relations given above are utilized by the ACE computer code[3] to solve the mass balances for chemical systems of interest.

Condensed Phase Surface Removal

Considerations to this point have been based on the assumption that all the mass leaving the ablating surface is in the gas phase, and this flux has been denoted by $(\rho v)_w$. However, for many ablation situations of interest, some of the material leaving the surface is in a condensed phase such as removal of a liquid layer. For the surface elemental balance, as shown in Fig. 1, there is an additional flux term leaving the surface. This results in the surface elemental balance [analogous to Eq. (3)] having the form

$$\tilde{j}_{k_w} + (\rho v)_w \tilde{K}_{k_w} + \sum_{\ell=1}^{L} \dot{m}_\ell \tilde{K}_{k_\ell} = m_c \tilde{K}_{k_c} \tag{14}$$

Adding all of elements yields the obvious net mass continuity relation

$$(\rho v)_w + \sum_{\ell=1}^{L} \dot{m}_\ell = \dot{m}_c \tag{15}$$

When "failing" occurs, an additional equilibrium relation is available for each condensed phase species ℓ for which $\dot{m}_\ell > 0$. Also, the surface temperature must be consistent with the assigned fail temperatures, i.e.,

$$T_w \leq T_{fail} \text{ for the surface species} \tag{16}$$

$$T_w \geq T_{fail} \text{ for any species for which } \dot{m}_\ell > 0 \tag{17}$$

These additional equilibrium relations, surface temperature restrictions, and assigned fail temperatures are sufficient to solve for the additional unknowns $\dot{m}_\ell$ introduced into elemental mass balance relations.

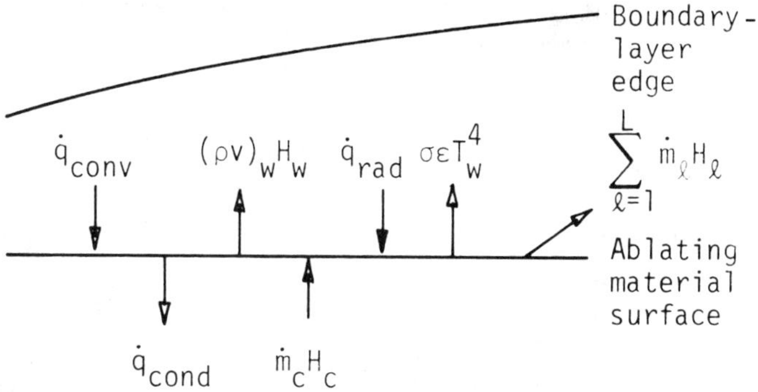

Fig. 2 Surface energy fluxes.

Surface Energy Balance

No Condensed Phase Surface Removal

Consider the fluxes of energy entering and leaving a control surface affixed to the ablating surface. For the case of unity Prandtl and Lewis number with no condensed phase removal, these fluxes may be illustrated as shown in Fig. 2 (neglecting the condensed phase removal term), where q_{conv} is the energy convected to the surface as a result of boundary-layer transport events. This term includes effects of both heat conduction due to a temperature gradient in the gas adjacent to the surface, and energy transport due to endothermic or exothermic chemical reactions at the surface. Conservation of energy at the surface requires that

$$q_{conv} + \dot{m}_c H_c - (\rho v)_w H_w + q_{rad} - \sigma\varepsilon T_w^4 = q_{cond} \qquad (18)$$

When all diffusion coefficients are equal, and if the Prandtl and Lewis numbers are unity, Lees[2] has shown that solutions to the boundary-layer energy equation can be conveniently expressed in terms of a dimensionless heat transfer coefficient C_H where

$$q_{conv} = \rho_e u_e C_H (H_e - H_w) \qquad (19)$$

If attention is limited to steady-state ablation, the conduction flux can be expressed by

$$q_{cond} = \dot{m}_c (H_c - H_\infty) = \dot{m}_c (H_c - \Delta H_f^0) \qquad (20)$$

where H_∞ is the enthalpy of the ablation material far from the ablating surface which is well represented by the heat of formation of the ablation material at the JANNAF 298 K reference state. Substituting Eq. (19) and Eq. (20) for the terms in Eq. (18) yields the steady-state surface energy equation when there is no condensed phase removal, i.e.,

$$\rho_e u_e C_H (H_e - H_w) - (\rho v)_w H_w + q_{rad} - \sigma\varepsilon T_w^4 + \dot{m}_c \Delta H_f^o = 0 \quad (21)$$

For the assumptions employed, the similarity of the boundary-layer elemental conservation and energy equations indicates that if the corresponding boundary conditions are also similar, then $C_M = C_H$. Equation (21) can be normalized by the mass transfer coefficient, $\rho_e u_e C_M$ and, utilizing the definitions of Eq. (4) and Eq. (8), becomes

$$H_e + (1 + B')H_w + q_{rad}/\rho_e u_e C_H - \sigma\varepsilon T_w^4/\rho_e u_e C_H + B'\Delta H_f^o = 0 \quad (22)$$

As discussed above, B' and H_w are related to T_w from solution of surface mass balances. Therefore, for a given set of environmental conditions, solutions to Eq. (22) can be obtained iteratively or graphically for the steady-state value of B', from which the steady-state recession rate is simply obtained.

Condensed Phase Surface Removal

When some of the material leaving the surface is in a condensed phase, as would be the case for liquid layer removal, the surface energy balance must be modified to account for the energy associated with the condensed phase removal. As illustrated in Fig. 2, the surface energy balance contains an additional energy flux term and, analogous to Eq. (18), may be written as

$$q_{conv} + \dot{m}_c H_c - (\rho v)_w H_w + q_{rad} - \sigma\varepsilon T_w^4 - \Sigma \dot{m}_\ell H_\ell = q_{cond} \quad (23)$$

where H_ℓ is taken as the enthalpy of each liquid phase "failing" at the surface. Utilizing Eqs. (19) and (20) as before, the steady-state surface energy equation with condensed phase removal becomes

$$\rho_e u_e C_H (H_e - H_w) - (\rho v)_w H_w + q_{rad} - \sigma\varepsilon T_w^4 - \Sigma \dot{m}_\ell H_\ell + \dot{m}_c \Delta H_f^o = 0 \quad (24)$$

Since the surface mass loss consists of both gaseous and condensed phase species, the normalized mass fluxes can be redefined as

$$B' = \dot{m}_c/\rho_e u_e C_M \tag{25}$$

$$B'_v = (\rho v)_w/\rho_e u_e C_M \tag{26}$$

$$B'_\ell = \dot{m}_\ell/\rho_e u_e C_M \tag{27}$$

where, from Eq. (15),

$$B' = B'_v + \Sigma B'_\ell \tag{28}$$

If a mass-averaged composite enthalpy is defined by

$$H_{comp} = (B'_v H_w + \Sigma B'_\ell H_\ell)/B' \tag{29}$$

then Eq. (24) can be normalized and rewritten as

$$H_e - H_w + q_{rad}/\rho_e u_e C_H - \sigma\epsilon T_w^4/\rho_e u_e C_H - B' H_{comp} + B' \Delta H_f^o = 0 \tag{30}$$

which is analogous to Eq. (22) for the steady-state ablation of a material with both gas and condensed phase mass loss. Again, B', H_w, and H_{comp} are related to T_w from the surface mass balance solutions.

Transpiration Effects

The heat and mass transfer coefficients in the above equations are "blown" values, i.e., they are the appropriate transfer coefficients in the presence of ablation. However, most boundary-layer analysis techniques evaluate the transfer coefficient for a nonablating surface. Likewise, experimental facilities, such as arc plasma jets, define heat transfer rates with various calorimetric techniques which can be used to define a nonablating heat transfer coefficient. The effects of foreign gas injection on boundary-layer heat and mass transfer has been studied by many investigators. However, the applicability of these results to chemically reacting boundary layers has been questioned.[4,5] In this paper, the nonablating transfer coefficient is corrected for mass transfer using the relation

$$C_H/C_{H_o} = \phi/(e^\phi - 1) \tag{31}$$

where

$$\phi = 2\lambda B'_o \tag{32}$$

and

$$B_0' = (\rho v)_w / \rho_e u_e C_{H_0} \tag{33}$$

The nonablating heat transfer coefficient, $\rho_e u_e C_{H_0}$, is defined from a nonablating boundary-layer solution or from experimental measurements. Efforts to correlate chemically reacting boundary-layer solutions[6,7] have demonstrated reasonable success if nonsimilar effects are negligible. The value of the correlation parameter, λ, was selected as 0.5 for all of the ablation predictions presented in this paper since this value

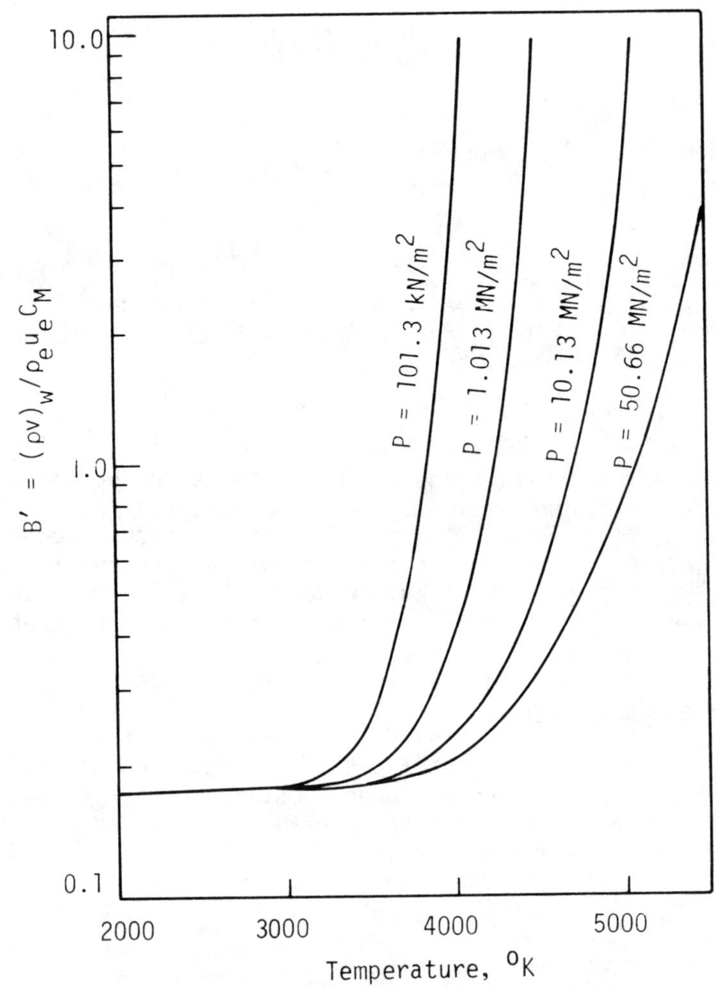

Fig. 3 B′ map for carbon in air.

has been used quite successfully to predict turbulent ablation of graphite and carbon-carbon. It should be noted that when there is condensed phase surface removal, only the gaseous mass flux is utilized in the correction to the nonablating heat transfer coefficient, in accordance with Eq. (33).

Surface Chemistry

Carbon

The classical equilibrium ablation of carbon in air is illustrated in Fig. 3 as a map of B' vs temperature and pressure. (These solutions are applicable to graphites or carbon-carbon composites.) The figure exhibits the well-known "diffusion limited" plateau in which surface mass loss is due to carbon oxidation according to the reaction

$$2C(s) + O_2(g) + 3.718N_2(g) \rightleftarrows 2CO(g) + 3.718N_2(g) \qquad (34)$$

where there are 3.718 moles of nitrogen for each mole of oxygen in air. The approximate value of B' for this reaction can be defined as the ratio of the mass of the solid reactants to the mass of the gaseous reactants, i.e.,

$$B' = \frac{\text{mass of carbon}}{\text{mass of air}} = \frac{2(12.011)}{[32.0 + (3.718)28.014]} = 0.176 \qquad (35)$$

At temperatures above 3000 K, the equilibrium ablation of carbon is characterized by the pressure-dependent sublimation of carbon as shown in Fig. 3. The calculations employed thermochemical data for 29 gaseous species and one condensed species from JANNAF.[8]

Tantalum Carbide

The ablation of tantalum carbide in oxidizing environments like air is, most likely, more complex than suggested by the analytical models developed by Ziering[9] and McDonald, Nickerson, and Buch.[10] The previous investigators selected to constrain the composition of surface melt layers in accordance with the binary phase diagram for the tantalum/carbon system.[11] However, at the heated surface of tantalum carbide ablating in air the surface chemistry must address the ternary system, tantalum/carbon/oxygen. An examination of the literature failed to identify a ternary phase diagram for this system. Actually, to be strictly correct, considerations should extend to the quaternary system, tantalum/carbon/oxygen/nitrogen. Since no nitrides have been identified through x-ray diffraction analysis of post-test tantalum loaded carbon-carbon specimens, the potential formation of tantalum nitrides has been neglected in this model.

The applicability of the binary phase diagram at the ablating surface is questionable. In fact, ternary phase diagrams for other systems demonstrate the formation of many phase compositions that are not observed for the binary system. Thus, the model presented here for the tantalum/carbon/oxygen system is not constrained by any predetermined phase composition rules. It should also be noted that the ACE[3] computer code does not handle condensed phase solution chemistry. The modeling allows several "failed" condensed species to be simultaneously removed at the surface. Their respective concentrations are determined from equilibrium considerations subject to the constraint of conservation of elemental mass.

Predictions of the normalized total mass loss rate, B′, of pure tantalum carbide heated in air are illustrated in Fig. 4 as a function of surface temperature and pressure. The "fail" temperatures of potential surface species are noted in the figure. It is seen that the predictions exhibit a significant dependency on pressure. For example, between 2058 K and 3258 K there is a diffusion limited plateau associated with the formation and fail of the condensed phase oxide described by

$$\begin{aligned} 2\mathrm{TaC(s)} + 7/2\ \mathrm{O_2(g)} + 7/2\ (3.718)\mathrm{N_2(g)} &\rightleftarrows \mathrm{Ta_2O_5(s)} \\ &+ 2\mathrm{CO(g)} + 7/2\ (3.718)\mathrm{N_2(g)} \end{aligned} \tag{36}$$

and

$$\mathrm{Ta_2O_5(s)} \rightleftarrows \mathrm{Ta_2O_5}(\ell) \tag{37}$$

where B′ is defined by the ratio of the mass of the condensed phase reactants to the mass of the freestream reactants, i.e.,

$$B'_{oxide} = \frac{\text{mass of TaC}}{\text{mass of air}} = 0.809 \tag{38}$$

However, at pressures below 1.013 MN/m^2 the mass loss rate is enhanced by the formation of gas phase oxides TaO and TaO_2. At the tantalum fail temperature, the mass loss rate increases until a new diffusion limited plateau is reached, i.e.,

$$2\mathrm{TaC(s)} + \mathrm{O_2(g)} + (3.718)\mathrm{N_2(g)} \rightleftarrows 2\mathrm{Ta(s)} + 2\mathrm{CO(g)} + (3.718)\mathrm{N\ (g)} \tag{39}$$

and

$$Ta(s) \rightleftarrows Ta(\ell) \tag{40}$$

where

$$B'_{Ta} = \frac{\text{mass of TaC}}{\text{mass of air}} = 2.83 \tag{41}$$

Note that at 10.13 MN/m^2, equilibrium considerations dictate that both the condensed phase oxide [$Ta_2O_5(\ell)$] and metal [$Ta(\ell)$] are removed simultaneously. At 50.66 MN/m^2, tantalum melt is totally suppressed. At the peritectic decomposition temperature of ditantalum carbide, the mass loss rate increases at constant temperature until a new diffusion limit is reached, i.e.,

$$2TaC(s) + 1/2O_2(g) + 1/2(3.718)N_2(g) \rightleftarrows Ta_2C(s) + CO(g) + 1/2(3.718)N_2(g) \tag{42}$$

and

$$Ta_2C(s) \rightleftarrows Ta_2C(\ell) \tag{43}$$

where

$$B'_{Ta_2C} = \frac{\text{mass of TaC}}{\text{mass of air}} = 5.66 \tag{44}$$

The B′ map illustrates that, above the peritectic decomposition temperature, mass loss rates increase until the tantalum carbide melt temperature is attained. At the tantalum carbide fail temperature, the mass loss rate can increase indefinitely as dictated by surface energy considerations as both tantalum carbide and ditantalum carbide are failing simultaneously.

In addition to the species considered for the analysis of carbon in air, this analysis considered three gas phase species (Ta, TaO, TaO_2) and four condensed phase species (Ta, Ta_2O_5, Ta_2C, TaC). Thermochemical data for Ta_2C was obtained from Schick[12] while the data for the other tantalum species is from JANNAF.[8] The "fail" temperatures assigned to each condensed phase specie correspond to their respective melt temperatures and are listed in Table 1.

Tantalum Carbide Loaded Carbon-Carbon Composites

The equilibrium thermochemical ablation of a TaC loaded carbon-carbon composite displays characteristics representative

Table 1 Fail temperature of condensed phase species

Species	Reference	Fail temperature, K
Ta_2O_5	8	2058
Ta	8	3258
Ta_2C	12	3773
TaC	8	4273

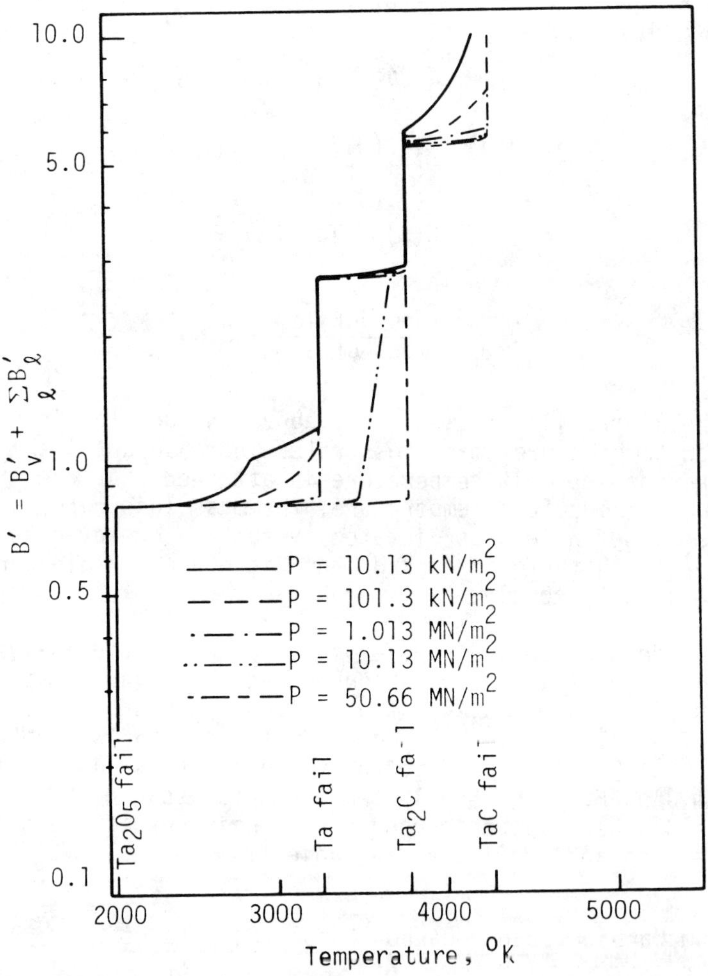

Fig. 4 B′ map for tantalum carbide in air.

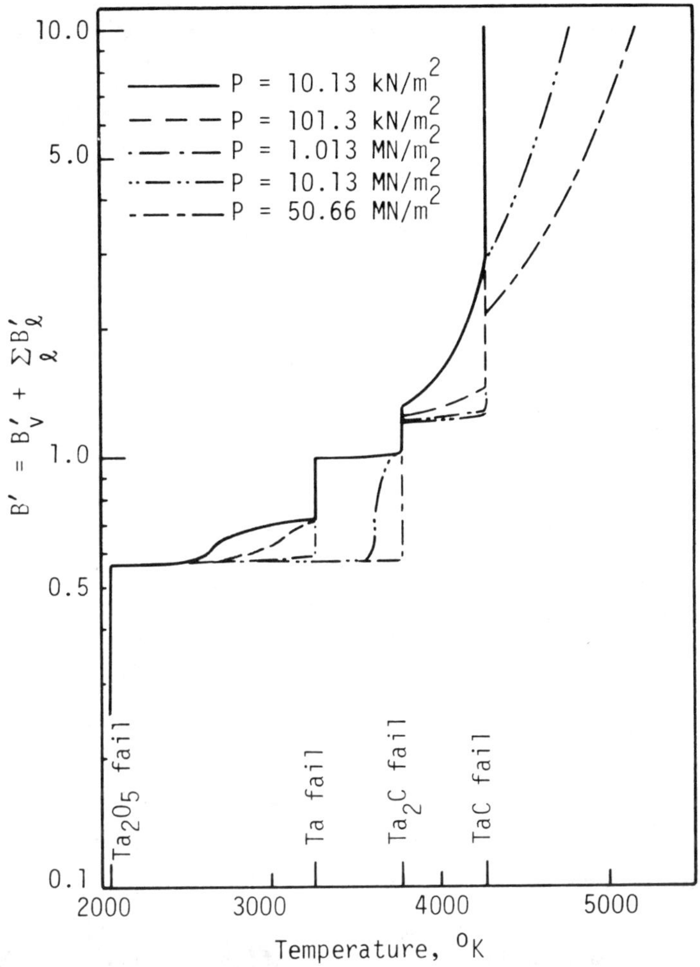

Fig. 5 B′ map for tantalum carbide loaded (50% by volume) carbon-carbon.

of the predicted performance of both pure carbon and pure tantalum carbide. Figure 5 illustrates the predicted normalized mass loss (B′) map for a TaC loaded (50% by volume) carbon-carbon in air. Similar maps are shown for a 23% and a 12% TaC loading (by volume) in Figs. 6 and 7, respectively. At temperatures below the melt temperature of tantalum carbide, the B′ map exhibits the same features as pure tantalum carbide but compressed over a smaller range of normalized mass loss rates. This is not surprising since it would be anticipated that the performance of a composite reflects that of its basic constituents. The results for the composite are controlled by the performance of the tantalum carbide constituent with the mass loss rates reflecting the reduced TaC concentration in the composite. This is simply demonstrated by evaluating the value of B′ for the diffusion limited oxide formation and fail. For the 50% by volume TaC loaded carbon–carbon, the surface reaction can be written [analogous to Eq. (36)] as

$$\begin{aligned} x_{TaC}TaC(s) &+ (1 - x_{TaC})C(s) + (5/4\ x_{TaC} + 1/2)O_2(g) \\ &+ (5/4\ x_{TaC} + 1/2)(3.718)N_2(g) \\ &\rightleftarrows (1/2\ x_{TaC})Ta_2O_5(s) + CO(g) \\ &+ (5/4\ x_{TaC} + 1/2)(3.718)N_2(g) \end{aligned} \tag{45}$$

and

$$Ta_2O_5(s) \rightleftarrows Ta_2O_5(\ell) \tag{46}$$

where x_{TaC} is the mole fraction of tantalum carbide in the composite. Using a density of 14.29 g/cm^3 for tantalum carbide and 1.90 g/cm^3 for the surrounding carbon-carbon, the TaC mole fraction for a 50% volume fraction is 0.319. The value of B′ is again defined as the ratio of the mass of the solid reactants to the mass of the freestream reactants, i.e.,

$$B'_{oxide} = \frac{\text{mass of TaC + mass of carbon}}{\text{mass of air}} = 0.57 \tag{47}$$

which, as shown in Fig. 5, agrees with the analysis results.

Above the melt temperature of tantalum carbide, the thermochemical ablation behavior of the composite resembles the sublimation characteristics of carbon. The total mass loss is comprised of a gas phase component due to volatilization of free carbon in the system and a condensed phase component due to "fail" of tantalum carbide. As would be expected, the fraction of the total mass loss attributable to condensed phase "fail" increases proportionately with increased TaC loading of the composite. It could be argued that the liquid layer formed due to melt of the tantalum carbide in the composite covers the surface and limits the surface temperature to the tantalum carbide melt temperature. This effect can be easily modeled with the current approach by "failing" condensed phase carbon at the tantalum carbide melt temperature. Such modeling would result in enhancement of the mass loss rate since no energy would be accommodated by the sublimation of carbon.

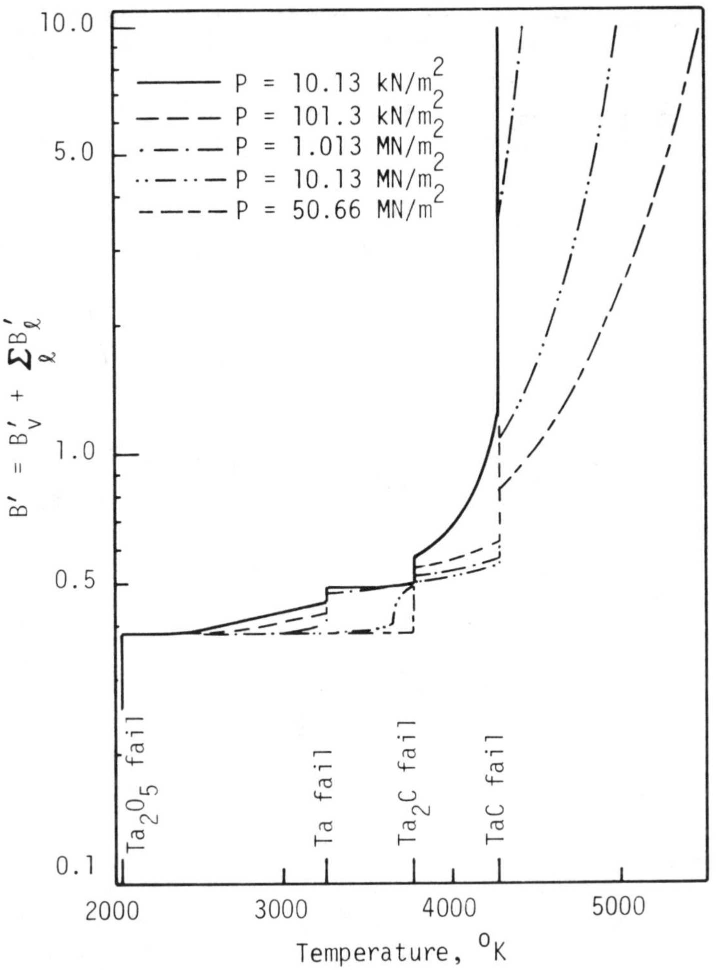

Fig. 6 B′ map for tantalum carbide loaded (23% by volume) carbon-carbon.

However, comparison of Figs. 3 and 4 illustrate that, in the region of the tantalum carbide melt temperature, the normalized mass loss rate of tantalum carbide is, in general, significantly larger than for carbon. This implies that the TaC component in a composite would undergo substantial subsurface melting. However, examination of post-test specimens from arc tests of three-dimensionally woven carbon-carbons with uniaxial TaC fibers demonstrates a relatively uniform surface. The absence of any differential ablation between the carbon fibers and matrix as compared to the TaC fibers sug-

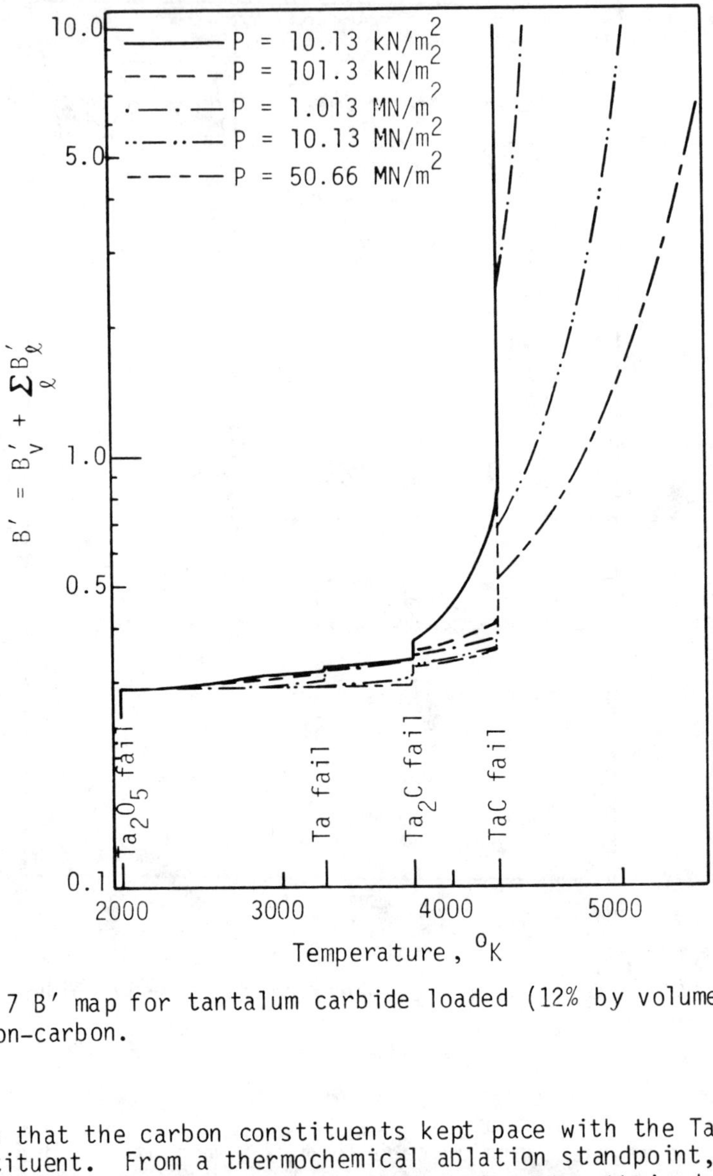

Fig. 7 B′ map for tantalum carbide loaded (12% by volume) carbon-carbon.

gests that the carbon constituents kept pace with the TaC constituent. From a thermochemical ablation standpoint, this behavior requires that the carbon constituents attained surface temperatures above the melt temperature of tantalum carbide.

To evaluate the recession rate of a TaC loaded carbon-carbon composite for a given set of environmental conditions requires simultaneous solution of surface mass and energy balances. The mass balance solutions obtained from the ACE

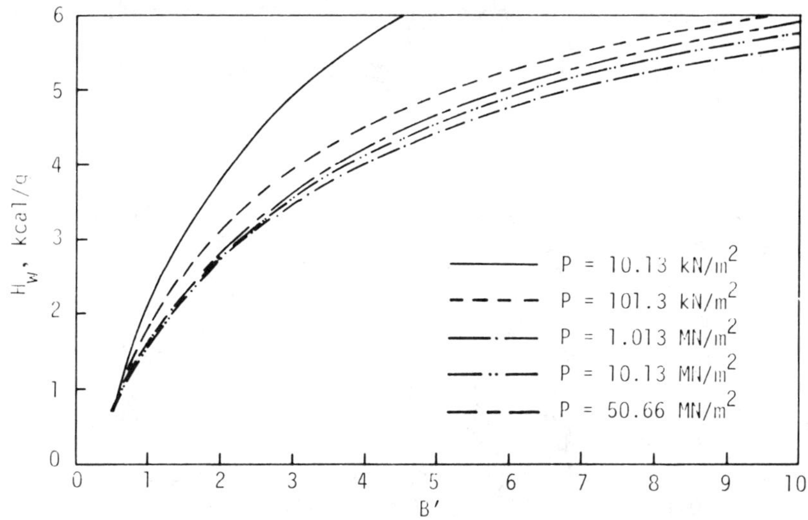

Fig. 8 Enthalpy of gaseous ablation products at the surface of a tantalum carbide loaded (23% by volume) carbon-carbon.

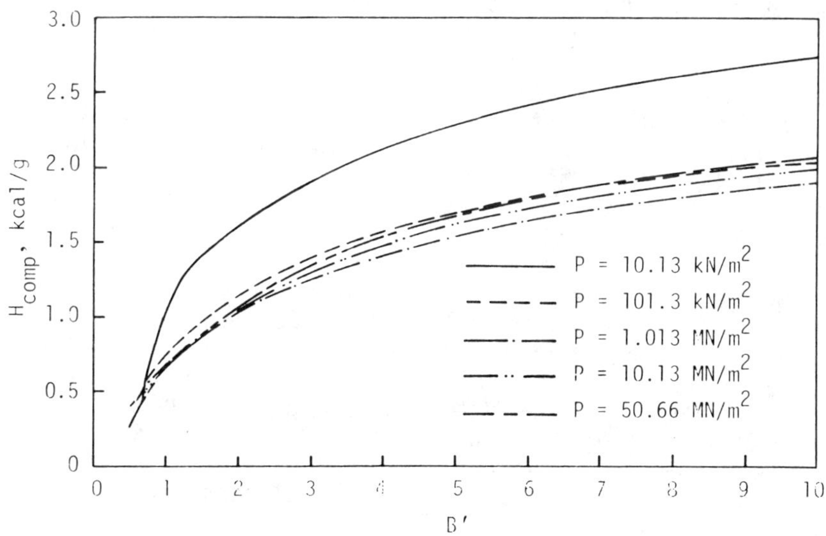

Fig. 9 Composite enthalpy of surface species for a tantalum carbide loaded (23% by volume) carbon-carbon.

computer code provide parametric data for T_w, H_w, and H_{comp} as a function of B′ and pressure. These data are used in solution of the surface energy equation. Values for H_w and

H_{comp} as a function of pressure and B′ are illustrated for a 23% TaC loading (by volume) in Figs. 8 and 9, respectively. Similar data for other compositions are best obtained from ACE mass balance equilibrium solutions.

Experimental Data

The usefulness of an analytical ablation model can only be evaluated by the accuracy with which actual material performance can be predicted. For conical configurations, these predictions are complicated by the effects of shape change on the inviscid flowfield and boundary layer as well as augmentation of turbulent convective heating due to interactions of the boundary layer with surface roughness. Some of the physical mechanisms involved in these coupled interactions are not yet fully understood, but reasonably accurate procedures for the analysis of coupled shape change have been developed employing empirical correlations.

Existing ablation data for TaC loaded carbon-carbons have primarily been obtained from tests in a high-pressure arc jet facility. The data base is not large and, in general, is limited to TaC volume fractions less than 40%. Unfortunately, the data are often presented without complete documentation of material construction. For example, the type of fiber, weave, weave spacing, and densification process have not been shown to have any significant effect on the steady-state turbulent recession rate of carbon-carbons. However, for a TaC loaded composite, the form in which the tantalum carbide is included may have an effect on performance. If the tantalum carbide is in the form of a uniformly distributed powder, potential subsurface melting can result in an enhancement of ablation rate in comparison to the theoretical values derived here, since the material performance would be more representative of a porous carbon. However, if the tantalum carbide is in the form of uniaxial fibers, which was the case for the data shown herein, indepth heat conduction should confine melting to the near surface region. The data included here were obtained from Ref. 10 which provides little, if any, information on measured surface temperature and material construction.

Figure 10 illustrates the steady-state turbulent recession rate data vs the volume percent of tantalum carbide in the composite. All of these tests were conducted at the same test condition, but the initial configuration of the models may have been different. The predictions from the model developed in Ref. 10 are shown for comparison and are in reasonable

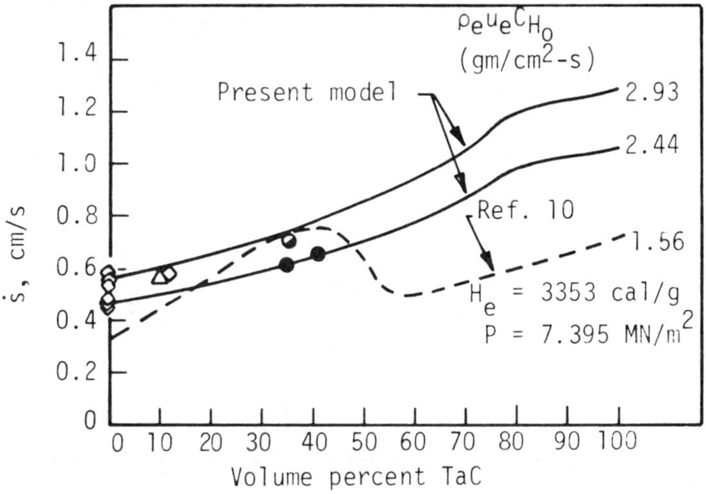

Fig. 10 Comparison of measured and computed steady-state recession rates for TaC loaded carbon-carbons in a high pressure arc.

agreement with the data for the TaC loaded composites although the inflections in predicted performance are somewhat disturbing. This behavior is due to the assumption that the melt layer obeys a temperature/composition rule prescribed by the binary phase diagram. However, the model presented in Ref. 10 does not predict the carbon-carbon data very well since the heat transfer coefficient employed in the calculations was based on smooth wall values accounting only for the nose radius of the stable turbulent shape. All current prediction procedures account for the effects of roughness augmentation on the convective heat transfer coefficient.

Calculations with the present model were made for a range of roughness augmentation factors from 1.56 to 1.87 which is typical of correlations of carbon-carbon performance. As expected, the predictions for a pure carbon-carbon are representative of the data. Furthermore, if the same augmentation factors are assumed to apply to the TaC loaded composites as well, the present model predicts the TaC loaded recession data quite well. It is apparent that if roughness augmentation effects were included in Ref. 10, the computed recession rates would significantly overpredict the data at the higher TaC volume fractions. Figure 11 compares the steady-state surface temperatures calculated with the present analysis with those presented in Ref. 10. The current predictions exhibit no sensitivity of surface temperature to heat transfer

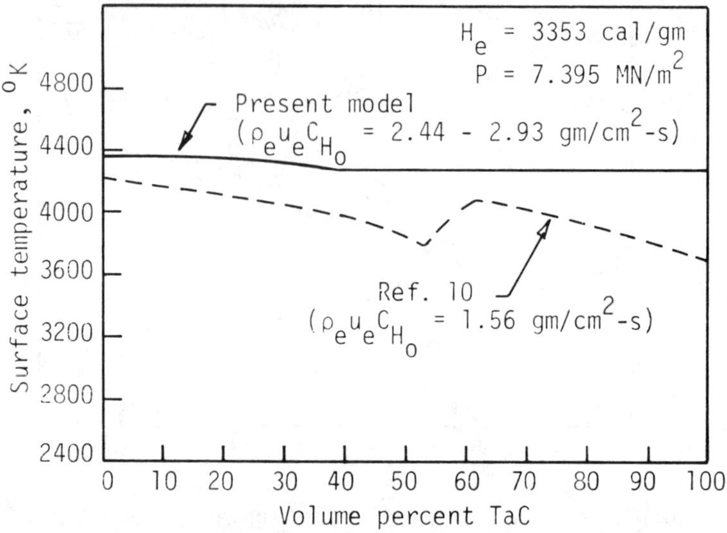

Fig. 11 Comparison of computed steady-state surface temperature for TaC loaded carbon-carbons in a high pressure arc.

coefficient for the range of values considered. As shown, surface temperature decreases gradually with increased TaC loading up to a 40% TaC volume fraction. At higher volume fractions, the surface temperature is predicted to be the tantalum carbide melt temperature. Again, the surface temperature predictions presented in Ref. 10 reflect the constraint of obeying the binary phase diagram.

The present model predicts the existing recession data well and is consistent with the trend of the data in contrast to the model presented in Ref. 10. However, verification of the model requires surface temperature data as well as tests at higher TaC volume fractions. Some ablation data on tantalum carbide was reported by Ziering,[9] but these data were obtained at low pressures in a facility where the free-stream oxygen concentration is uncertain due to consumption of the carbon cathode and is, therefore, considered unreliable.

Acknowledgment

This work was performed by Acurex Corporation with partial support from the Defense Nuclear Agency, the USAF Space and Missile Systems Organization, and the Air Force Materials Laboratory.

References

[1]Kendall, R. M., Rindal, R. A., and Bartlett, E. P., "A Multicomponent Boundary Layer Chemically Coupled to an Ablating Surface," AIAA Journal, Vol. 5, June 1967, pp. 1063-1071.

[2]Lees, L., "Convective Heat Transfer With Mass Addition and Chemical Reactions," Third AGARD Colloquium on Combustion and Propulsion, Pergamon Press, New York, 1959.

[3]Powars, C. A. and Kendall, R. M., "Aerotherm Chemical Equilibrium (ACE) Computer Program," Aerotherm Corporation, Mountain View, Calif., UM-69-7, May 1969.

[4]Rosner, D. E., "Convective Heat Transfer With Chemical Reaction," Aerochem Research Laboratories, Inc., Princeton, N. J., ARL 99, Aug. 1961.

[5]Spalding, D. B., Convective Mass Transfer, An Introduction, McGraw-Hill, New York, 1963.

[6]Bartlett, E. P. and Putz, K. E., "Heat- and Mass-Transfer Blowing Correction Correlations for Charring Ablators, Part 1, Equal Diffusion Coefficients," Sandia Laboratories, Albuquerque, N. M., SC-RR-710260, November 1971.

[7]Bartlett, E. P. and Grose, R. D., "An Evaluation of a Transfer Coefficient Approach for Unequal Diffusion Coefficients," Aerotherm Corporation, Mountain View, Calif., FR-69-50, June 1969.

[8]JANNAF Thermochemical Tables, Dow Chemical Company, Midland, Michigan, including Supplement No. 41, June 1974.

[9]Ziering, M. B., "Thermochemical Ablation of Ceramic Heat Shields," AIAA Journal, Vol. 13, May 1975, pp. 610-616.

[10]McDonald, J. E., Nickerson, G. R., and Buch, J. D., private communication, Prototype Development Associates, Inc., Santa Ana, Calif., Feb. 1976.

[11]Hansen, M., Constitution of Binary Alloys, 2nd ed., McGraw-Hill, New York, 1958.

[12]Schick, H. L., Thermodynamics of Certain Refractory Compounds, Vol. II, Academic Press, New York, 1966.

CATALYTIC RECOMBINATION OF NITROGEN AND OXYGEN ON HIGH-TEMPERATURE REUSABLE SURFACE INSULATION

Carl D. Scott*

NASA Lyndon B. Johnson Space Center, Houston, Tex.

Abstract

The energy-transfer catalytic recombination coefficient for nitrogen and oxygen recombination on the surface coating of high-temperature reusable surface insulation (HRSI) is inferred from stagnation point heat flux measurements in a high-temperature dissociated arc jet flow. The resulting catalytic recombination coefficients are correlated with an Arrhenius model for convenience, and these expressions may be used to account for catalytic recombination effects in predictions of the heat flux on the HRSI thermal protection system of the Space Shuttle Orbiter during reentry flight. Analysis of stagnation point pressure and total heat balance enthalpy measurements indicates that the arc heater reservoir conditions are not in chemical equilibrium. This is contrary to what is usually assumed for arc jet analysis and indicates the need for suitable diagnostics and analyses, especially when dealing with chemical reaction phenomena such as catalytic recombination heat-transfer effects.

Nomenclature

A_{eff} = area of inviscid core of nozzle flow at the nozzle exit
h = enthalpy
h_T = total enthalpy
h_N = enthalpy of dissociation of nitrogen
h_O = enthalpy of dissociation of oxygen
k = Boltzmann constant
Le = Lewis number

Presented as Paper 80-1477 at the AIAA 15th Thermophysics Conference, Snowmass, Colo., July 14-16, 1980.

*Research Engineer.

m = mass of atom or molecule
$\dot{m}$ = arc heater gas flow rate
p = pressure
q = heat flux
R_{eff} = effective body radius
T = temperature
V = velocity
α = atom mass fraction
β = chemical energy accommodation coefficient
β' = stagnation point velocity gradient parameter
γ = energy-transfer catalytic recombination coefficient
γ' = catalytic atom recombination coefficient
ε = emittance
ρ = density
σ = Stefan-Boltzmann constant
ϕ = catalytic parameter defined by Eq. (2)
μ = viscosity

Subscripts:

FC = fully catalytic
N = nitrogen
O = oxygen
TC = thermocouple
w = wall
∞ = freestream
Ni = nickel
HRSI = high-temperature reusable surface insulation

Introduction

A large area of the thermal protection system of the Space Shuttle Orbiter consists of high-temperature reusable surface insulation (HRSI) that is coated with a reaction cured glass (RCG) material having a high emittance to provide thermal radiation relief. The coating also provides some degree of structural durability and forms a seal against moisture penetration. Because this coating has a high silica content, it is expected to be fairly noncatalytic with respect to nitrogen and oxygen recombination.[1,2] This is consistent with measurements in dissociated arc jet flow where the temperature of the HRSI surface is several hundred degrees lower than would be predicted based on the heat flux measured with a calorimeter having a highly catalytic surface. This temperature difference has been attributed to a low catalytic recombination coefficient on the HRSI. Because the shock layer and boundary-layer air is dissociated during Orbiter entry, the heat flux to the surface due to recombination also would be reduced during flight.

The fraction of the available chemical (dissociation) energy transferred to a surface depends on the rate at which atoms recombine and give up their dissociation energy on the surface. The net energy transferred depends on the product of the recombination coefficient γ' and the chemical energy accommodation coefficient β. In general, this product γ is a function of wall temperature, and until now, there have been no published measurements of this important factor for nitrogen and oxygen recombination on the RCG coating of HRSI.

This paper describes measurements made in the NASA Lyndon B. Johnson Space Center (JSC) 10-MW arc tunnel facility and analyses used to determine the energy-transfer coefficient for catalytic recombination of nitrogen and oxygen on the RCG coating over the wide range of temperatures anticipated during entry. The implications of these measurements to Orbiter heating rates are discussed briefly.

Experiments

Two sets of experiments were conducted - one using a 10 MW constricted arc heater to produce partially dissociated flow of nitrogen and the other using a 5 MW heater to produce a flow of partially dissociated air. The range of enthalpy of nitrogen was 5.5-30.8 MJ/kg and that of air was 20.1-37.5 MJ/kg. To achieve the lower enthalpy nitrogen cases, nitrogen was introduced downstream of the arc column in the "reservoir" (Fig. 1). It was assumed that there was adequate mixing to ensure uniform inviscid flow properties. The flow then expanded out a conical nozzle having a 15 deg half-angle.

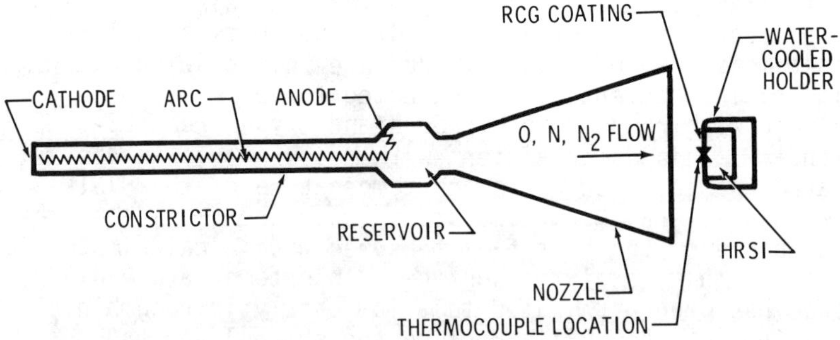

Fig. 1 Schematic view of sample in arc jet flow.

Samples of RCG-coated HRSI material (10.16 cm diam × 5.08 cm thick) were supported in a water-cooled holder (12.7 cm o.d. and inserted into the flow with the flat face normal to the flow (Fig. 1). Each sample was instrumented with platinum/platinum-10% rhodium (Pt/Pt-10% Rh) thermocouples located in the surface coating at the center of the sample (the stagnation point). One sample was instrumented radially and in depth to obtain an estimate of the significance of conduction. The radial conduction was negligible and the conduction normal to the surface was less than 3% of the incident heat flux.

To serve as a relatively highly catalytic reference, the heat flux was measured by a nickel-coated slug calorimeter mounted in a fixture having the same geometry as the HRSI sample and holder. In addition, the stagnation point pressure was measured at the same location in the flow as the heat flux by a 1.27 cm diam pitot pressure probe. All measurements were made 11.6 cm downstream of the nozzle exit.

Analysis

The energy-transfer recombination coefficient $\gamma = \gamma' \beta$ was determined using the stagnation point heat-transfer theory of Goulard.[3] Goulard's theory does not account for the recent refinements of the incomplete energy accommodation and quenching theory developed by Rosner and Feng.[4] This more developed theory was not used in the present work because the γ' and β are not known separately, nor are the molecular excited states into which the unaccommodated energy goes. Likewise, the wall and gas phase quenching rates are not known. Therefore, these results may be biased, yielding effective recombination coefficients that do not differentiate quenching of excited molecules from direct energy accommodation.

The use of the energy-transfer catalytic recombination coefficient γ rather than the recombination coefficient γ', which pertains to the number of recombined atoms, is suggested by the results of Breen et al.[5] In those measurements on a coating similar to the one investigated here, it was found that γ' determined from atom concentration measurements in a duct flow reactor was much greater than expected from heat-transfer measurements in arc jets. These two measurements are consistent only if a fraction of the available recombination energy is actually transferred to the wall; i.e., $\beta < 1$. This phenomenon is seen in other cases; e.g., γ' and β have been measured directly for several metals in dissociated nitrogen by Halpern and Rosner[6] and in oxygen by Melin and Madix.[7]

To use Goulard's theory,[3] it is necessary to assume that γ' can be replaced by $\gamma = \gamma'\beta$. Actually, this is true if $\beta = 1$ or if $\gamma' \ll 1$, as can be seen from the Goulard relation rederived to take into account incomplete energy accommodation and two atomic species.

$$\frac{q}{q_{FC}} = \frac{1 + \left(Le^{2/3}\ \phi_N - 1\right)\dfrac{h_N\alpha_N}{h_T} + \left(Le^{2/3}\ \phi_O - 1\right)\dfrac{h_O\alpha_O}{h_T}}{1 + \left(Le^{2/3} - 1\right)\dfrac{h_N\alpha_N}{h_T} + \left(Le^{2/3} - 1\right)\dfrac{h_O\alpha_O}{h_T}} \tag{1}$$

where

$$\phi = \beta \Bigg/ \left(1 + \frac{0.47\ Sc^{-2/3}\ \sqrt{2\beta'\mu_s\rho_s}}{\rho_w\gamma'}\sqrt{\frac{2\pi m}{kT}}\right) \tag{2}$$

The catalytic recombination rate constant k_w often found in the literature is related to the recombination coefficient γ' by[8]

$$k_w = \sqrt{kT/2\pi m}\ \gamma' \tag{3}$$

However, since most of the "measurements" of k_w are based on energy flux measurements, the reported values also have β implicitly included in the definition of k_w.

The velocity gradient parameter for the stagnation point of a sphere is

$$\beta' = V_\infty/R_{eff}\left[\rho_\infty/\rho_s(2 - \rho_\infty/\rho_s)\right]^{1/2} \tag{4}$$

The effective radius R_{eff} for a flat-faced cylinder was obtained from Bade and Yos[9] and was based on the work of Boison and Curtiss.[10]

The values of the flowfield properties are obtained by the simple flow conservation equations for an inviscid stream tube.

Enthalpy: $h_T = 1/2\ V_\infty^2 + \alpha_N h_N + \alpha_O h_O + C_p T_\infty$ (5)

Momentum (pressure): $p_{T_2} = \rho_\infty V_\infty^2 + p_\infty$ (6)

Mass: $\dot{m} = \rho_\infty V_\infty A_{eff}$ (7)

The total quantities on the left side of Eqs. (5-7) and p_∞ are measured, and these equations and the frozen gas equation of state are solved simultaneously to obtain ρ_∞, V_∞, α_N, and T_∞. In the air cases, the oxygen is assumed to be fully dissociated since it dissociates at a low temperature. The expanding nozzle flow is assumed to be a chemically frozen ideal gas. This assumption is borne out in nozzle flow calculations in Ref. 9 that include the chemical rate equations. Unfortunately, those calculations assume chemical equilibrium in the reservoir as the initial condition. It appears that chemical equilibrium does not exist in the reservoir of the JSC arc jets and the results of the nonequilibrium arc tunnel analysis (NATA) calculation[9] were only used to obtain an estimate of the boundary-layer displacement thickness in the nozzle (hence, A_{eff}) and to establish the validity of the frozen flow assumption.

The properties at the stagnation point behind the normal shock were obtained from normal shock relations using the frozen ratio of specific heats γ, and the shock layer flow was also assumed to be chemically frozen. This assumption is supported by the reacting shock layer calculations of Refs. 11 and 12. Transport properties were calculated for the nonequilibrium mixture using cross-section data and equations developed by Hansen.[13]

Results

For the nitrogen cases, the JSC 10-MW arc jet was run at various test conditions and gas injection configurations that resulted in a wide range of enthalpies and HRSI sample temperatures. For the air cases, the JSC 5-MW arc heater was used. A summary of the test conditions is given in Table 1. The basic measured quantities are gas flow rate, arc current and voltage, and coolant water flow rate and temperature rise, all of which are used to calculate the heat balance total enthalpy h_T. Other measured quantities

Table 1 Arc jet test conditions

Measured					Calculated free stream properties				
Gas flow rate, kg/s	Arc current, A	Pitot pressure, N/m^2	Heat flux Ni calorimeters, W/cm^2	Heat flux RCG coating, W/cm^2	Enthalpy, MJ/kg	N atom mass fraction	Velocity, m/s	Density, g/m3	Temperature, K
Arc jet nitrogen test									
0.0454	300	752	37.0	21.8	20.1	0.45	2800	0.099	500
.0454	600	951	53.0	32.0	27.3	.61	3300	.086	590
.0454	750	965	73.0	38.9	32.2	.77	3300	.089	500
.0908	300	1190	36.1	18.8	13.0	.28	2300	.217	50
.0908	450	1390	54.0	29.3	16.9	.37	2700	.190	140
.182	425	1910	21.0	8.38	5.5	.066	2000	.497	44
.182	1600	2610	60.0	39.6	12.7	.24	2670	.367	60
.0454	300	427	10.9	4.43	7.07	.12	1660	.155	340
.0454	400	465	14.3	5.85	8.62	.16	1780	.147	370
.0454	500	503	17.7	7.00	9.68	.19	1900	.140	410
.0908	300	827	12.9	5.64	5.45	.078	1710	.284	150
.0908	400	896	16.8	7.60	6.70	.11	1810	.273	150
Arc jet air test, $\alpha_0 = 0.23$									
0.0227	250	1170	31.6	20.1	12.0	0.17	2900	0.14	420
.0277	400	1320	41.0	26.7	15.7	.27	3200	.13	440
.0277	800	1510	56.5	37.5	22.4	.47	3500	.12	500

are stagnation point (pitot) pressure, heat flux to the nickel-coated slug calorimeter, and the surface temperature of the HRSI samples. The heat flux to the HRSI sample was obtained by the time-dependent numerical curve fit procedure of Williams and Curry.[14] These results indicated that since the thermocouple is 0.38 mm beneath the coating surface, the simple relation $q = \varepsilon\sigma T_{TC}^4$ underpredicts the heat flux by about 6%. Therefore, all heating rates obtained from this relation were increased by 6%.

Using the measured flow properties, Eqs. (5-7) were solved to obtain the flow properties presented in Table 1. For the nitrogen cases, the nitrogen atom mass fraction α_N is plotted in Fig. 2 as a function of total enthalpy and is only slightly dependent on the arc jet gas flow rate $\dot{m}$. However, α_N corresponding to $\dot{m}$ = 0.182 kg/s is slightly lower than that for lower flow rates at the same enthalpy. This is probably due to the higher pressure in the arc heater reservoir, which drives the flow toward chemical equilibrium before the gas expands out the nozzle.

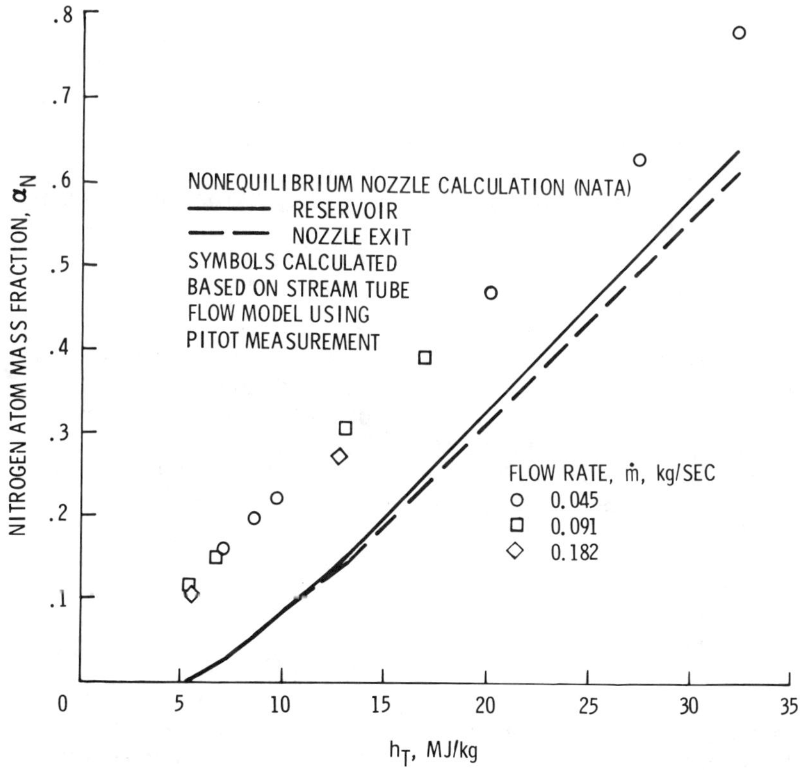

Fig. 2 Nitrogen atom mass fraction in arc jet flow.

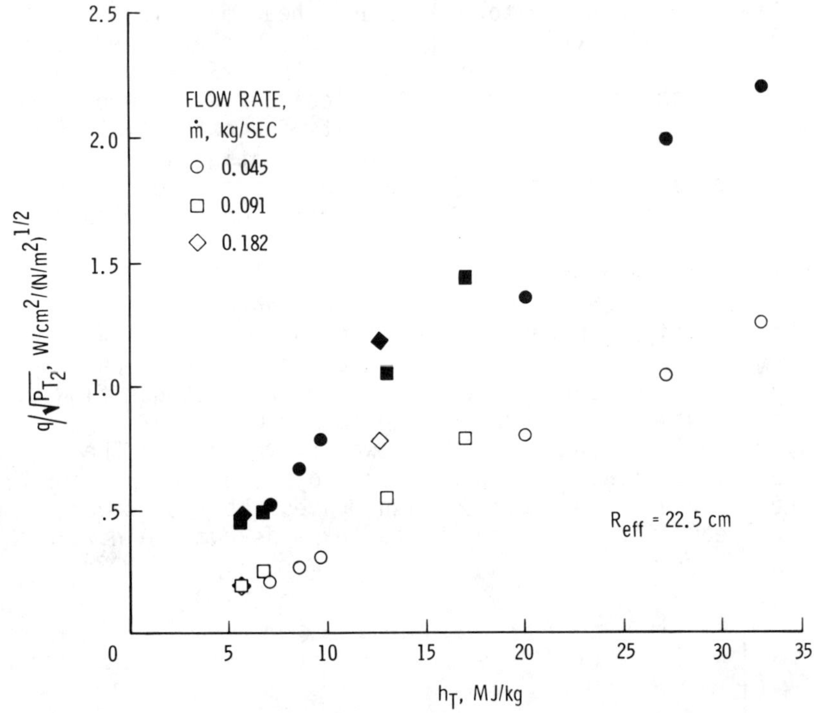

Fig. 3 Heat flux to nickel-coated calorimeter (solid symbols) and RCG coating (open symbols) normalized by the square root of the stagnation point pressure (nitrogen flow).

As mentioned previously, chemical equilibrium does not appear to be achieved in the arc heater reservoir. This is indicated by two observations. One is a comparison (Fig. 2) of the atom fraction calculated above and that computed by the NATA code.[9] The other is the difference in heating rates to the nickel-coated slug calorimeter and to the HRSI. It can be seen in Fig. 3 that the HRSI heat flux is much lower than the calorimeter heat flux even at very low enthalpies where very little difference should exist because of a lack of nitrogen dissociation if the flow were in chemical equilibrium. The difference in heat flux due to wall temperature differences between the calorimeter and the HRSI is not nearly enough to account for the observed differences in heating. (Compare Fig. 3 with Fig. 2.) The fact that the heat flux to the HRSI in the low enthalpy cases is only about one-half that to the nickel-coated calorimeter is consistent with the assertion that HRSI has a relatively low catalytic energy accommodation coefficient γ and that the flow is out of chemical equilibrium in the nozzle reservoir. The lack of chemical equilibrium in the reservoir is probably due to the characteristic reaction time being long compared with the resident time of the flow in the reservoir. The relatively high degree of dissociation probably results from the high enthalpy core gas in the arc column not equilibrating with the cooler gas added in the reservoir. Because of the small diameter of the reservoir, the velocity in the reservoir is high enough that the flow does not reach recombination equilibrium before it freezes in the nozzle expansion. Therefore, one cannot use an equilibrium reservoir starting condition for a nozzle flow calculation such as NATA.

The NATA code assumes the reservoir properties are in chemical equilibrium and uses equilibrium properties as initial conditions for the nozzle flow calculation. Note in Fig. 2 that the atom fraction in the expanded flow at the nozzle exit predicted by the NATA code is very nearly equal to the reservoir values. This implies that the flow is chemically frozen at approximately the reservoir composition and justifies assuming the flow is frozen throughout the nozzle.

Stagnation point boundary-layer edge properties were obtained from the ideal gas normal shock relations[15] using a frozen ratio of specific heats and assuming the gas then compresses isentropically to zero velocity. These values were then used in calculating the heat flux ratio q/q_{FC} as functions of the catalytic recombination coefficient, now interpreted as the catalytic energy accommodation coefficient γ. The results of the calculations for nitrogen are given in Fig. 4 for each test condition.

The value of γ_N is obtained from the curves of Fig. 4 using the measured heat flux ratio q/q_{Ni} adjusted for wall temperature differences and the fact that nickel is not fully catalytic according to the following relation.

$$\frac{q}{q_{FC}} = \left.\frac{q}{q_{Ni}}\right|_{meas} \cdot \frac{h_T - h_{w_{Ni}}}{h_T - h_{w_{HRSI}}} \cdot \left.\frac{q_{Ni}}{q_{FC}}\right|_{calc} \tag{8}$$

The factor $q_{Ni}/q_{FC}|_{calc}$ is calculated from Eq. (1) and requires knowledge of the recombination coefficient for the reference surface, nickel. However, there is a rather large

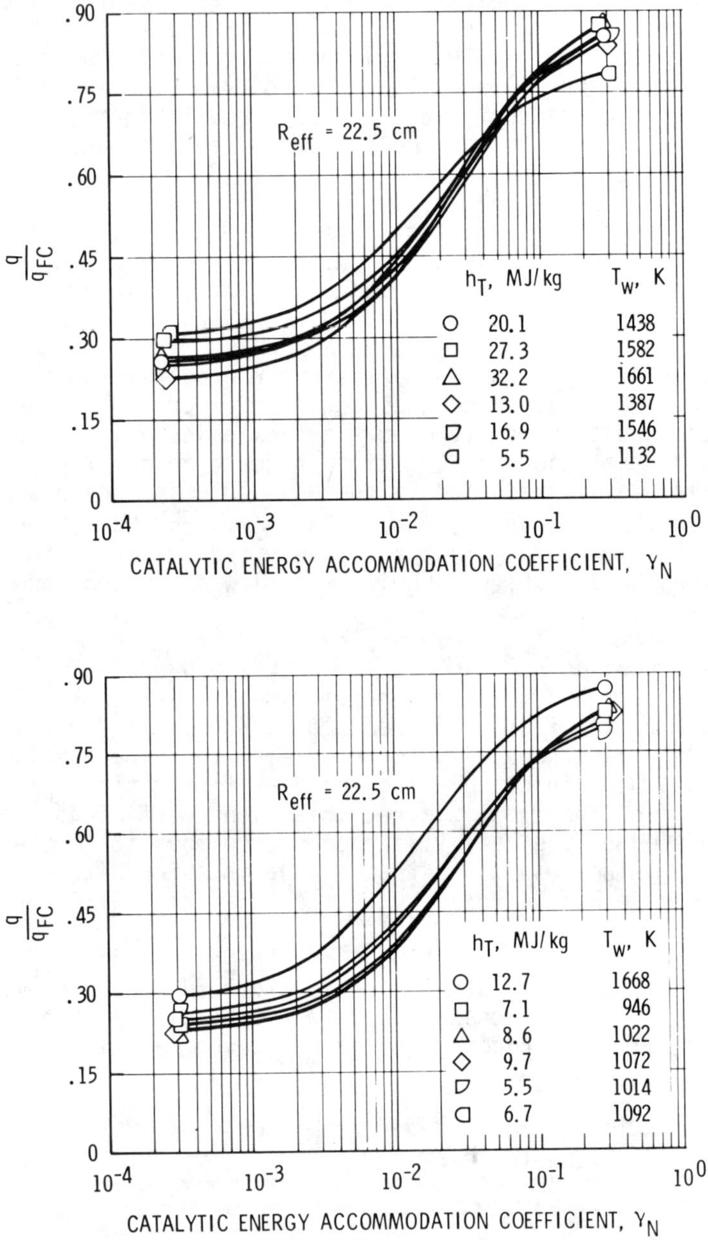

Fig. 4 Relative heat flux to stagnation point of flat-faced cylinder for various nitrogen test conditions (Eq. (1)).

uncertainty in the recombination coefficient for nickel that carries over to the RCG results. The recombination coefficient for nitrogen on nickel has been measured by a number of investigators. Some have used flow reactors and/or side arm methods,[1-3,16] whereas others have used arc jet heat flux methods[17,18] similar to the method used here. Oxygen recombination on nickel and nickel oxide has been investigated using the flow reactor/side arm technique.[5,19,20] Melin and Madix[7] measured γ' and β separately for oxygen recombination on nickel (oxide). The product $\gamma'\beta$ should be directly applicable to heat transfer in the approximation used here. Myerson[21] measured the catalytic accommodation coefficient γ_O of oxygen on nickel in a flow reactor. There have been no measurements of γ_O for pure oxygen on nickel in arc jets.

The reference value of $\gamma_N = 0.1$ for nitrogen recombination on nickel at 350 K that is used here is the mean of the γ_N and γ'_N values reported in Refs. 1-3 and 16-18. The reference value of $\gamma_O = 0.0085$ for oxygen recombination at room temperature was taken from Myerson[21] because his measurements were made at pressures closer to those experienced in the present work than the results of Melin and Madix[7] ($\gamma_O = 0.0027$). The pressure can affect the apparent γ if excited molecules quench on the wall or in the gas near the wall. This effect, ignored in the present analysis, is the subject of further investigation.

Although separate measurements of γ' and β are not available for nitrogen recombination on nickel, there is some arc jet evidence[17] that supports a value of $\gamma'\beta \approx 0.1$. However, there also exists some evidence[18] to support a lower value. A summary of published recombination coefficients is given in Table 2.

Likewise, the value of γ_O at three test conditions in air is obtained from the similar curves in Fig. 5, where the value of γ_N found from the pure nitrogen experiments is used in Eq. (1). The resulting catalytic recombination energy accommodation coefficients γ_N and γ_O are plotted in Figs. 6 and 7, respectively. Although there is considerable scatter, they appear to have an Arrhenius dependence on temperature; i.e., the rate appears to be controlled by an activation energy mechanism. The detailed mechanism and energetics of the recombination events will not be discussed in this paper because the energy states of the N_2 and O_2 molecules produced are not known.

Table 2 Reported recombination coefficient

Gas/surfaces species	γ, γ'	k_w, cm/s (a)	$\gamma'\beta$	Reference temperature, K	Technique	Reference
N/Ni	0.000065	(1.12)		300	Side arm	Rahman and Linnett[1]
N/Ni	.022	(400)		350		Rojinsky as reported by Goulard[3]
N/Ni	.1	(1900-2700)		400-800	Atom balance flow tube	Wood and Wise[16]
N/NiO	.014	(258)		358	Side arm	Dickens and Sutcliffe[2]
N/Ni(NiO)	.1	2200		500	Arc jet heat flux	Anderson[17]
	.4	1200		1000		
N/Ni	.029	520		350	Arc jet heat flux	Pope[18]
O/Ni	.0085	(150)			Flow reactor flux heat flux	Myerson[21]
O/Ni	.028	(500)		298	Side arm	Greaves and Linnett[19]
O/NiO	.0077	(138)		298	Side arm	Greaves and Linnett[19]
O/NiO	.0089	(160)			Side arm	Greaves and Linnett[20]
O/NiO	.01	(180)		300	Flow tube reactor	Breen et al.[5]
O/Ni	.017	(50)	0.0027		Flow tube reactor heat flux	Melin and Madix[7]
Air/Ni	.09	1700		400	Arc jet heat flux	Anderson[17]
Air/Ni	.017	300		350	Arc jet heat flux	Pope[18]

[a]Parentheses indicate values calculated from reported values.

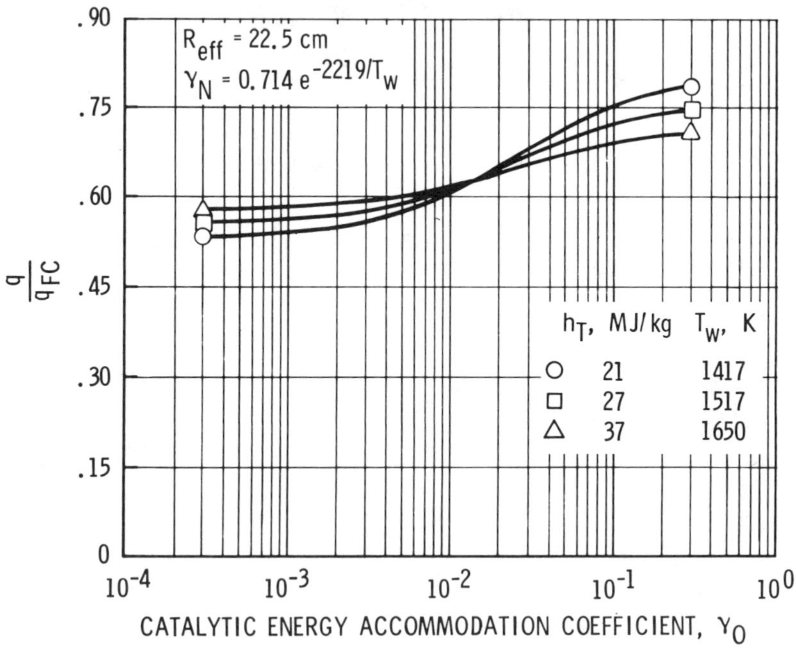

Fig. 5 Relative heat flux to stagnation point of flat-faced cylinder for various air test conditions. (Eq. (1))

Fitting the γ data to an Arrhenius model provides the convenient analytical expressions. For nitrogen

$$\gamma_N = 0.0714e^{-2219/T_w} \qquad 1670 < T_w < 950 \text{ K} \tag{9}$$

and for oxygen

$$\gamma_O = 16e^{-10271/T_w} \qquad 1650 < T_w < 1400 \text{ K} \tag{10}$$

The scatter in the data seen in Figs. 6 and 7 results from uncertainties in measurements and the assumptions and approximations required to calculate the flow properties and heat flux ratios. The uncertainties are expected to be larger for small γ where q/q_{FC} is a very slowly varying function of γ.

The error bars in Figs. 6 and 7 correspond to the extremes of the assumed nickel reference recombination coef-

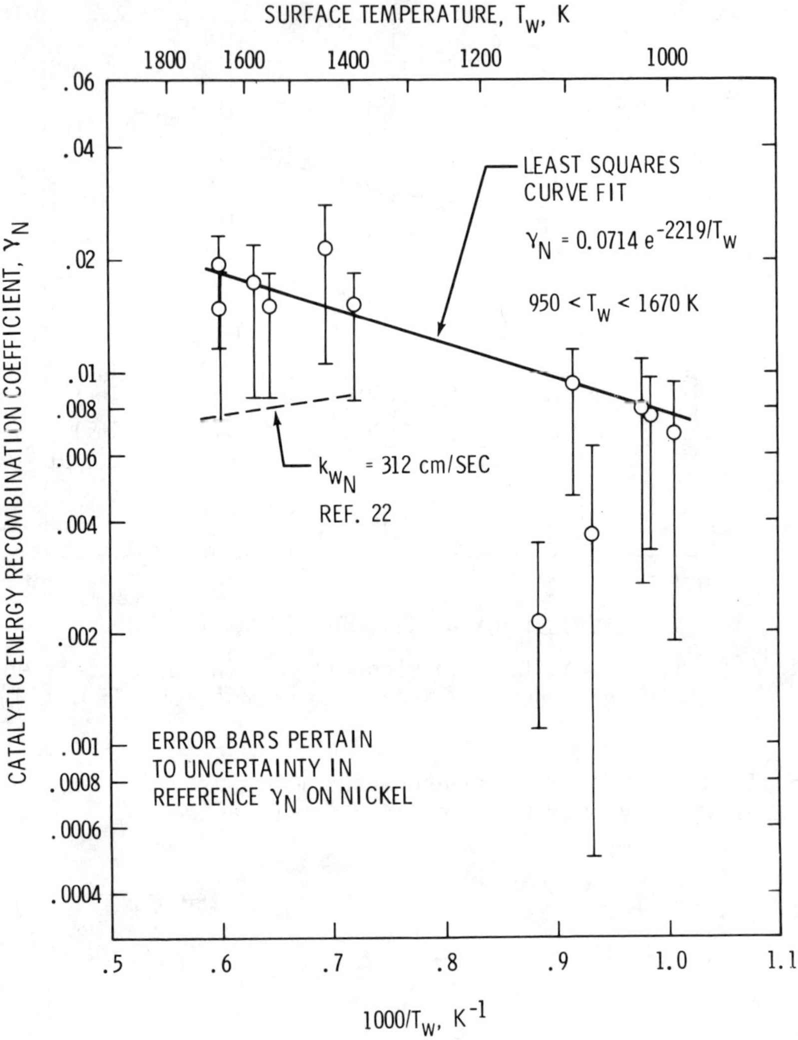

Fig. 6 Energy-transfer catalytic recombination coefficient for nitrogen on RCG-coated HRSI.

ficients. This indicates the need for a better understanding of recombination energy accommodation phenomena on even simple materials like nickel.

It is possible to determine the approximate influence of γ_O and γ_N on the Orbiter heating rates by comparing the present results with previous estimates. Rakich and Lanfranco[23] calculated the heat flux on the windward centerline of the Orbiter using a technique that combined a

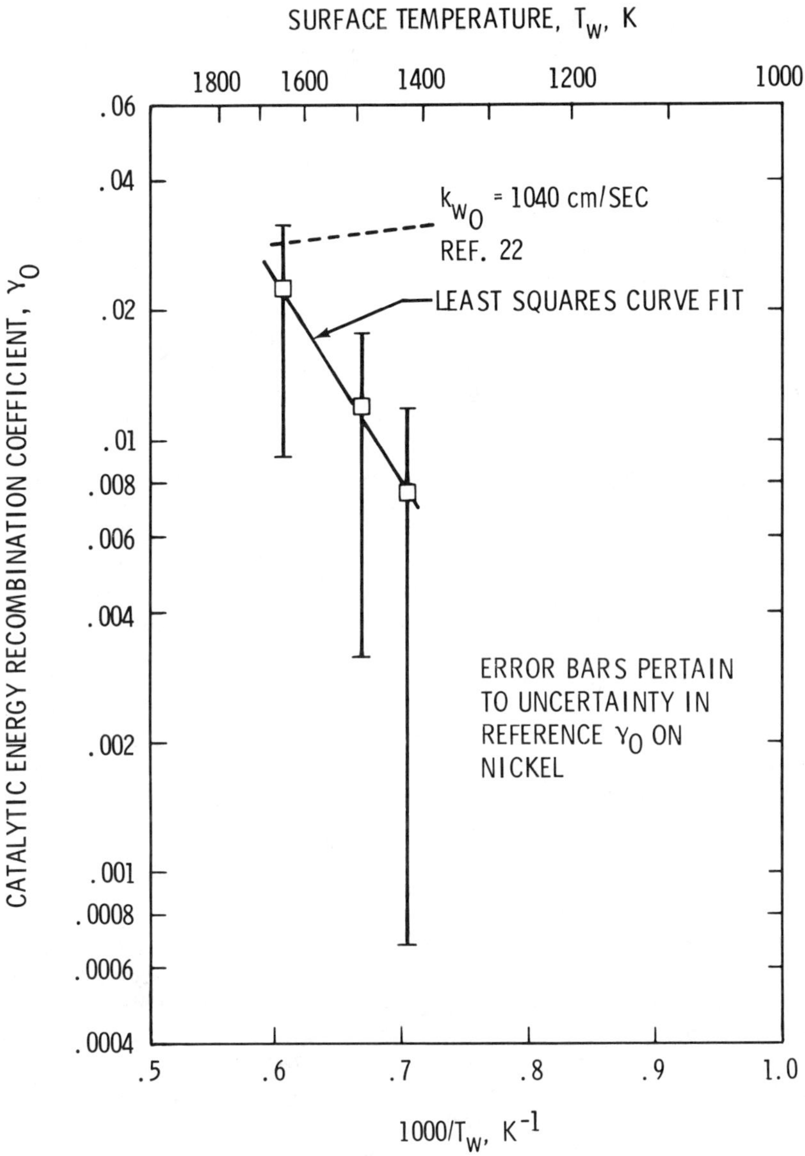

Fig. 7 Fnergy-transfer catalytic recombination coefficient for oxygen on RCG-coated HRSI.

reacting inviscid flowfield solution with a quasi-three-dimensional boundary-layer analysis. The catalytic recombination rate constants k_{wN} = 312 cm/s and k_{wO} = 1040 cm/s of Tong, Buckingham, and Curry[22] and Tong, Morse, and Curry[24] were used in this analysis. These rates were inferred from a series of arc jet tests[16] for various materials and surface temperatures and can only be considered as rates representative of thermal protection materials.

These rates are compared with the present results in Fig. 6. To compare directly, it was necessary to convert k_W to γ at some unspecified temperature. Therefore, the k_W of Tong et al. are shown as curves in Figs. 6 and 7 over the temperature range that the measurements were probably made. Their nitrogen recombination rate k_{W_N} is lower than the present results by as much as a factor of 2 over the temperature range of interest, whereas their k_{W_O} is larger by as much as a factor of 3. The results of Rakich and Lanfranco indicate that the heat flux ratio q/q_{FC} at the stagnation point may be 0.79 to 0.90 during the peak heating portion of the Orbiter entry. On the windward centerline downstream of the stagnation point where HRSI is used, this factor is somewhat nearer unity. Since the γ_O of Tong et al. is larger than the present results, the predictions of Rakich and Lanfranco[23] should overpredict the heat flux early in the trajectory and late in the trajectory where very little nitrogen is dissociated. Where a large amount of nitrogen is also dissociated, their predictions may underpredict the true heat flux. Based on Figs. 3 and 4 of Ref. 23, it appears that q/q_{FC} at an altitude of 68.88 km may be in the range from 0.90 to 0.95. A more accurate estimate of the heating rate ratio on the Orbiter during entry will require using the present results directly in a calculation such as the one in Ref. 23.

Conclusions

This work is the first systematic attempt to determine the high-temperature energy-transfer catalytic recombination coefficient on a refractory material. Although the study did not address the particular interaction mechanism and energy states of the recombined molecules, the results should be of value in engineering calculations of the heat flux to the RCG coated thermal protection material HRSI in dissociated environments such as arc jet tests and re-entry flight.

The following specific conclusions can be drawn from the results.

1. The arc jet flow in the reservoir of the JSC arc tunnel facility is not in chemical equilibrium and remains frozen throughout the nozzle. This indicates the need for

improved diagnostics and higher fidelity analyses than have been used previously in arc jet facilities. It may be desirable to modify the arc heaters by increasing the reservoir cross-sectional area and decreasing the throat area to be assured of longer residence times so that the flow is initially in equilibrium. Otherwise, computation of the reacting nonuniform flow in the arc column and reservoir is necessary to start a nozzle flow calculation more sophisticated than the simple method used in the present work.

2. The reaction cured glass coating on high-temperature reusable surface insulation is relatively nonaccommodating to the recombination energy as compared with nickel, resulting in reduced heat flux in dissociated flows. Knowing the energy accommodation recombination coefficients as functions of temperature, one can compute the heat flux to the Space Shuttle Orbiter during re-entry with a higher accuracy than previously possible and thus remove some of the conservatism in the current heating rate predictions for a fully catalytic surface. This may increase the operational limits of the vehicle.

3. The energy accommodation recombination coefficient for both oxygen and nitrogen appears to have an Arrhenius behavior with temperature; however, additional work is needed to develop a gas/surface interaction model that accounts for this behavior in detail.

4. The amount of uncertainty in γ_O and γ_N results to a large extent from the lack of accurate reference values (e.g., on nickel). Basic recombination studies should be done to understand the phenomena that are overlooked or incorrectly interpreted that result in the large variation of reported results. These studies should include the effects of excited molecule production and quenching as well as a more accurate understanding of the flow used in various techniques such as flow reactors and arc jets.

Acknowledgment

The author is grateful to S. D. Williams for carrying out the heat flux calculations from the thermocouple temperature response.

References

[1]Rahman, M. L. and Linnett, J. W., "Recombination of Atoms at Surfaces, Part 10 - Nitrogen Atoms at Pyrex Surfaces," Trans-

actions of the Faraday Society, Vol. 67, Part 1, 1971, pp. 170-198.

[2]Dickens, P. G. and Sutcliffe, M. B., "Recombination of Oxygen Atoms on Oxide Surfaces, Part 1 - Activation Energies of Recombination," Transactions of the Faraday Society, Vol. 60, 1964, pp. 1185-2308.

[3]Goulard, R., "On Catalytic Recombination Rates in Hypersonic Stagnation Heat Transfer," Jet Propulsion, Vol. 28, Nov. 1958, pp. 737-745.

[4]Rosner, D. E. and Feng, H., "Energy Transfer Effects of Excited Molecule Production by Surface-Catalyzed Atom Recombination," Journal of the Chemical Society, Faraday Transactions I, Vol. 70, 1974, pp. 884-907.

[5]Breen, J., Rosner, D. E., Delgass, W. N., et al., "Catalysis Study for Space Shuttle Vehicle Thermal Protection Systems," NASA CR-134124, 1973.

[6]Halpern, B. and Rosner, D. E., "Chemical Energy Accommodation at Catalyst Surfaces," Chemical Society of London, Faraday Transactions I, Physical Chemistry, Vol. 74, Part 8, 1978, pp. 1883-1912.

[7]Melin, G. A. and Madix, R. J., "Energy Accommodation During Oxygen Atom Recombination on Metal Surfaces," Transactions of the Faraday Society, Vol. 67, 1971, pp. 198-211.

[8]Scott, C. D., "Wall Boundary Equations with Slip and Catalysis for Multicomponent Non-Equilibrium Gas Flows," NASA TM X-58111, Dec. 1973.

[9]Bade, W. L. and Yos, J. M., The NATA Code - Theory and Analysis, Vol. 1, NASA CR-2547, June 1975.

[10]Boison, J. C. and Curtiss, H. A., "An Experimental Investigation of Blunt Body Stagnation Point Velocity Gradient," ARS Journal, Vol. 29, 1959, pp. 130-135.

[11]Scott, C. D., "An Experimental and Analytical Study of Slip and Catalytic Boundary Conditions Applied to Spheres in Low Reynolds Number Arc Jet Flows," Rarefied Gas Dynamics, edited by M. Becker and M. Fiebig, 1974, pp. D.14-1 - D.14-12.

[12]Scott, C. D., "Reacting Shock Layers with Slip and Catalytic Boundary Conditions," AIAA Journal, Vol. 13, No. 10, Oct. 1975. pp. 1271-1278.

[13]Hansen, D. F., "Approximations for the Thermo-dynamic and Transport Properties of High Temperature Air," NASA TR R-50, 1959.

[14]Williams, S. D. and Curry, D. M., "An Analytical and Experimental Study for Surface Heat Flux Determination," Journal of Spacecraft and Rockets, Vol. 14, No. 10, Oct. 1977, pp. 632-637

[15]Ames Research Staff, "Equations, Tables and Charts for Compressible Flow," NACA Report 1135, 1953.

[16]Wood, B. J. and Wise, H., "The Interaction of Atoms with Solid Surfaces," Rarefied Gas Dynamics, Supplement I, edited by L. Talbot, Academic Press, New York, 1961, pp. 51-59.

[17]Anderson, L. A., "Effect of Surface Catalytic Activity on Stagnation-Point Heat-Transfer Rates," AIAA Journal, Vol. 11, No. 5, May 1973, pp. 649-656.

[18]Pope, R. B., "Stagnation-Point Convective Heat Transfer in Frozen Boundary Layers," AIAA Journal, Vol. 6, 1968, pp. 619-626.

[19]Greaves, J. C. and Linnett, J. W., "The Recombination of Oxygen Atoms at Surfaces," Transactions of the Faraday Society, Vol. 54, July 1958, pp. 1323-1330.

[20]Greaves, J. C. and Linnett, J. W., "Recombination of Atoms at Surfaces, Part 5 - Oxygen Atomes at Oxide Surfaces," Transactions of the Faraday Society, Vol. 55, Part 8, Aug. 1959, pp. 1346-1354.

[21]Myerson, A. L., "Exposure-Dependent Surface Recombination Efficiencies of Atomic Oxygen," Journal of Chemical Physics, Vol. 50, 1969, pp. 1228-1234.

[22]Tong, H., Buckingham, A. C., and Curry, D. M., "Computational Procedure for Evaluation of Space Shuttle TPS Requirements," AIAA Paper 74-518, Palo Alto, Calif., June 1974.

[23]Rakich, John V. and Lanfranco, Martin J., "Numerical Computation of Space Shuttle Laminar Heating and Surface Streamlines," Journal of Spacecraft and Rockets, Vol. 14, May 1977, pp. pp. 265-272.

[24]Tong, H., Morse, H. L., and Curry, D. M., "Non-Equilibrium Viscous-Layer Computations of Space Shuttle TPS Requirements," Journal of Spacecraft and Rockets, Vol. 12, Dec. 1975, pp. 739-743.

PARTICLE ACCELERATION USING A HELIUM ARC HEATER

J. H. Painter*

McDonnell Douglas Astronautics Company-St. Louis Division
St. Louis, Mo.

Abstract

The feasibility of using arc heated helium to accelerate 80 μm ATJ graphite particles to velocities of 3 km/s has been demonstrated. A long accelerator nozzle was used to expand helium from reservoir conditions of 10 MPa and 17 MJ/kg to velocities > 4 km/s. The arc heated helium flow accelerated 80 μm ATJ graphite particles to velocities > 3 km/s, as measured by a laser velocimeter. Accelerator effluent diagnostics revealed a Mach 2.24 stream, with a severe degradation in total pressure. In spite of a significant energy loss to the accelerator walls, probe measurements indicate the exit centerline enthalpy to be comparable to the bulk enthalpy prior to expansion.

Nomenclature

A = nozzle cross-sectional area, m^2
C = heat-transfer correlation coefficient
C_d = drag coefficient
C_p = helium specific heat at constant pressure, J/(kgK)
d = nozzle diameter, m
H = enthalpy, MJ/kg
h = heat transfer coefficient, W/(m^2K)
L = accelerator length, m
M = Mach number

Presented as Paper 80-1478 at the AIAA 15th Thermophysics Conference, Snowmass, Colo., July 14-16, 1980.

*Senior Technical Specialist.

P = pressure, MPa
Pr = Prandtl number
Q = power loss, kW
$\dot{q}$ = local nozzle heat flux, kW/cm^2
Re = Reynolds number
r = calorimeter radius, cm
r_c = nozzle radius of curvature, m
T = temperature, K
V = velocity, m/s
x = axial position in nozzle, m
α = nozzle divergence angle, deg
β = boundary-layer property correction factor
γ = ratio of specific heats
δ_* = boundary-layer displacement thickness, m
θ = angle between wedge and flow, deg
μ = viscosity, Pa·s
ρ = density, kg/m^3
σ = stress, MPa
τ = shear stress, MPa
ϕ = angle between shock and flow, deg
ω = viscosity-temperature exponent

Subscripts

A = allowable value
ad = adiabatic wall value
e = elastic value
e = accelerator exit value
g = gas-property or gas-side value
i = impact value
o = reservoir value
r = reference value
s = stagnation value
w = wall value
x = based on axial position
* = nozzle throat value

Introduction

The ablation of thermal protection systems (TPS) can be accelerated by erosion in a flight environment that contains rain, ice, and/or dust particles. Measurements of stagnation region heat-transfer rates in hypersonic particle erosion facilities have indicated values that may exceed those in a clear environment.[1] Numerous computer codes have been developed to predict the behavior of thermostructural systems in a particle-erosion flight environment;[1,2] all these codes require erosion coefficients that must be determined empirically.

The currently available hypervelocity erosion facilities do not adequately simulate the real flight environment. Some of the methods used to obtain erosion data include: 1) subjecting the preheated specimen to hypervelocity particles accelerated by a light gas gun or shock tunnel; 2) firing a specimen down a ballistic range equipped to introduce particles in the test section; 3) mounting the model in a hypersonic wind tunnel into which dust particles are injected; 4) impacting single/multiple particles on the model using various accelerating devices; and 5) utilizing hypersonic arc heated air facilities. These simulation techniques have one or more of the following deficiencies: low velocities, short exposure time, inadequate particle dispersion, inadequate particle and/or model size, and lack of simultaneous heating and erosion simulation. The last problem is of particular significance since erosion phenomena of TPS materials are coupled to the heating environment.

In the technique outlined in Ref. 3, the ablation would be obtained with a high-pressure arc heater operating on air, and for erosion, particles accelerated to hypervelocities in a separate light gas nozzle would be introduced into the ablation jet by impingement of the two flows. However, the predicted performance of the nozzle using this technique indicates particle velocities of less than 1.4 km/s for a bulk helium enthalpy of 5 MJ/kg.

MDC has demonstrated that an arc heater can provide both high-pressure and high-temperature reservoir conditions on hydrogen, helium, and air.[6] The objective of the present work was to design an accelerator nozzle compatible with the arc heater and to demonstrate its particle acceleration capability. Helium was chosen for these tests since it is inert, offers higher gas velocities than air, and interacts less with the primary air test stream in a combined ablation erosion facility.

Results from this work will be used to develop a particle acceleration system that will operate in conjunction with an arc air heater for simulating a combined ablation/erosion RV flight environment. The particle acceleration technique used for these tests involved operating an arc heater on helium at pressures to 10 MPa with an enthalpy of approximately 17 MJ/kg. Graphite particles were introduced into the gas stream and accelerated to hypervelocities (3 km/s).

Particle Acceleration System

Arc Heater

The arc heater used in these tests was a nominal 250 kW Huels-type, high voltage unit. It was mounted and tested in the Plasma Arc Tunnel (PAT) facility as shown in Fig. 1. Two tandem, cylindrical electrodes were separated by a threaded cylindrical insulator. Both electrodes and the center chamber were water-cooled. Four tubular gas injectors were used. The gas was injected tangential to the heater walls, and the resultant radial pressure gradient stabilized the arc on the heater centerline. A magnetic field coil was used on the rear electrode to prevent arc transfer to the rear plug. The nozzle module developed for previous high-pressure hydrogen and helium tests was used in the preliminary system tests of this program.

Figures 1 and 2 show the arc heater and particle accelerator installations. Each component was individually water cooled and instrumented for power losses using differential temperature sensors and turbine flowmeters mounted on the test stand. The accelerator exhausted into a 2 m diam, 9 m long tank that could be evacuated between tests.

A 3 Ω ballast resistor was necessary to provide the immediate voltage reserve required by an arc excursion without the inherent time delay of the power supply. The resistor consisted of 60 water-cooled, stainless steel, 6.4 mm diam tubes with 0.5 mm wall thickness; each tube was 3 m long. The tubes were manifolded in two banks of 30, and each bank was separately cooled. The maximum current capability was in excess of 800 A.

Cooling water was supplied by a three-stage centrifugal pump capable of 76 kg/s at 4 MPa. The normal flow rates were 1.6 kg/s to the accelerator, 1.5 kg/s to the nozzle, 3.0 kg/s to the cathode body, and 2.3 kg/s to the anode. The ballast resistor required 6 kg/s, and the spin coil, 126 g/s.

Power was supplied by two, three-phase, full-wave rectifier modules with saturable reactor current controls. The units were connected in series for an open circuit voltage of 10.2 kV. The maximum power capablity of these two units was 5 MW for 6 min.

The gas supply system included three storage tanks having a 1.6 m^3 volume at 24.5 MPa. The gas flow was regulated at 18

Fig. 1 Particle accelerator installation.

Fig. 2 Particle acceleration system installation.

MPa using a microflute valve. Final control was accomplished using a dome loaded regulator. The arc heater injectors were calibrated at a subsonic pressure ratio corresponding to normal operation by using a critical-orifice flowmeter whose calibration is traceable to the National Bureau of Standards.

Instrumentation was used to monitor arc power input, power losses, gas pressures, helium flow rate, and particle flow rate.

It was found that an arc current of 420 A was required to achieve helium temperatures of 3200 K. The arc current was limited to this value so that electrode erosion would not be a major factor. Although the arc heater nominally operates at 250 kW, it successfully reached power levels near 1 MW. The voltage at 10 MPa and 3200 K was 2100 V for a power input of 0.88 MW.

Particle Accelerator Design

The analyses of Refs. 3, 4, and 5 were used to develop a particle acceleration code (PARTACC), and estimates were made of the particle velocities that can be achieved in a high-temperature helium expansion. The drag coefficient was made an input with provision for adding a correction. Estimates of the slip Reynolds number based on properties behind the normal shock, combined with drag data from Refs. 3 and 4, indicated a range of drag coefficients from 0.71 to 1.5. A one-dimensional isentropic core flow was assumed.

A base helium reservoir condition of 10 MPa and 3800 K was subjected to individual variations of several input parameters to determine the corresponding effects on particle velocity. Table 1 shows the calculated particle velocity at the exit of a Mach 4 nozzle (atmospheric exhaust from 10 MPa) for the range of conditions examined. These estimates indicate that it is feasible to achieve a velocity of 3 km/s with 100 μm particles. Smaller particles (25 μm) can be accelerated to 4 km/s. A 40% increase in stagnation temperature results in only an 8% increase in particle velocity. The effect of pressure was also examined; doubling the reservoir pressure increased the final particle velocity by only 17%.

Design of the particle accelerator nozzle maximized the exit particle velocity within the limitations imposed by the arc heater reservoir conditions, by fabrication feasibility and cost effectiveness, by the desired test core size, and by heat-transfer and stress considerations.

Particle Acceleration. The HEARC arc heater computer code[6] was used to predict the available gas reservoir conditions for various nozzle throat diameters. The arc current was varied from 200 to 500 A, and the arc pressure was varied from 8 to 12 MPa. A power limitation of 1 MW was imposed on the heater to eliminate electrode erosion problems.

The PARTACC code was used to determine the exit particle velocities, using the appropriate available reservoir conditions for each nozzle. It was determined that the maximum particle velocities could be obtained with the smallest nozzle throat size. Since the nozzle diverger length was scaled linearly with throat diameter, this choice also resulted in the shortest nozzle.

The gas velocity at the nozzle exit was limited in part by the ratio of reservoir to exhaust pressures. For an atmospheric exhaust and 10 MPa reservoir pressure, the maximum Mach number of the gas would be 4 with proper expansion, while overexpansion without separation might yield a maximum of 5. The effect of exit Mach number on final particle velocity was examined. In general, there was a maximum particle velocity near Mach 3; furthermore, the minimum nozzle length for a given particle velocity was obtained at Mach 3.

Table 2 is a PARTACC output showing the acclerator characteristics for the selected reservoir conditions and nozzle configuration. The particles are 80 μm ATJ graphite ($\rho_p = 1.73\ g/cm^3$) with an injection velocity of 250 m/s. The predictions for drag coefficients of 0.7 and 1.0 are shown. The actual coefficients were expected to be between these two values. Thus, PARTACC predicted an exit particle velocity in excess of 3 km/s.

Length and Boundary Layer. The effect of nozzle length (expansion angle) was examined for the optimum reservoir conditions and Mach number. A 10% increase in particle velocity required a 14% increase in nozzle length. The nozzle diverger length (660 mm) was selected to yield the required particle velocity at the lowest anticipated drag coefficient.

The long, small expansion angle diverger creates a thick boundary layer which could approach the centerline of the nozple. Therefore, the nozzle walls were moved outward a standard displacement thickness determined by

$$\delta^*/x = 0.0362\ Re_x^{-0.14} \qquad (1)$$

Table 1 Parameter effects on particle velocity in a helium accelerator

Parameter				
Drag coefficient	0.75	1.00[1]	1.25	1.50
Final velocity (m/s)	2890	3210	3400	3540
Particle size (μm)	25	50	75	100[1]
Final velocity (m/s)	4260	3770	3450	3210
Total temperature (K)	3000	3400	3800[1]	4200
Final velocity (m/s)	3030	3130	3210	3450
Injection velocity (m/s)	250[1]	500	750	1000
Final velocity (m/s)	3210	3220	3230	3240
Particle material density (g/cm^3)	1.25	1.75[1]	2.25	2.75
Final velocity (m/s)	3480	3210	3000	2830
Total pressure (MPa)	5	10[1]	15	20
Final velocity (m/s)	2640	3210	3540	3770

[1]Base conditions;
[2]Additional base conditions: exit Mach No. = 4.0; specific heat ratio = 1.67.

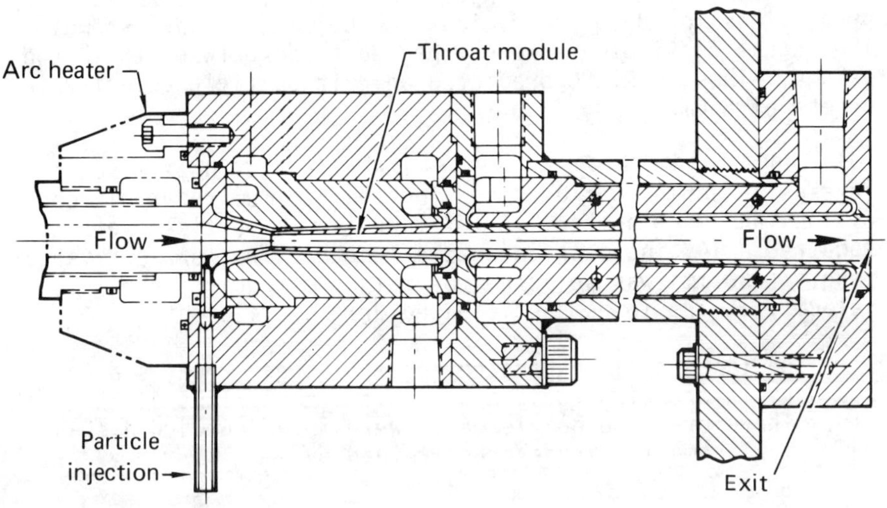

Fig. 3 Particle accelerator nozzle.

Table 2 PARTACC-predicted particle velocities[1]

x/L	x, m	V_p/V_g	V_p, m/s	V_g, m/s	M	C_d
0.00	0.000	0.09	250	2885	1.00	1.00
0.10	0.066	0.46	1727	3784	1.51	1.00
0.20	0.132	0.53	2175	4100	1.75	1.00
0.30	0.198	0.57	2456	4315	1.95	1.00
0.40	0.264	0.59	2661	4476	2.13	1.00
0.50	0.331	0.61	2821	4604	2.30	1.00
0.60	0.398	0.63	2950	4710	2.46	1.00
0.70	0.463	0.64	3056	4796	2.60	1.00
0.80	0.529	0.65	3145	4871	2.74	1.00
0.90	0.596	0.65	3224	4936	2.87	1.00
1.00	0.662	0.66	3291	4992	3.00	1.00
0.00	0.000	0.09	250	2885	1.00	0.70
0.10	0.066	0.41	1538	3784	1.51	0.70
0.20	0.132	0.48	1963	4101	1.75	0.70
0.30	0.199	0.52	2235	4316	1.96	0.70
0.40	0.264	0.54	2432	4476	2.13	0.70
0.50	0.331	0.56	2586	4604	2.30	0.70
0.60	0.397	0.58	2712	4709	2.45	0.70
0.70	0.463	0.59	2817	4797	2.60	0.70
0.80	0.529	0.60	2906	4871	2.74	0.70
0.90	0.596	0.60	2983	4936	2.87	0.70
1.00	0.663	0.61	3051	4993	3.01	0.70

[1]Helium: γ = 1.67; mole wt = 4.0 g/mole; nozzle: M_e = 3.00; A_c/A^* = 2.99; L = 0.660 m; reservoir: P = 10 MPa; T = 3200 K; particle: d_p = 80 μm; ρ_p = 1.73 g/cm³; V_{inj} = 250 m/s.

where

$$Re_x = V_g \rho_r x \mu_r^{-1}$$

Figure 3 depicts the particle accelerator nozzle assembly. The accelerator was made in two modules to reduce the axial throat stress and to prevent a total loss if the throat should fail.

$$\dot{q} = h_g (T_{ad} - T_w) \tag{2}$$

Accelerator Heat Transfer. The accelerator heat flux was severe in the throat and diverger sections. The local heat flux was computed using the technique developed in Ref. 7: where

$$h_g = \left[(C/d_*^{0.2})\ (\mu^{0.2}\ C_p/Pr^{0.6})\right.$$

$$\left.(\rho_*\ V_*)^{0.8}\ (d_*/r_c)^{0.1}\right]\ (A_*/A)^{0.9}\ \beta \tag{3}$$

$$\beta = \left\{\left[(T_w/2T_o)\ (1 + M^2(\gamma-1)/2) + 0.5\right]^{0.8-\omega/5}\right.$$

$$\left.(1 + M^2(\gamma-1)/2)^{\omega/5}\right\}^{-1} \tag{4}$$

$$T_{ad} = T_g\left[1 + Pr^{1/3}\ M^2(\gamma-1)/2)\right] \tag{5}$$

For reservoir conditions of 10 MPa and 3200 K, the nozzle throat heat flux was 8.2 kW/cm^2 requiring a coolant velocity of 80 m/s to maintain a water side wall temperature < 500 K. Above 500 K there would have been insufficient elastic wall thickness to withstand the throat hoop stress. A coolant flow rate of 1.9 kg/s through an annulus 1 mm wide was calculated to be sufficient for proper heat transfer; the actual flow rate was 1.6 kg/s with an inlet pressure of 3.3 MPa. No erosion or nozzle damage was encountered until particle flow was initiated.

Integration of the heat flux [Eq. (2)] over the length of the diverger section yields the total diverger energy loss:

$$Q = 2.5\ \pi\ (\dot{q}_*\ d_*^2/\sin\alpha)\left[(A/A^*)^{0.1} - 1\right] \tag{6}$$

The computed loss for the design nozzle was 237 kW, equivalent to a stagnation temperature drop in the gas of nearly 1000 K. This loss was not considered in the PARTACC code.

Nozzle Stress. The maximum octahedral shearing stress, τ_w, at the nozzle throat was computed using the technique outlined in Ref. 8:

$$\tau_W = 1/3 \left[(\sigma_1 - \sigma_2)^2 + (\sigma_2 - \sigma_3)^2 + (\sigma_3 - \sigma_1)^2 \right]^{1/2} \quad (7)$$

where σ_1 is the net axial stress from the opposing cooling water pressure and gas thrust on the nozzle, σ_2 is the radial stress, and σ_3 is the combined circumferential and thermal stress. The allowable octahedral stress, τ_A, was assumed to be[8]

$$\tau_A = \sigma_e \sqrt{2}/3 \quad (8)$$

The nozzle wall was fabricated from OFHC copper which has a low tensile strength above 590 K. Arc heater experience at pressures to 25 MPa had shown that the combined mechanical and thermal stress relations were valid only in a thin elastic layer near the cooling water where temperatures were below 590 K. The remainder of the wall behaved essentially in a plastic manner.

The STRESS computer code was used to optimize the heat flux [Eq. (2)] and combined stress [Eq. (7)] capability of the nozzle. The code computed the water side wall temperature of the nozzle for the design wall heat flux and cooling scheme.

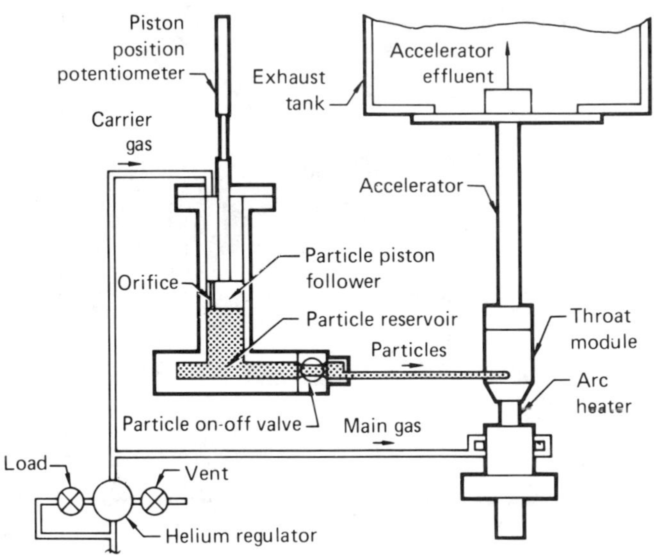

Fig. 4 Modified particle feed system.

The elastic portion of the wall was then used to compute the octahedral shear stress which was compared with the allowable values to determine the factors-of-safety. For the design selected, the computed factor-of-safety at the nozzle throat was 1.9 at the inner elastic radius and 2.8 at the outer elastic radius. These margins could be realized only by using two modules to decrease the downstream flange size of the throat section, thus reducing the longitudinal tensile stress at the nozzle throat by a factor of 2.6.

It was necessary to decrease the coolant pressure at the downstream flange of the diverger section in order to reduce the combined stresses near the diverger inlet retaining flange. The inlet pressure was adjusted to maintain the outlet pressure at 1.8 MPa, the minimum allowable without local boiling at the nozzle wall. With this pressure reduction, the diverger section could be fabricated in one piece with a safety factor of more than 1.5 for the worst conditions.

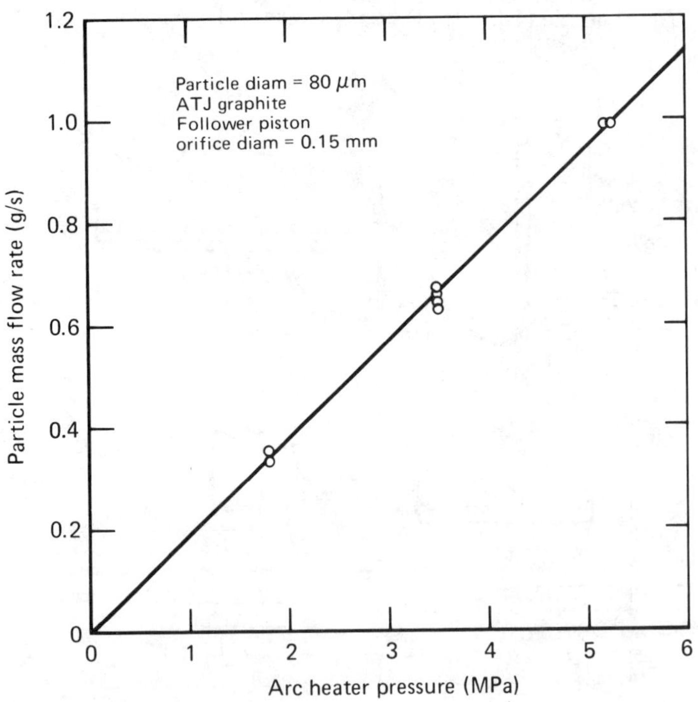

Fig. 5 In-situ particle feed system calibration.

Particle Feed System

Figure 4 shows the particle feed system as it was connected to the arc heater. Carrier gas was provided from the arc heater injector manifold and controlled by an orifice in the follower piston. The pressure differential on the piston maintained contact between the particles and piston allowing a direct measure of the volume change in stored particles via a linear potentiometer. An on-off ball valve was used to control particle flow during tests.

The particle feed system was calibrated in situ using an orifice plate between the accelerator modules to simulate hot flow pressure levels. Figure 5 illustrates the results of that calibration. For an arc heater pressure of 5 MPa, the particle flow rate was approximately 1 g/s. Higher flow rates could be achieved by enlarging the orifice in the particle follower piston.

The particles used in these tests were nominally 80 μm diam ATJ graphite. The particle size distribution was analyzed by Coulter Electronics, Inc. Less than 10% of the particulates (by volume) were smaller than 65 μm before testing. The geometric mean particle diameter was 80 μm.

The output of the linear potentiometer was recorded throughout the tests. When the desired arc heater condition was reached, the ball valve between the particle reservoir and

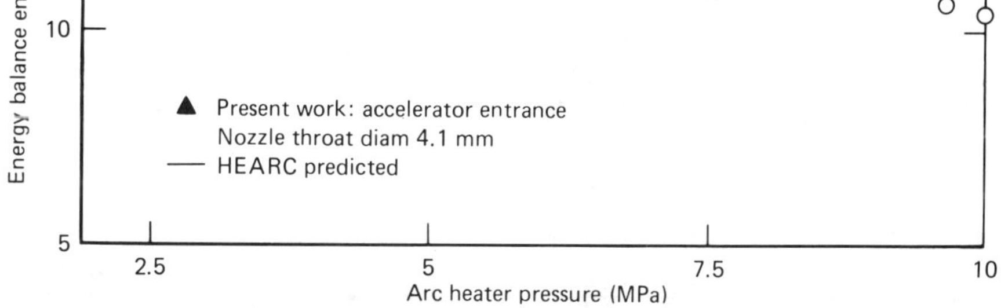

Fig. 6 Helium enthalpy performance of arc heater.

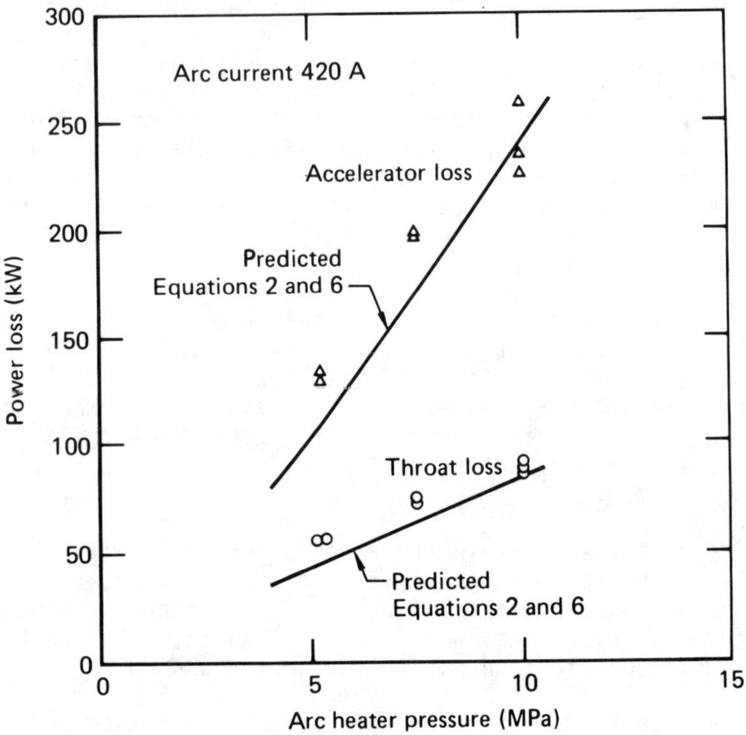

Fig. 7 Accelerator and throat power losses.

Table 3 Energy balance accelerator enthalpies

Test	Pressure, MPa	Current A	Gas accelerator enthalpy MJ/kg Entrance	Exit
23	5	423.8	17.6	10.3
24	5	414.3	15.6	8.4
25	5	424.0	15.8	8.5
26	10	424.3	16.1	9.8
27	10	419.3	17.1	10.2
28	10	428.5	17.9	10.0
29	10	435.7	17.5	10.5
30	7.5	426.1	18.1	10.5

the accelerator entrance was opened long enough to observe particle flow and record velocity data (usually 10 s).

Arc Heater Performance

The performance of the arc heater operating on helium with a 4 mm diam throat has been documented over a wide range of arc current and pressure.[6] These data were obtained at arc currents from 240 to 400 A and arc pressures from 2.5 to 10 MPa. In the present program, addition of the accelerator to the arc heater reduced the exit enthalpy considerably, but the other parameters did not change significantly.

The enthalpy pressure characteristics of the arc heater operating on helium are shown in Fig. 6 compared with those predicted by HEARC.[6] The enthalpies shown are energy balance determined values, and the pressures were measured on the side wall of the arc chamber.

The enthalpies shown can be converted to temperature by multiplying the enthalpy in MJ/kg by 192 to obtain temperature in Kelvin. Helium is not ionized below 8000 K, and ideal gas relations are applicable in the gas temperature range of these tests, 3000 to 3500 K.

The power losses to the accelerator and nozzle throat module are shown in Fig. 7 compared with those predicted using Eq. (6). The power loss is linear with arc pressure (helium flow rate) and reaches 250 kW at 10 MPa. The combined nozzle throat and accelerator losses (Fig. 7) are more than one-third of the total arc power input. These losses are inherent in this particle acceleration technique. Table 3 compares the helium enthalpies at the entrance and exit of the accelerator.

Accelerator Performance

Evaluation of the accelerator performance is obtained primarily from the velocities of the particles emerging from the accelerator exit. Additional performance parameters are the exit gas velocity, particle spatial density, and particle flux uniformity.

The initial tests of the accelerator flow were made without particles, with the objective of measuring the Mach number, impact pressure, and stagnation enthalpy of the gas stream. From these measurements, the accelerator effluent velocity was calculated and the particle velocity was estimated.

In the final tests, 80 μm ATJ graphite particles were added to the gas flow, and measurements of the particle flow rate and exit velocity were made. The spatial density was estimated from particle flow rates plus particle and gas velocities.

Test Stream Diagnostics

Figure 8 shows the effluent from the arc heated helium accelerator. The bright, conically shaped central core

Fig. 8 Accelerator effluent without particles (P_o = 5 MPa).

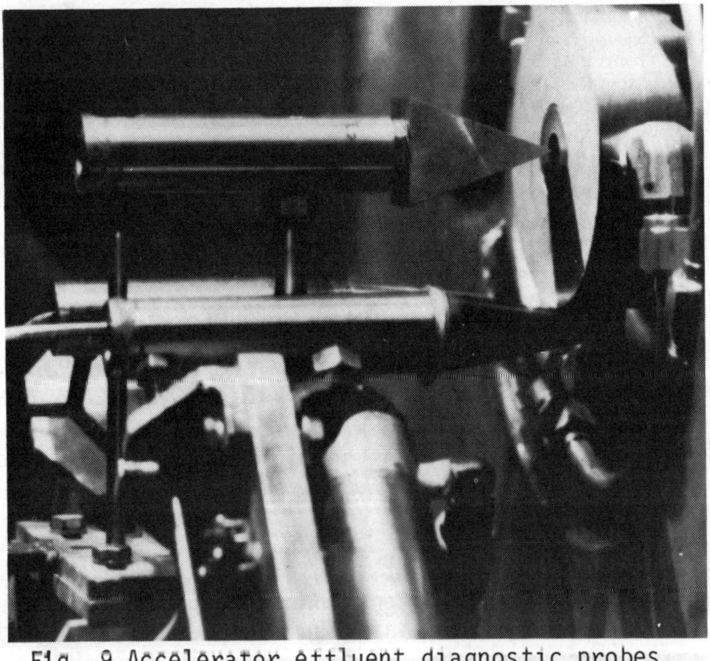

Fig. 9 Accelerator effluent diagnostic probes.

continues to expand and entrains the cold gas in the exhaust tank, forming a jet that extends nearly 30 cm downstream before becoming invisible. The effluent was relatively quiescent with brightness variations related to the downstream arc-termination location. The jet was longer and less luminous at 10 MPa than it was at 5 MPa.

Three probes were used to determine the accelerator effluent characteristics. They were mounted 45 deg apart as shown in Fig. 9 on individual linear actuators. The actuator moved the probes rapidly to the centerline of the stream where they remained for a sufficient time for each sensor to yield adequate data. The probes were sequenced to avoid interference.

Mach Wedge Measurements. The Mach number of the accelerator effluent was determined using a 20 deg half-angle wedge inserted in the stream. Figure 10 shows a typical photograph of the wedge in the stream; the shock angle is apparent and yields the stream Mach number by use of the relation for a two-dimensional wedge is supersonic flow:

$$\tan \theta/\cot \phi = 2(M_1^2 \sin^2 \phi - 1)/M_1^2(\gamma + \cos 2\phi) + 2 \quad (9)$$

Fig. 10 Mach wedge in stream.

where θ is the angle between the wedge surface and the flow direction, i.e., 20 deg, and ϕ is the angle between the shock wave and the flow direction. The measured ϕ is 72 deg $\pm$ 1 deg from Fig. 10 and the calculated Mach number is 2.24 $\pm$0.05. This Mach number is lower than the design Mach 3 which was based on displacing the nozzle wall from an isentropic inviscid core by the viscous boundary-layer displacement thickness. Evidently the degradation in total pressure increases the boundary-layer thickness faster than predicted by Eq. (1). The lower Mach number did not reduce the predicted particle acceleration capability significantly. The predicted gas velocity was lowered by less than 10%, but the predicted particle velocity was reduced by less than 2%.

Impact Pressure Measurements. The accelerator effluent impact pressure was measured at two arc heater pressure levels, 5 and 10 MPa. The results of these measurements are shown in Table 4. The initial measurements were made with a hemispherical tungsten probe having an orifice at the stagnation point. The measured pressures were lower than anticipated. A second, smaller probe was tested to reduce possible flow blockage. Both probes used a pressure tansducer (CEC 4-326-12), and the line volume was filled with silicone oil to reduce response time. The probes were inserted in the stream 25 ms and measured steady centerline values for at least 90 ms. The measurements (Table 4) show good agreement between the two different size probes (7 vs 11 and 12 vs 14).

The measured impact pressures indicate a severe degradation in total pressure in the accelerator. At Mach 2.24, the impact to total pressure ratio for helium is 0.67, which combined with the measured impact pressures indicates that the

Table 4 Impact pressure measurements

Test	Probe diameter, mm	Arc heater pressure, MPa	Impact pressure, MPa
2	7.6	5.0	0.54
3	7.6	5.1	0.60
4	7.6	5.1	0.52
5	7.6	5.1	0.61
7	7.6	5.1	0.61
11	3.8	5.1	0.60
12	3.8	9.8	1.20
14	7.6	10.0	1.23

stream total pressure decreased from 5 MPa to 0.9 MPa, and from 10 MPa to 1.8 MPa. The severe energy loss in the accelerator (one-third of the total power input) was in good agreement with turbulent flow predictions (Fig. 7), and was caused by the high-temperature differential between the gas and wall, and by the high heat-transfer coefficients. Assuming the validity of Reynolds analogy, the high heat-transfer coefficients also mean high friction coefficients and significant momentum losses.

Calorimeter Measurements. Calorimetric measurements were made in the accelerator effluent to supplement the energy balance enthalpy measurements made on the arc heater and accelerator. The measured stagnation point heating rates were combined with the impact pressure to calculate the effluent stagnation enthalpy.

The calorimeter used was a hemisphere with a null point sensor. These calorimeters (Acurex Corporation) use a chromel-alumel thermocouple located on the sensor centerline 0.25 mm from the front surface. The sensor diameter was 2.4 mm. The calorimeter thermocouple output was recorded on an oscillograph (CEC 5-124) at a paper speed of 1.5 m/s. The response time of the calorimeter was quoted by the manufacturer as 0.5 to 0.7 ms. The calorimeter entered the stream center in less than 10 ms and was removed after 70 ms. Temperatures in excess of 600 K were reached in this time period.

Figure 11 depicts the measured stagnation point heating rates in the accelerator effluent for arc heater pressures of 5 and 10 MPa (impact pressures of 615 and 1230 kPa respectively). The 5 MPa condition was repeated five times using two different calorimeters, resulting in identical time-temperature traces within the accuracy of measurement. The average heating rate at 5 MPa was 2.86 kW/cm^2, and at 10 MPa, 4.58 kW/cm^2.

Calculated Gas Velocity. The measured Mach number, impact pressure, and heat flux were combined to calculate the accelerator effluent velocity. An empirical stagnation point heat-transfer relation in several gases at high enthalpy levels has been developed by Zoby.[9] Interpolation of the coefficients for hydrogen, air, argon, and nitrogen yields the following equation for helium:

$$\dot{q}_s = 0.17\,(H_s - H_w) \cdot \sqrt{P_i/r} \tag{10}$$

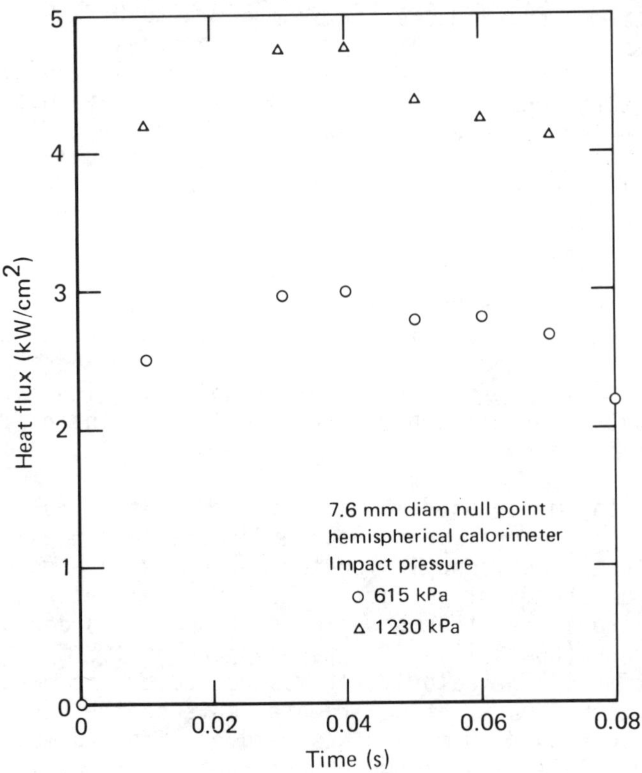

Fig. 11 Accelerator effluent heat flux.

Equation (10) and the measurements of $\dot{q}_s$, P_i, and M yield a calcualted gas velocity of 4.4 km/s at 5 MPa heater pressure and 4.7 km/s at 10 MPa. The calculated stagnation enthalpies were 15.9 and 17.9 MJ/kg at 5 and 10 MPa, respectively, which are close to the measured energy balance enthalpies upstream of the accelerator. The average, or bulk, enthalpies at the accelerator exit were 8.5 and 10 MJ/kg, respectively. It is not unusual in the arc heated test stream for the centerline enthalpy to be as much as twice the average value. The energy losses occur predominately from the gas near the cooled accelerator wall. Additional radial profiles of heating rate and pressure would be required to specify the exit enthalpy distribution.

All diagnostic probes were positioned near the accelerator exit, whereas the subsequent velocity measurements were made 247 mm downstream. The effluent was underexpanded at 10

MPa. The calculated static pressure at the nozzle exit, based on M = 2.24 and an impact pressure of 1.23 MPa, was 0.155 MPa. Further expansion of the flow to the 0.1 MPa ambient pressure would increase the test stream Mach number to 2.6 which would result in a 5% increase in the gas velocity. It was concluded that the increased residence time in the higher velocity jet would further accelerate the particles between the exit and the measuring station.

Particle Acceleration Tests

A total of 15 tests were made where nominal 80 μm diam ATJ graphite particles were accelerated by arc heated helium. Two primary arc heater conditions were tested, 5 and 10 MPa, at an arc current of 420 A. A third pressure, 8 MPa, was used in the final test. The normal procedure was to stabilize the arc heated flow without particles, and then activate the particle flow for 10 s while velocity data were required. The particle flow period was minimized to reduce accelerator erosion.

All of the tests were made with the effluent exhausting into a large tank which had a 15 cm diam valve open at the end opposite the arc heater to allow the gas to escape. Most of the larger particles were captured inside the tank and later removed by a high air mass flow entering through a side window and exhausting through a steam ejector system. The tank was equipped with centerline windows which allowed the operator to view, and the laser-beam to traverse the accelerator effluent.

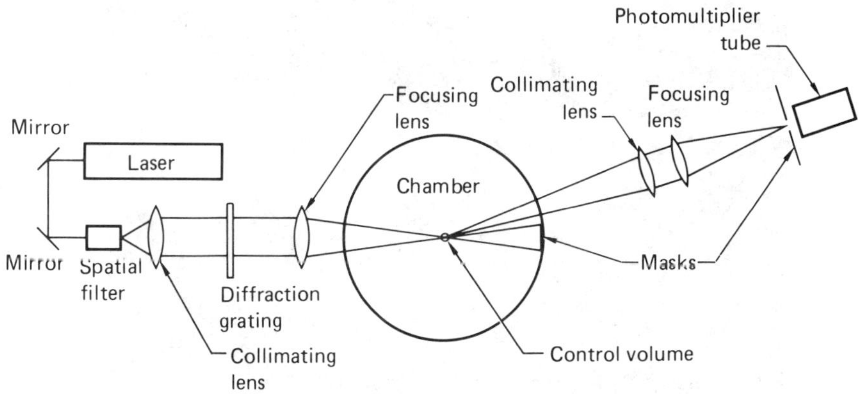

Fig. 12 Laser velocimeter system.

Particle Velocity Measurements. A particle-sizing interferometer was modified extensively by Dr. James Trolinger of Spectron Development Laboratories and by AFFDL to convert from a laser Doppler system to a laser grating system.[10] The change was necessitated by the high-frequency signals anticipated from particles moving at 3 km/s.

The resulting system (shown in Fig. 12) utilized an argon ion laser (Spectra Physics Model 164-08) operating at a wavelength of 514.5 nm as the source of light. The laser beam was passed through a spatial filter and collimated before passing through a diffraction grating. The light was then focused normal to the accelerator axis 25 cm downstream of the exit plane. The laser radiation scattered by particles in the flow was focused upon a photomultiplier tube (PMT) entrance slit by a dual receiving lens. The axis of these receiving optics was vertically offset from horizontal by about 15 deg to avoid focusing the laser beam directly upon the PMT. The PMT output signal had a frequency on the order of 1 MHz. The signal was intended to be processed by a particle size analyzer (Spectron Development Laboratories) which outputs a digitized frequency into a microcomputer (Applied Digital Communications Model 70-131-161) with a flexible disk drive. The microcomputer was intended to use these data to generate velocity histograms of the particles in the accelerator effluent. Early in the testing it was found that the PMT

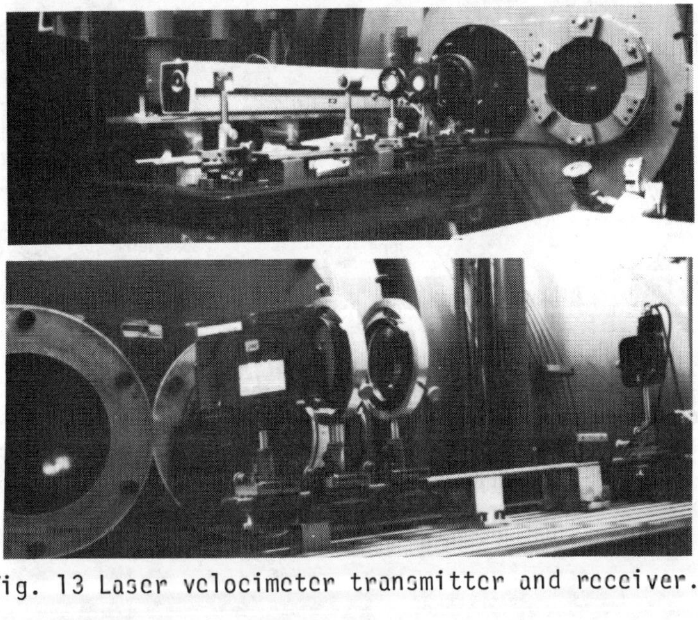

Fig. 13 Laser velocimeter transmitter and receiver.

Fig. 14 Laser velocimeter alignment.

output exceeded the 1.5 V maximum signal level of the analyzer, and thus other techniques were required to statistically analyze the data. A data type recorder (Bell and Howell Model VR3700B) with a 2 MHz frequency response was used in the remaining tests to record the analog PMT output signal for future processing. The PMT signals were also monitored by using a storage oscilloscope (Tektronix Model 7633).

Figure 13a shows the laser velocimeter transmitter on one side of the test tank, and Fig. 13b shows the receiver apparatus on the other side of the tank. The diffracted laser beam formed multiple control volumes on the accelerator axis with the three center volumes of primary intensity as shown in Fig. 14. The beams were aligned perpendicular to the accelerator centerline using a second laser and small orifice

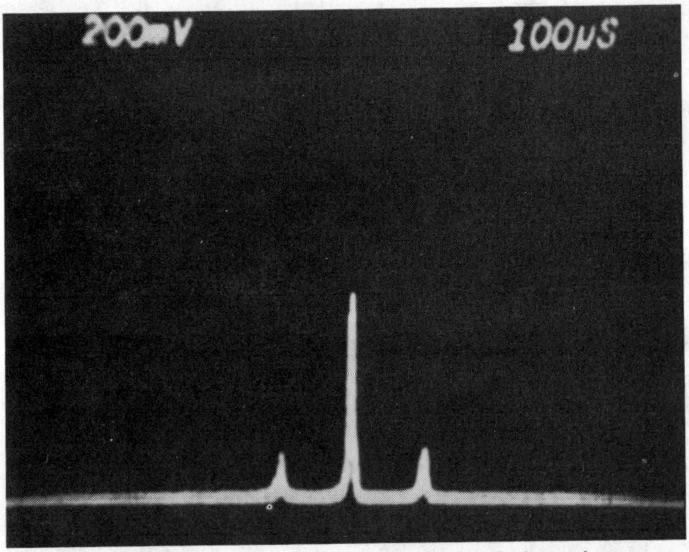

Fig. 15 Laser velocimeter calibration.

plugs as shown in Fig. 14. The distance between control volume centers was 3.05 mm. Thus the time required for a particle traveling at 3048 m/s (10,000 ft/s) to pass two adjacent beams was 1.0 μs and the frequency was 1 MHz.

A rotating disk with a human hair attached perpendicular to the periphery was used to calibrate the system. The peripheral speed could be varied to a maximum of 29 m/s. The output of the PMT with the calibrator at top speed is shown in Fig. 15. The intensity of the center volume is 3.5 x the intensity of the adjacent volumes, which assists in distinguishing the signal of a single particle from two particles in coincidence on the oscilloscope photographs.

Figure 16 and 17 present photographs of the PMT output via the oscilloscope and tape recorder for two tests. The photographs were taken from reruns of the taped output. The occurrence of events similar to the calibration signal was evident in many tests.

Two predominant velocities appear in the data, and they are in reasonable agreement with the calculated gas and particle velocities as shown in Fig. 18. The centerline gas velocity was calculated using the measured impact pressure, stagnation point heat flux, and wedge determined Mach number. Virtually identical centerline gas velocities were computed using PARTACC with the assumption that the centerline stagnation temperature was not degraded in the expansion. This assumption is based on the measured centerline stagnation point heating rates and impact pressures which indicate that the accelerator energy loss occurred primarily in the flow near the wall. PARTACC predicted particle velocities are also shown in Fig. 18. The lower band of measured velocities

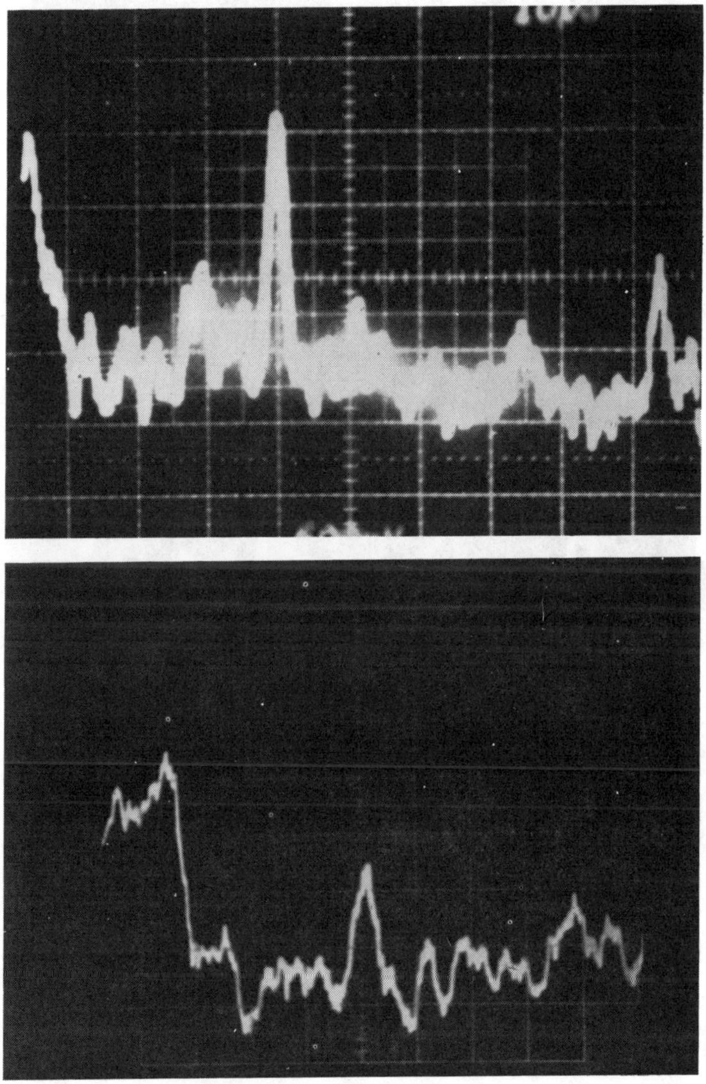

Fig. 16 Laser velocimeter output signals, test 29: 1.25 μs/cm (top), 0.625 μs/cm (bottom).

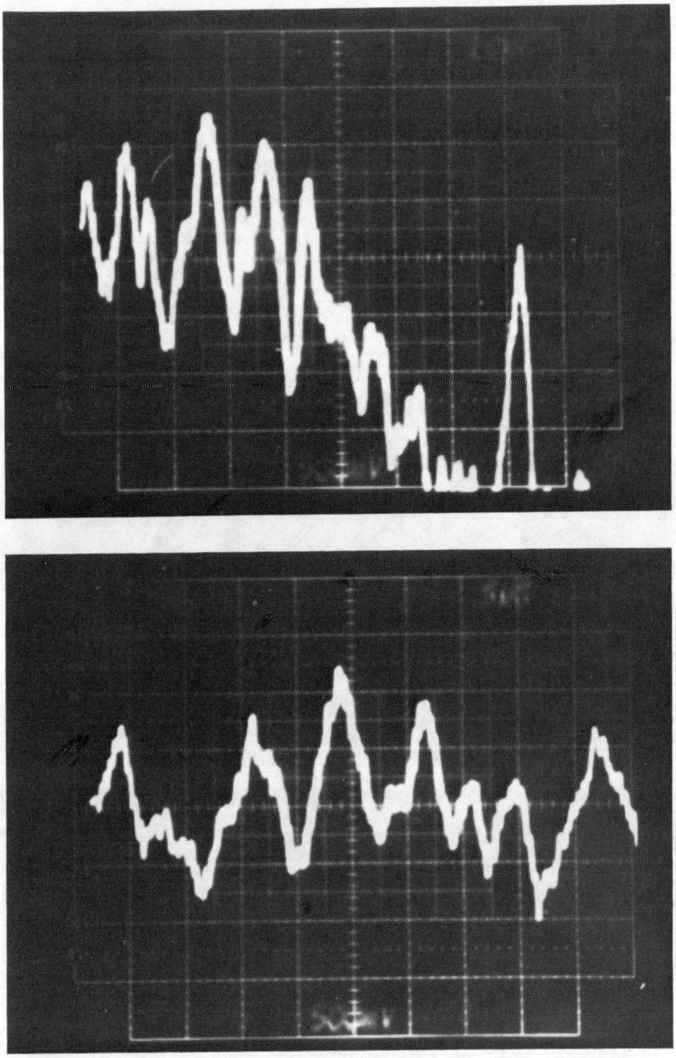

Fig. 17 Laser velocimeter output signals, test 30: 1.25 μs/cm (top), 0.625 μs/cm (bottom).

(large particles) agree with prediction to ± 15% at 5 MPa and ± 6% at 10 MPa. PARTACC does not model the frictional effects in the core flow and thus does not predict the strong degradation of total pressure measured with the probes.

The PARTACC prediction in Fig. 18 was for 80 μm diam ATJ graphite particles. Evidently the higher measured band of velocities represents smaller particles. A microscopic analysis of the particles in the collection tank after acceleration showed particles sized from 1 μm to 80 μm with the predominant size being the latter. The smaller particles followed the gas velocity closely. Evidently the particle injection technique generated smaller particles, or they were impact initiated. The particles did not appear to have ablated but rather retained rough characteristics, as if they were fractured. The presence of smaller particles adds to the complexity of analyzing the measured velocities.

Accelerator Erosion. The in situ particle flow calibrations caused severe erosion of the copper nozzle throat and accelerator liners, particularly at the minimum diameter and in the area downstream of the calibrator orifice. The throat was eroded, and the accelerator wall was eroded to the water passage. The accelerator liner was repaired by electron beam welding a new entrance region to the remaining expansion section. These liners were used for tests 15 through 22 with no further problems.

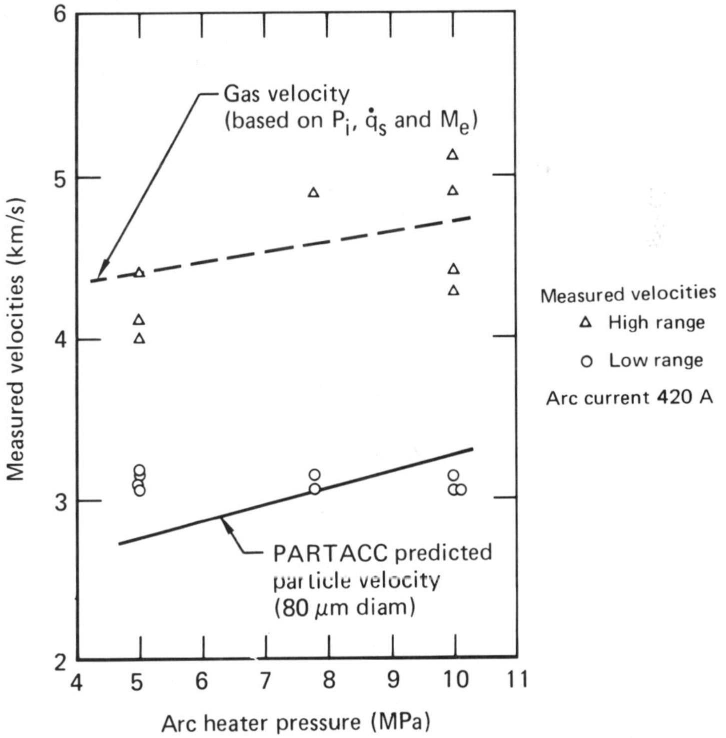

Fig. 18 Measured particle velocities.

New liners were installed in the nozzle throat and accelerator starting with test 23. At the end of test 30, the nozzle and accelerator liners had eroded only slightly. The liners apparently would have had a significant life time before requiring replacement. The increased throat area increased the helium flow rate linearly for a given arc pressure and slightly reduced the exit Mach number: however, the gas and particle velocities at the exit were not affected significantly by erosion.

Conclusions

The feasibility of using arc heated helium to accelerate 80 μm particles to velocities of 3 km/s has been demonstrated. A small Huels-type arc heater operating on helium at pressures to 10 MPa and enthalpies to 18 MJ/kg has been shown to be adequate as a hot gas source. A water-cooled, thin wall copper accelerator nozzle sustained heating rates to 8 kW/cm^2 without damage. Particle velocities in the accelerator effluent were measured using a laser velocimeter (LV). Time of flight measurements using the LV photomultiplier output need to be fully statistically analyzed. However, preliminary data indicate two velocity ranges that compare favorably with calculated gas and particle velocities at the accelerator exit. Inferred particle velocities exceed 3 km/s, and inferred gas velocities exceed 4 km/s.

Acknowledgments

This work was supported by the United States Air Force Flight Dynamics Laboratory (AFFDL) under Contract F33615-77-C-3069; Project Engineer, K. R. Cramer.

References

[1]Dunbar, L. E., and Courtney, J. F., "Minuteman Hot Structures Heating Augmentation Study, Volume I - Mechanisms and Analyses," SAMSO TR 73-272, Aug. 1973.

[2]Migotsky, E., "Gasdynamic Particle Accelerators - An Optimization," McDonnell Douglas Report MDC G4163, March 1973.

[3]Fewell, K. P., and Kessel, P. A., "Analysis, Design and Test of Components of a Combined Ablation/Erosion Nozzle," AEDC TR-75-154, March 1976.

[4]Korkan, K. D., Petrie, S. L., and Bodonyi, R. J., "Particle Concentration in High Mach Number Two Phase Flows," USAF ARL 74-0102, July 1974.

[5]Korkan, K. D., Petrie, S. L., Rice, T. S., and McCovey, E., "Advanced Two-Phase Flow Investigation in an Erosion-Ablation Facility Including Diagnostics," AFFDL TR-77-78, Aug. 1977.

[6]Painter, J. H., and Shaeffer, J. F., "Performance Characteristics of a Huels-Type Arc Heater Operating on Hydrogen, Helium or Air," AEDC TR-76-25, March 1976.

[7]Van Kuren, J. T., "The Mechanical Design of a High Temperature Mach Number 4 Axisymmetric Wind Tunnel Nozzle," USAF ASD TDR 62-445, June 1962.

[8]Lewis, H. F., and Horn, D. D., "Some Considerations in the Design of Water-Cooled Nozzle Throat Sections for High Temperature, High Pressure Operation," AEDC TR-64-51, April 1964.

[9]Zoby, E. V., "Empirical Stagnation-Point Heat Transfer Relation in Several Gas Mixtures at High Enthalpy Levels," NASA TND-4799, 1968.

[10]Trolinger, J. D., "Development and Application of a Laser Velocimeter to Measure Very High Speed Particle Velocities," AFFDL TR-79-3067, June 1979.

TEMPERATURE AND ABLATION OPTICAL SENSOR

Joseph J. Cassaing,* Patrice M. Lelièvre,+
Pierre Durrenberger,≠ and Daniel L. Balageas§

ONERA, Châtillon, France

Abstract

For heat-shield recession measurement a novel technique, different from current radiation and acoustic measurement methods, has been ground and flight tested. A combined ablation/temperature gage, using an optical view of the radiation from a sensor embedded within the heat shield is described. Flight measurements are compared with those of thermocouples, and the passage of the ablation front is observed. Numerical simulation of ablation and heat diffusion into the instrumented ablator is possible, allowing accurate evaluation of errors due to the presence of the gage.

Nomenclature

D	= diameter
I	= current intensity of photodiode
L	= length of TAOS cavity
r,z	= cylindrical coordinates
T	= temperature
T_A	= ablation temperature
$\overline{T}_{TAOS}$, $\overline{T}_i$	= internal mean temperature of the flat bottom of TAOS
$\overline{T}_0$	= external mean temperature of the flat bottom of TAOS
t	= time
t_A	= ablation front passage time in the ablator
t_O	= time of cavity opening

Presented as Paper 80-1479 at the AIAA 15th Thermophysics Conference, Snowmass, Colo., July 14-16, 1980.

*Research Engineer, Systems Department.
+Research Engineer, Materials Department.
≠Research Engineer, Physics Department.
§Head, Thermophysics Division, Systems Department.

z_A = axial position of ablation front
z_{TAOS} = axial position of internal surface of cavity bottom
δ_i = intrinsic error

Introduction

The problem of measuring the shield recession during re-entry may be solved by various techniques: gamma backscatter systems, direct gamma systems, and ultrasonic systems. These measurements, coupled with conventional temperature gage measurements (thermocouples), allow the flight evaluation of ballistic re-entry vehicle heat shields. Unfortunately, each technique has its limitations, particularly thermocouples which cannot deliver information for the temperature beyond 2500 K. Since 1977, ONERA has developed and flight tested a new and original sensor which allows the measurement of the internal temperature and the detection of the ablation front passage.[1,2] This optical sensor TAOS (Temperature and Ablation Optical Sensor) is essentially made of a long, narrow hollow blind cylinder and of a photodiode receiving, possibly through an optical fiber, the radiation emitted by the cylindrical cavity (see Fig. 1). The hollow cylinder is made of tight vitreous carbon and its typical dimensions are 2 mm diam and several tens of millimeters long. The diameter of the inner cylinder constituting the cavity is 1 mm or less. This gage is adjusted in a blind hole bored in the ablator, the axis of the cavity being parallel to the heat flux direction. When the material is subjected to intense heat, the cavity bottom is the hottest part of the sensor and a strong thermal gradient is generated along the cavity axis. In such conditions, the radiations received by the photodiode are issued mainly from the practically isothermal bottom of the cavity. Thus the temperature of this cavity bottom can be related to the current delivered by the photodiode which acts as a thermometric sensor. When the ablation front reaches the bottom of the cavity, the radiation received by the sensor drops abruptly, allowing the determination of the ablation time relative to this location. In these conditions, TAOS acts as an ablation gage giving only one ablation measurement. However, an important advantage is to be noted: the sensor allows the measurement of the ablation temperature, data that no other gage can deliver during flight.

TAOS as Temperature Gage

Various types of photodiodes can be used. A silicon photodiode was chosen in conjunction with silica-glass fibers, giving an effective spectral range of 400-1000 nm. Thus the temperature range for the measurement begins near 1600 or 1700 K and extends up to the highest temperatures reached by the ablator (4000 K in the case of carbon-carbon materials). This temperature range is particulary interesting, as its lower part may be explored by thermocouples, thus giving a possibility of comparison.

The first devices were cavities bored directly into the ablator. It appeared that the sensor output was highly dependent on the optical properties of the cavity, such as shape and roughness, so that a calibration was necessary for every cavity. Such calibrations were not possible. When using TAOS in charring ablators another problem was encountered: the pyrolysis gases generated by heating entered the cavity, introducing important perturbations in the signal. The solution was found by making the cavity in a cylinder of tight, homogeneous, and reproducible material. The cylinder was adjusted in a hole bored into the ablator. Vitreous carbon was

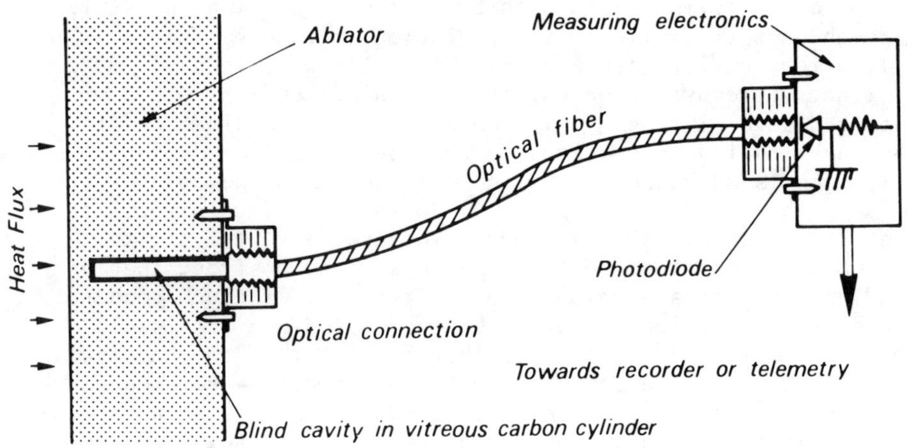

Fig. 1 - Schematic diagram of TAOS.

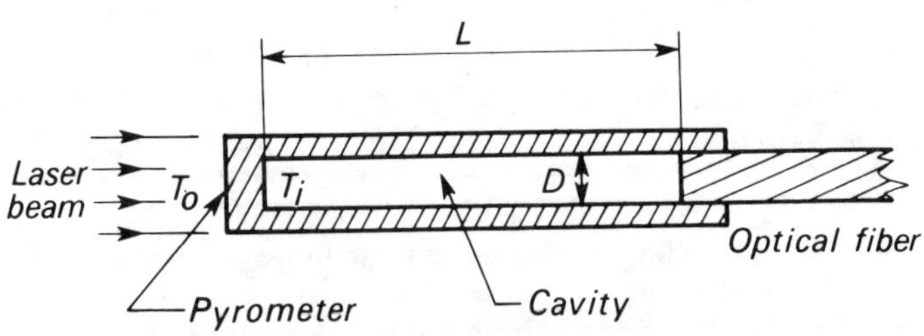

Fig. 2 Schematic view of TAOS during calibration tests.

selected and was found satisfactory. Due to the fact that practically all of the energy reaching the diode is emitted by the hot flat bottom of the cavity and that the colder lateral surfaces act as a poor reflector, the bottom of the cavity must be in view of the photodiode. So it is absolutely necessary to drill the cavity perfectly rectilinear, even in the case of very long devices. This is possible only by using an ultrasonic machining technique. This technique was chosen after comparison tests with conventional drilling and electro-erosion techniques.

The calibration of TAOS is achieved in the laboratory prior to its insertion into the ablator. The TAOS is submitted in vacumm to a continuous-wave (cw) CO_2 laser delivering a homogeneous heat flux upon the bottom of the cavity (see Fig. 2).

Under these conditions, temperatures similar to those encountered during flight are obtained. The time history of the outer mean temperature $\overline{T}_o$ is measured by an optical pyrometer, and the inner mean temperature $\overline{T}_i$ is derived using a two-dimensional finite-difference numerical simulation of the heat diffusion into the TAOS (see Fig. 3). The calibration curve relates the output current to the inner temperature. Beyond 2000 K, this current is roughly proportional to the eighth power of the temperature:

$$I = K\,\overline{T}_i^{\,8}$$

The coefficient K depends on the optical properties of vitreous carbon, the quality of the cavity machining, the geometry of the cavity (see Fig. 4), and the diameter of the optical fiber.

It is particularly interesting to study the length dependence of k. Experimental results obtained with a 1.35 mm diam cavity are presented in Fig. 5. By successively cutting this cavity, the length varied from 98 to 41 mm. Successive calibrations allowed the determination of the law k = k(L). For L greater than 76 mm, k is almost proportional to L^{-2}, indicating that only direct rays are received by the photodiode, the reflected rays being too much attenuated by the multiple reflections on the lateral surface of the cavity. On the contrary, the light reflection on the cylindrical surface of the cavity takes an increasing importance when L decreases: for L = 41 mm, the slope of the curve is only - 1.1.

The repeatability of TAOS as a temperature gage is demonstrated by successive calibration tests. Two successive calibration curves, T = T(I), are presented in Fig. 6. The

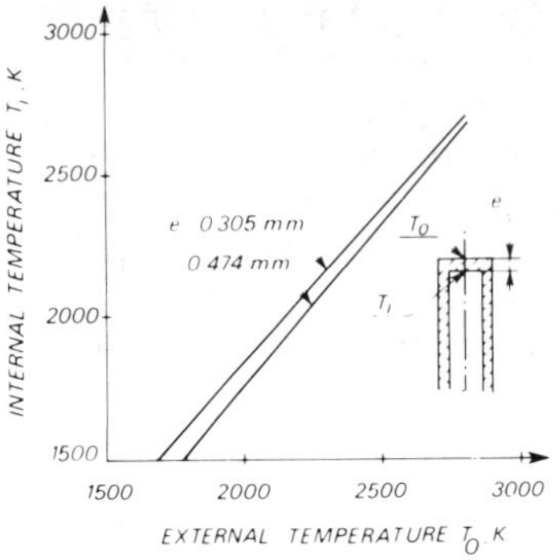

Fig. 3 Evaluation of TAOS bottom internal mean temperature $\overline{T}_i$ during calibration test, using two-dimensional numerical simulation of heat diffusion and experimental external mean temperature $\overline{T}_0$. Results for two TAOS with bottom thickness e = 0.305 mm and 0.474 mm and laser intensity I = 540 W x cm^{-2}.

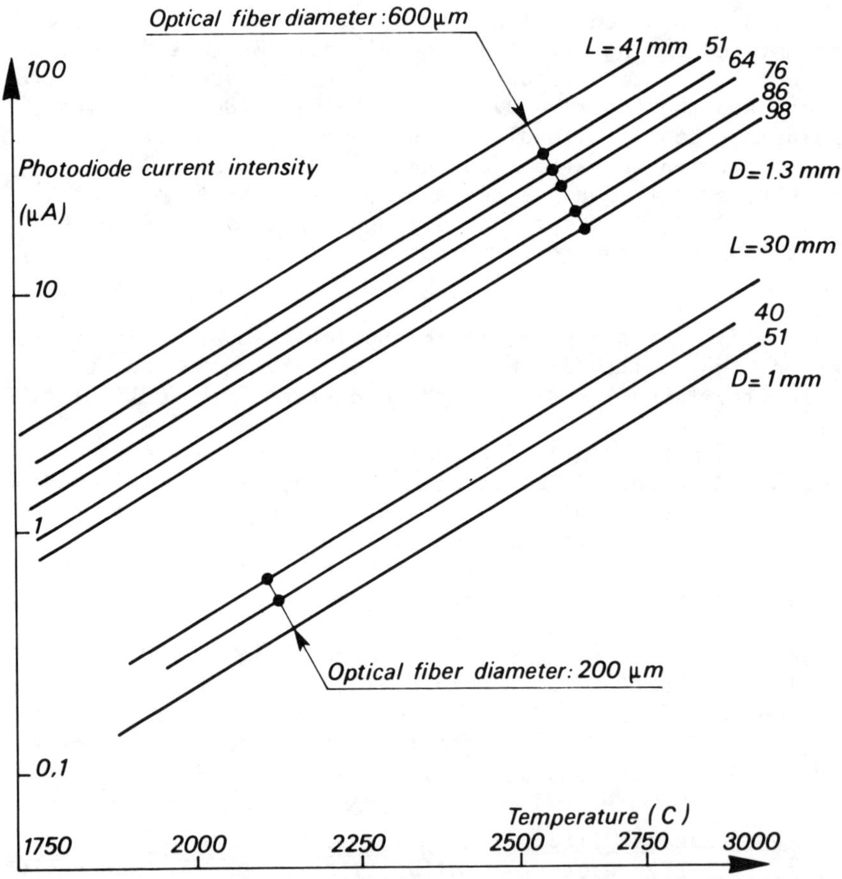

Fig. 4 Laboratory calibration curves.

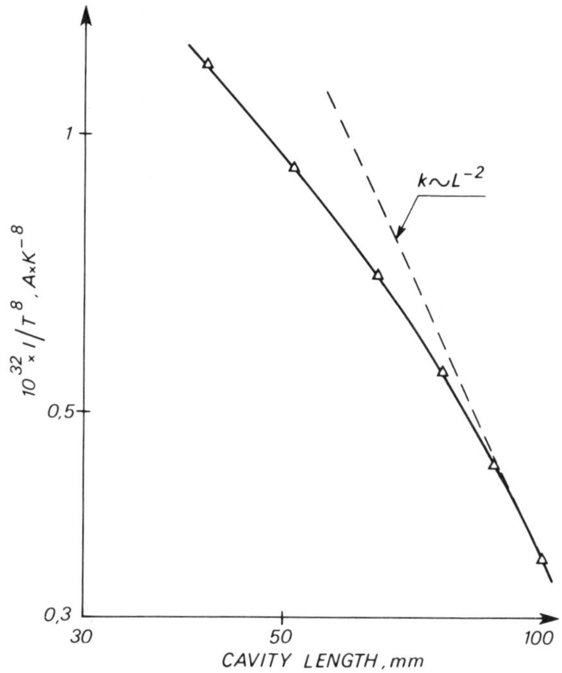

Fig. 5 Cavity length dependence of current intensity of the photodiode. The cavity diameter is 1.35 mm.

difference in temperature for the same photodiode intensity, for two calibrations, is less than 50 K for the whole temperature range explored.

A numerical simulation of heat diffusion into the instrumented ablator and into the sensor was developed. This modeling includes ablation phenomena. Thus the temperature measurement error due to the presence of the hollow cylinder may be evaluated with good accuracy. The correction depends on the thermophysical properties of the ablator and the re-entry conditions. This point will be examined in more detail later.

TAOS as Ablation Gage

TAOS were ground tested using two simulation facilities: a 5 MW Aerospatiale arc heater, and an ONERA facility coupling a 10 kW cw CO_2 laser and a Mach 2 wind tunnel. For these tests, TAOS were implanted in three-dimensional carbon-carbon materials and in three-dimensional reinforced charring ablators. The TAOS signal drop time was compared to time information

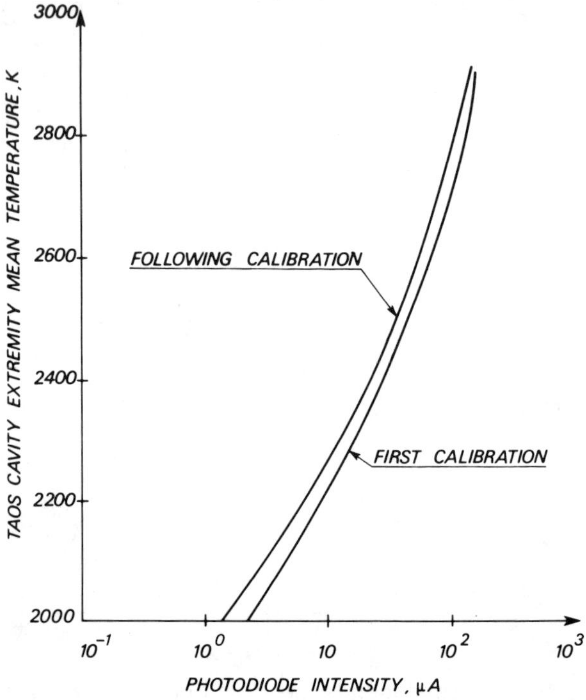

Fig. 6 Repeatability demonstration of TAOS by two successive calibration tests.

resulting from the examination of motion pictures of the ablating surface showing the opening of the cavity extremity. In all these experiments, the simultaneity of the two phenomena was verified with a precision better than 0.1 s, the signal drop being very sharp (see Fig. 7). It was concluded that this type of sensor was adequate and sufficiently safe for flight testing.

Flight Experiment

Four TAOS were mounted on conical parts of an experimental re-entry vehicle launched in 1979. The TAOS were inserted within a silica fiber reinforced resin ablator (SFRR). Two sensors indicated ablation, giving the cavity opening time with an uncertainty (current drop duration) of less than 0.1 s (see Fig. 8). The third sensor was not ablated but delivered a signal, and the last did not operate.

The perturbations due to the insertion of TAOS within the ablator were evaluated using a two-dimensional model of heat

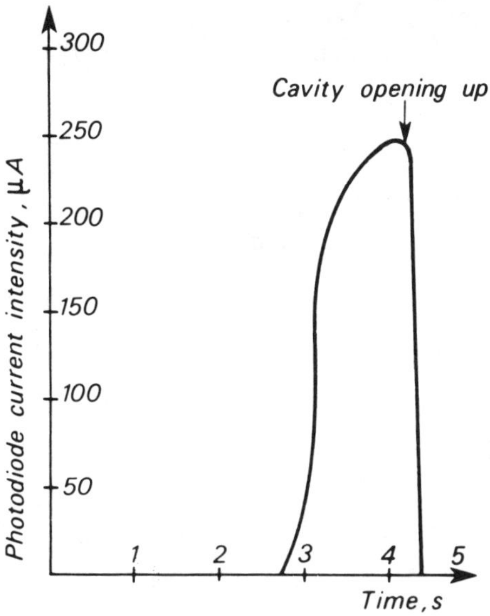

Fig. 7 Three-dimensional carbon-carbon material ablation observed by TAOS on the cw CO_2 laser facility, with laser intensity $I = 5000 \ W \times cm^{-2}$.

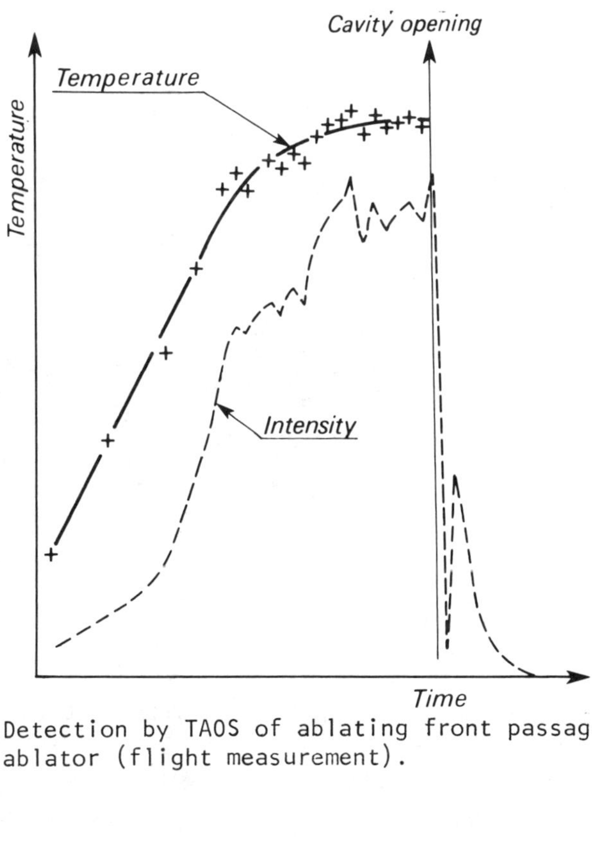

Fig. 8 Detection by TAOS of ablating front passage in a charring ablator (flight measurement).

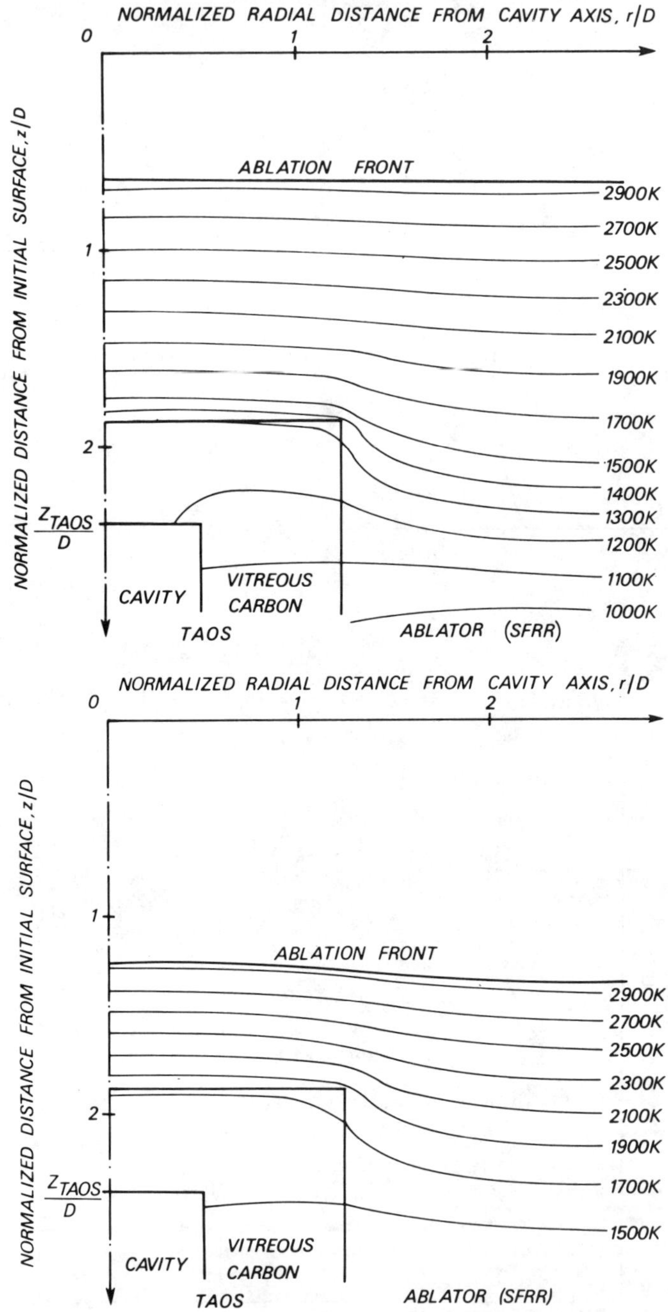

Fig. 9 Isotherms and ablation front position at four successive moments in TAOS region: numerical simulation results for SFRR ablator during flight test.

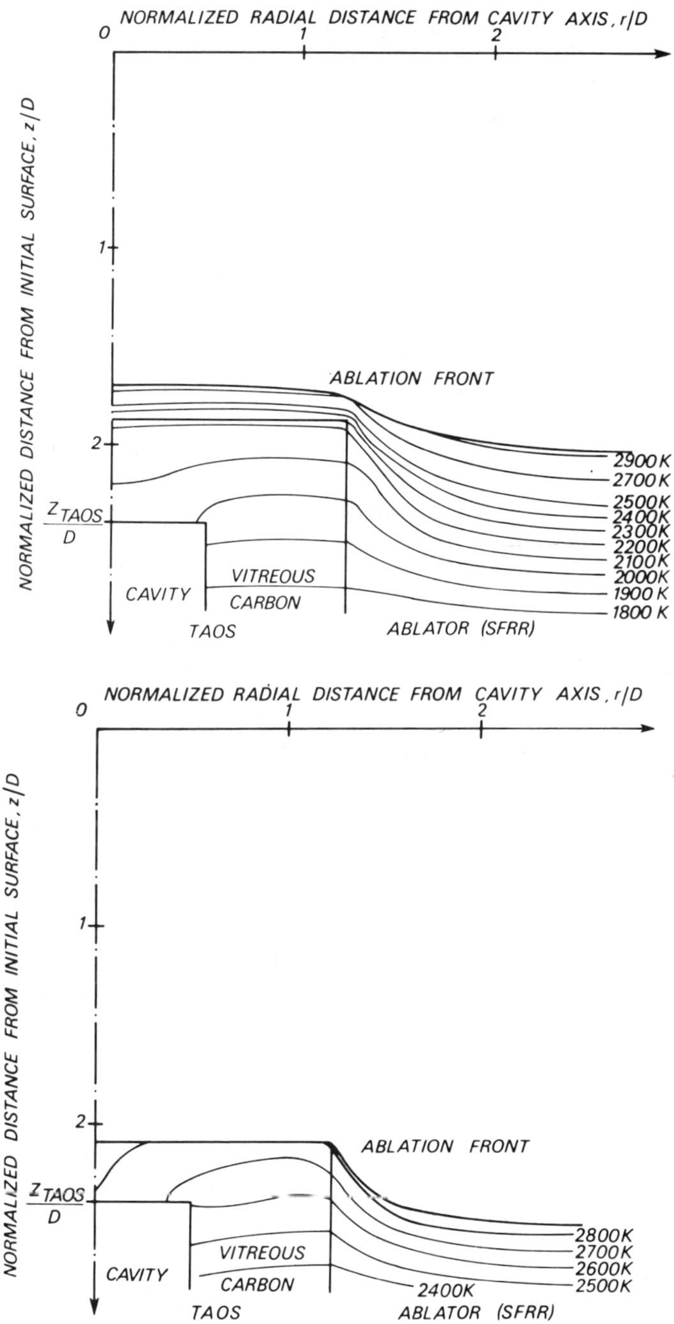

Fig. 9 (cont'd) Isotherms and ablation front position at four successive moments in TAOS region: numerical simulation results for SFRR ablator during flight test.

diffusion and ablation, applied to each sensor in the flight configuration. The isotherms and the ablation front positions around a sensor are presented in Fig. 9 at four successive times; this sensor was eventually ablated and delivered the cavity opening signal.

The distortion of the internal heat distribution is important, as will be seen in the following example (Fig. 10).

z_{TAOS} = the depth of the internal face of the sensor bottom

t_A = the time at which the ablation front reaches this depth, far from the sensor

t_0 = the sensor opening time

$\delta_i t_A$ = $t_0 - t_A$, the intrinsic error in ablation time

$\overline{T}_{TAOS}$ = the mean temperature of the internal face of the sensor

T_{SFRR} = the temperature at depth z_{TAOS}, far from the sensor

$\delta_i T$ = $\overline{T}_{TAOS} - T_{SFRR}$, the intrinsic error in internal temperature

T_A = the ablation temperature far from the sensor

$\delta_i T_A$ = $\overline{T}_{TAOS} - T_A$, intrinsic error in ablation temperature

The simulation provided the following, as well as the intrinsic error in ablation time $\delta_i t_A$:

1) The intrinsic error $\delta_i T$ of the instrument considered as an internal temperature sensor (Fig. 11), this error expressing the difference between the indicated temperature and that at the same depth within the ablating material. It obviously ends at time t_A.

2) The intrinsic error $\delta_i T_A$ of the instrument considered as an ablation temperature sensor (Fig. 12), this error expressing the difference between the indicated temperature and that of the ablation front of the ablator. It ends at time t_0.

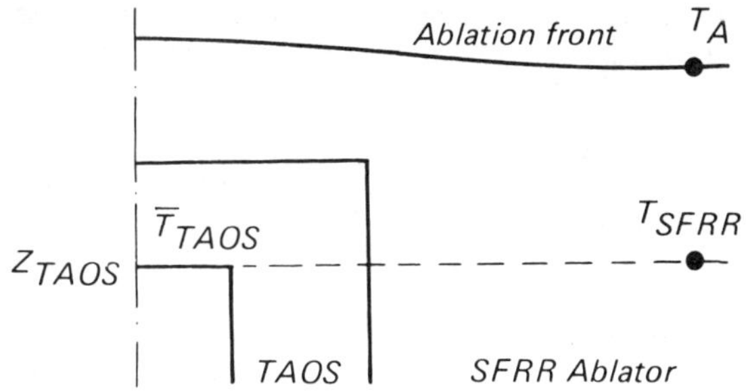

Fig. 10 Different temperatures considered in the simulation for correction of the TAOS rough data.

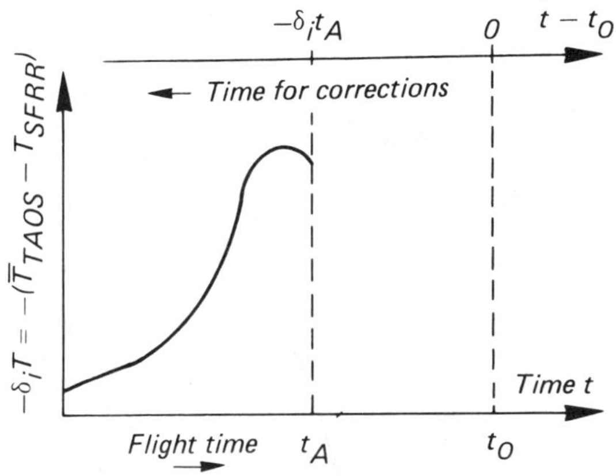

Fig. 11 Evolution of the intrinsic error of TAOS considered as an internal temperature sensor.

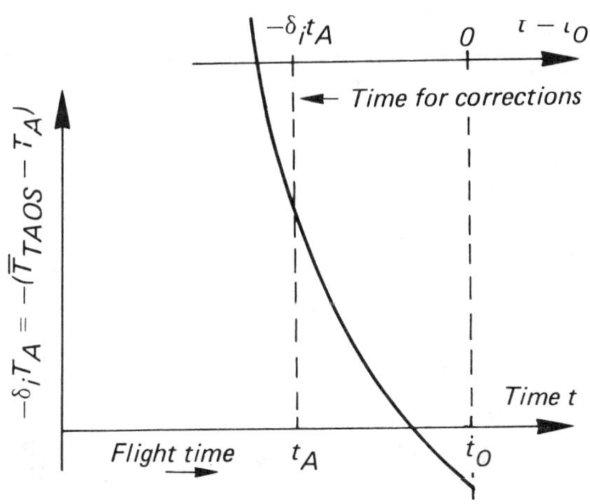

Fig. 12 Evolution of the intrinsic error of TAOS considered as an ablation temperature sensor.

From the two laws of Figs. 11 and 12 and from the rough experimental temperatures, corrected flight data may be generated (see Fig. 13).

These intrinsic errors may be important. In the case of the TAOS presented in Fig. 8, they reach

$$\delta_i t_A = 1.5 \text{ s}$$

$$\delta_i T_A \ (t = t_O) = 110 \text{ K}$$

In fact, these errors depend on the whole thermal history of TAOS and they would be different for other trajectories or other insertion depths.

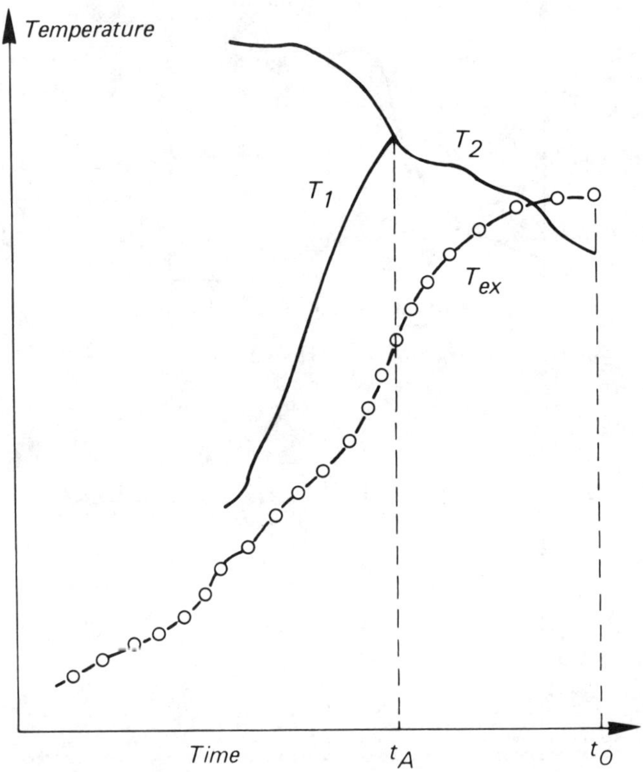

Fig. 13 Correction of TAOS flight data resulting from numerical simulations. —o—o— TAOS rough data T_{ex}; ——Intrinsic error corrected TAOS data; SFRR internal temperature at depth $z = z_{TAOS}$, $T_1 = T_{ex} - \delta_i T$; SFRR ablation temperature at TAOS location, $T_2 = T_{ex} - \delta_i T_A$.

The importance of intrinsic errors may certainly be attributed to the difference in the thermal and ablative properties of the TAOS material (vitreous carbon) and ablator material (silica fiber reinforced resin). The thermal diffusivity of vitreous carbon is compared to thermal diffusivities of standard ablators (see Fig. 14). From this comparison, and considering the fact that SFRR is the only material which chemically differs from the TAOS material, it is obvious that insertion of such a gage in SFRR is the worst configuration. Vitreous carbon TAOS used in conjunction with pure carbon ablators (CC, CFRR,GFRR) will give less perturbed measurements. In this case, intrinsic errors in ablation time are the reverse of the previous case (SFRR). In the present case, the pessimistic conclusion must be greatly tempered. First, because the silica-carbon chemical reactivity is ignored by the model and because these reactions, that certainly take place around the TAOS and on the top surface, must hasten the TAOS opening, thus decreasing the intrinsic error in ablation time $\delta_i t_A$. Secondly, because even in this unfavorable case (vitreous carbon TAOS embedded in SFRR), the comparison between corrected TAOS data and thermocouple

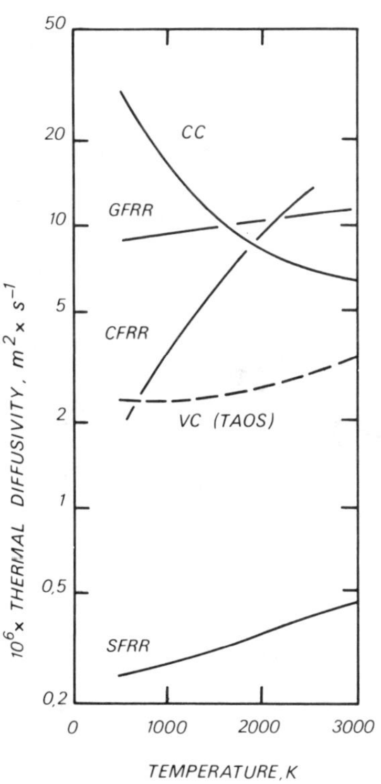

Fig. 14 Thermal diffusivities of TAOS material and standard ablators.

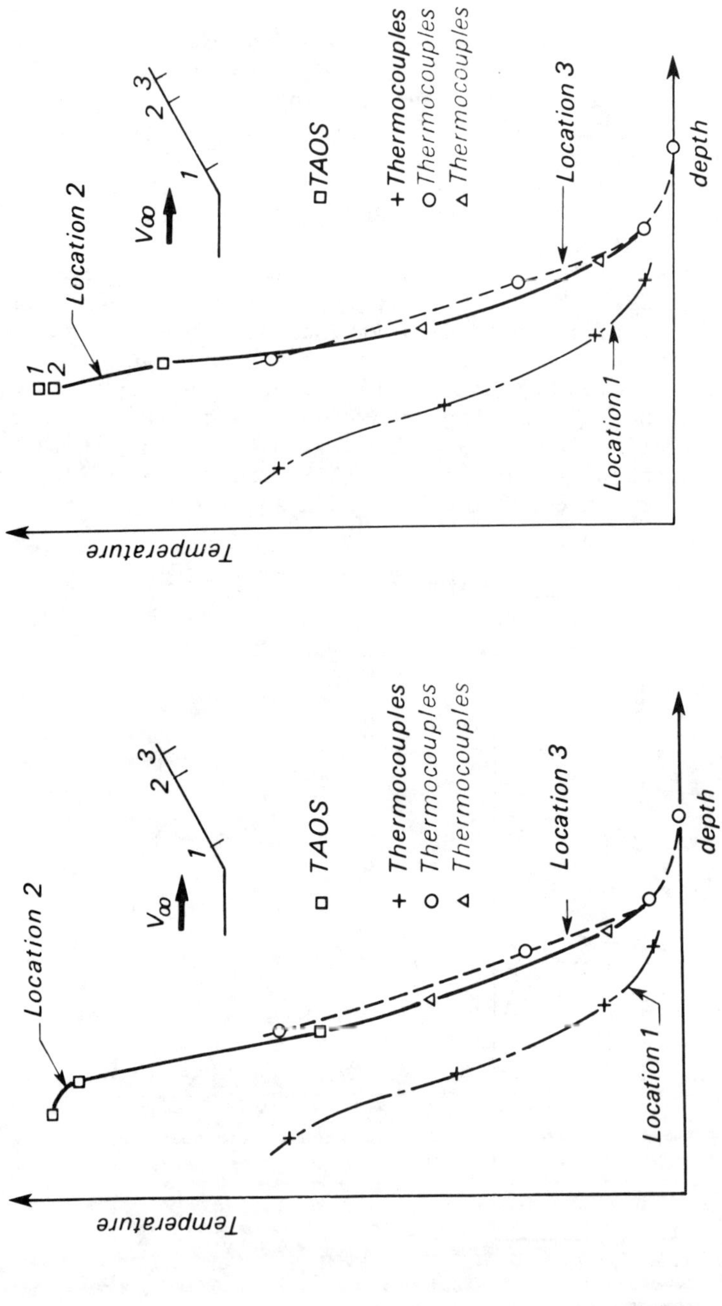

Fig. 15 Flight experiment: comparison between thermocouples and TAOS temperature measurements: a) TAOS rough data; b) intrinsic error corrected TAOS data.

measurements is encouraging (see Fig. 15). In Fig. 15a, TAOS data are rough values obtained directly from the photodiode intensity. In Fig. 15b, TAOS data are intrinsic error corrected using two-dimensional simulation results. The corrected temperature of TAOS 1 and 2 are related to the SFRR ablation front, this front having yet reached the depths $z = z_{TAOS\ 1}$ and $z = z_{TAOS\ 2}$. The corrected temperature of TAOS 3 is related to the inner temperature of the ablator at depth $z = z_{TAOS\ 3}$. At this time, the position of the ablation front z_A is intermediate between $z_{TAOS\ 2}$ and $z_{TAOS\ 3}$, and TAOS 1 is just before opening. The geometrical configuration of the three TAOS (all at station 2) relative to the SFRR ablation front is presented in Fig. 16. This figure gives information on the correction calculation. The corrected ablation temperature values given by TAOS 1 and 2, for which $z_{TAOS} < z_A$, must be identical. In fact, they differ by only 20 K.

Technological Evolution

More compact, rugged, and safe TAOS are under development where integration of the three components (cavity, optical fiber, photodiode) is realized. A photograph of the most recent type of gage is presented on Fig. 17. The laboratory production cost of such a gage, including calibration, is estimated to be $1000 each. This price is comparable to that of a high-performance thermocouple gage.

Another field of research consists of choosing and testing new tight and refractory TAOS materials for better adaptation of the gage to each type of ablator.

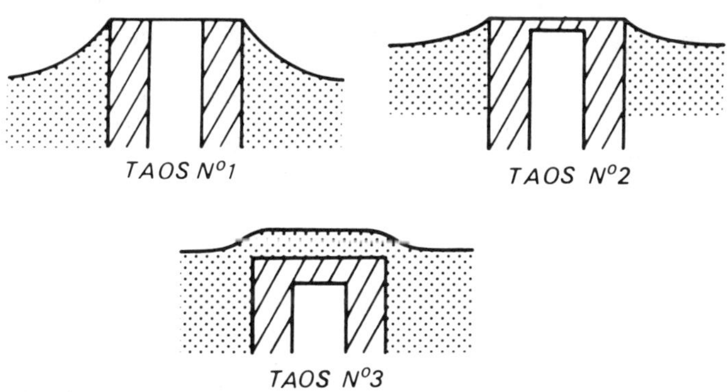

Fig. 16 TAOS configuration relative to ablation front at location 2 and time corresponding to data of Fig. 15.

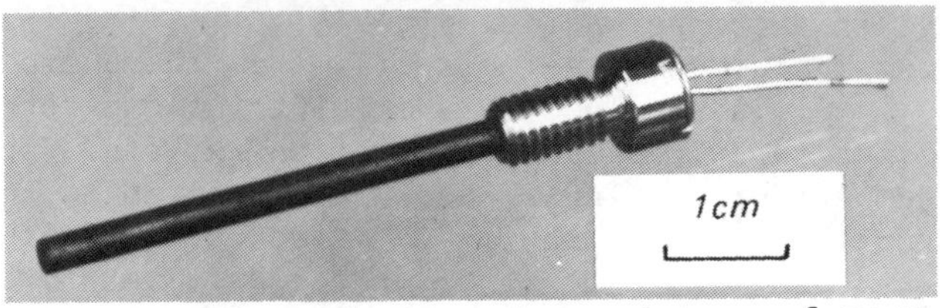

Fig. 17 View of most advanced type of TAOS.

Conclusions

Ground tests and flight experiments demonstrated the interest and reliability of a new type of sensor acting at the same time as both a temperature and ablation gage. The main characteristics of this temperature and ablation optical sensor (TAOS), which must be embedded into the ablator parallel to the heat flux direction, are:

1) Cylindrical shape with typical dimensions: 2 mm o.d. and some centimeters in length.

2) Temperature measurement range: from 1600 to more than 4000 K (temperature of the hotter end of the gage).

3) Temperature of the colder end of the gage must be lower than 900 K.

4) Temperature uncertainty is equilvalent to that of thermocouples, considering the fact that an accurate evaluation of errors due to the presence of the sensor in the ablator is possible. These intrinsic errors depend mainly on the nature of the ablator. For SFRR, intrinsic errors in ablation time and ablation temperature are near 1 s and 100 K, respectively. For a given material, these errors depend on the whole thermal history of the gage (heat flux time evolution, sensor location depth). In the case of pure carbon ablators, errors will be reversed. In fact, it would be desirable to adjust the nature of the TAOS material to the thermophysical properties of the ablator.

The possibility of measuring ablation temperature in flight is demonstrated. This information is available only by means of such a detector, and such data have not been obtained until now.

The use of TAOS for every part of a re-entry vehicle is dependent on the possibility of boring very long and rectilinear cavities into the ablators for gage insertion, and into the TAOS material for cavity realization. Cavities up to 50 mm long have been obtained and research is being done for longer cavity boring. TAOS will be particularly interesting for carbon-carbon nosetips where even high-performance thermocouples, whose highest operating temperatures do not exceed 2500 K, cannot give thermal information concerning important zones of the structure.

More rugged and more compact gages are being developed and will be flight tested shortly. The production costs of these improved gages are comparable to those of high-performance thermocouples.

Acknowledgment

The support of this research under CEA grant is gratefully acknowledged.

References

[1]Durrenberger, P., "Capteur Optique de Temperature et d'Ablation (COTA). Essais préliminaires," ONERA Report CR 02/3319 SHP, February 1977.

[2]"Capteur Optique de Température et d'Ablation (COTA)," ONERA Technical Sheet 1979-27, 1979.

WIND-TUNNEL STUDY OF ASCENT HEATING OF MULTIPLE RE-ENTRY VEHICLE CONFIGURATIONS

R.J. Berman* and D.E. Nestler+

General Electric Company, Philadelphia, Pa.

Abstract

A wind-tunnel study was conducted at Mach 8 to provide design information for the multiple re-entry vehicle ascent heating problem. The phase change paint technique identified shock impingement hot spots having heating rates as high as 30X the undisturbed values. Evidence of bistable shock structure was obtained from shadowgraphs. Evaluation of the data provided a simple flow model and interaction heating prediction technique for use in design analysis.

Nomenclature

c = specific heat
H = heat-transfer coefficient based on a recovery temperature of 0.9 T_0
H_R = reference heat-transfer coefficient (stagnation point value for nosetip R_N) based on method of Fay and Riddell[1]
k = thermal conductivity
M_∞ = freestream Mach number
n = exponent in Eq. (5)
P = pressure
$\dot{q}$ = heat flux
R_B = body radius
R_N = nose radius
Re_∞ = freestream Reynolds number

Presented as Paper 80-1456 at the AIAA 15th Thermophysics Conference, Snowmass, Colo., July 14-16, 1980.

*Supervising Engineer, Re-entry Systems Division.
+Staff Consultant, Re-entry Systems Division.

T = temperature
t = time
X = distance along base plate skirt (Fig. 1)
α = angle of attack
δ_s = bow shock standoff distance
ρ = density
ϕ = peripheral angle (Fig. 1)

Subscripts

aw = adiabatic wall value
i = initial value
max = maximum value
o = reservoir or total value
pc = phase change
s = stagnation value
U = undisturbed value
w = wall value

Introduction

The range and payload capability of advanced missile systems are affected by the re-entry system weight, aerodynamics, and shroud removal altitude. Current operational systems plan for shroud removal at an altitude of about 300,000 ft. Previous analyses have shown that significant range benefits can be realized by removing the shroud at about 150,000 ft, even though axial force coefficients are estimated to increase by about a factor of 5 when the shroud is removed at lower altitudes. Removal at this relatively low altitude, however, subjects the re-entry vehicles and deployment module to high heating and high aerodynamic loads due to multishock interactions. Preliminary estimates of the temperatures at the periphery of the deployment module indicated temperatures which would require additional insulation material. As a result, it was felt that existing semiempirical prediction techniques required verification and extension for design studies of this concept.

This paper presents the results of a wind-tunnel study which was conducted to address shock/boundary-layer interaction heating associated with multiple conical bodies mounted on a circular base. Evaluation of the heat-transfer data and shadowgraph shock patterns will be shown to provide a simple flow model and interaction heating prediction technique for use in design analysis.

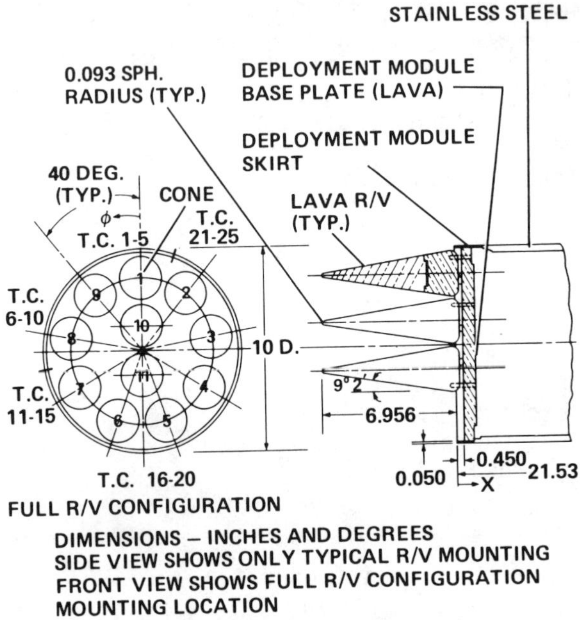

Fig. 1 Wind-tunnel model.

Experimental Program

A wind-tunnel test was conducted in the Arnold Engineering Development Center, Tunnel B, at a nominal Mach number of 8 and freestream Reynolds numbers of 0.5, 0.8, and 1.0 million/ft. This section will describe the model, test conditions, and measurement techniques.

Model

A drawing of the model with major external dimensions is shown in Fig. 1. The primary configuration tested included 11 cones on a circular base plate as shown in the front view. The second configuration tested was a single cone mounted on the base plate centerline, while the final configuration was simply the base plate with no cones. A photograph of the full configuration mounted on the base plate is shown in Fig. 2. The base plate and all cones were fabricated from lava (aluminum silicate) for use with phase change paint. The aft portion of the model was fabricated from a 400 series stainless steel with the forward skirt of 0.450 in. length machined down to a thickness of 0.050 in. and instrumented with Chromel® - Alumel® thermocouples (Fig. 3). Thermocouples were used on the base plate skirt because the scaled model skirt thickness was too thin to obtain valid heat-transfer data by the paint technique.

Test Conditions

Tests were conducted in Tunnel B of the Arnold Engineering Development Center. A summary of the nominal test conditions is given in Table 1. These Reynolds number conditions, when scaled to flight using cone model length, provide a proper Mach number/altitude simulation of ascent flight conditions.

Measurement Techniques

The primary heat-transfer data were obtained by the phase change paint technique. Prior to each run, the model was cleaned and cooled with alcohol and then spray painted with a phase change paint (Tempilaq ®). All painting was performed in the injection tank. The model initial surface temperature was measured and the model was then injected into the airstream for 20 s. During this time, the model surface temperature rise produced isotherm melt lines. The progress of the melt lines was photographed with two 70 mm sequence cameras at the standard rate of 1 frame/s. One sequence camera was

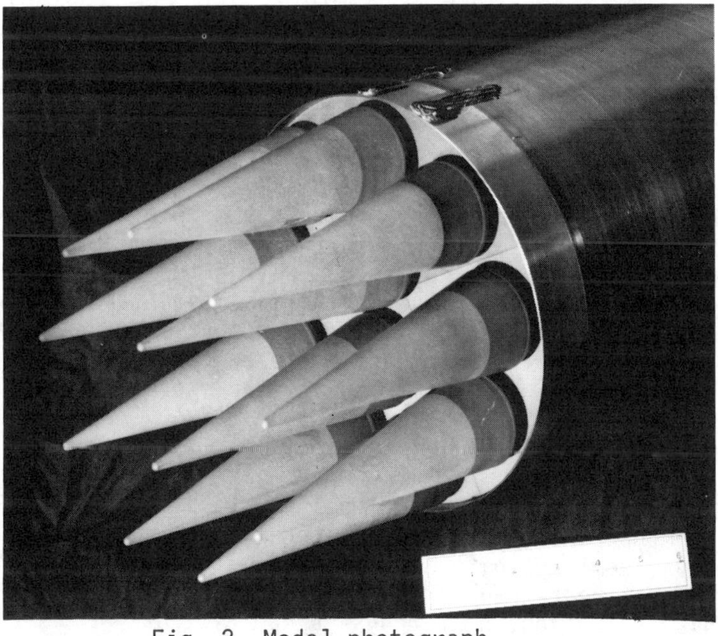

Fig. 2 Model photograph.

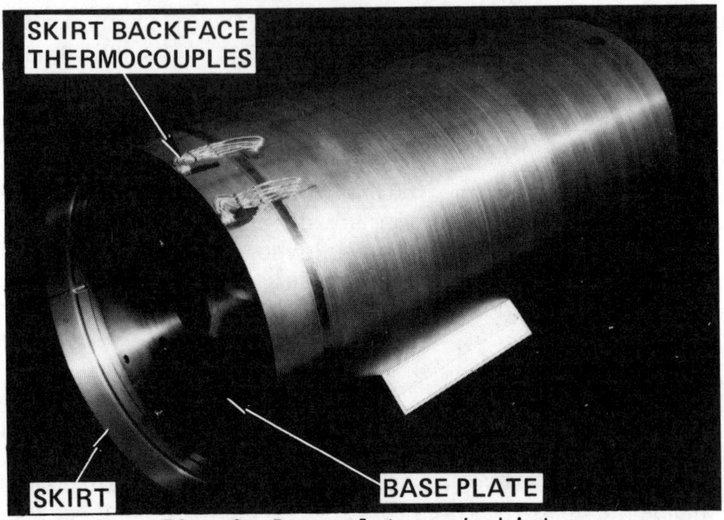

Fig. 3 Base plate and skirt.

Table 1. Summary of nominal test conditions

M_∞	P_0, psia	T_0, °R	Re_∞/ft x 10^{-6}	α, deg	Equivalent altitude, kft
7.94	200	1250	1.0	0	145
7.92	155	1220	0.8	0	155
7.89	90	1180	0.5	0 - 12	170

placed in an overhead window at an oblique angle to record the developing melt lines on the deployment module base plate. The other sequence camera viewed the model from a side window to record data on individual cones within the field of view. A third camera was installed in the injection tank to provide additional base plate melt line coverage after each test run was complete. This camera was triggered before and after each injection with the model pitched to zero incidence. In addition to the melt line photographs, optical coverage included a shadowgraph of each model injection.

Reduction of phase change paint data is based on the assumption that the surface temperature history is that of a homogeneous, semi-infinite slab subjected to an instantaneous and constant heat-transfer coefficient (Jones and Hunt[2]). For this evaluation, heat-transfer coefficients were based on assumed adiabatic wall temperatures of 0.9 T_0.

The lumped thermophysical property $\sqrt{\rho ck}$ was assumed to be constant with temperature, and a value of 0.125 Btu/ft^2-s$^{\frac{1}{2}}$ ^{o}R was selected from material literature for lava. The data were normalized by the cone stagnation point heat-transfer coefficient using a reference radius of 0.00775 ft, which corresponded to an individual cone nose radius.

The skin thermocouple data were recorded at the rate of about 15 loops/s starting at model liftoff and continuing throughout the injection cycle. The data were reduced for a period of time after the model reached the tunnel centerline dependent on the absolute heating rate level. Tunnel conditions and model attitudes were recorded and reduced at the time the model reached the centerline of the tunnel.

Results and Evaluation

Phase Change Paint Data

Typical photographs of the phase change paint melt lines at different times during a test are given in Figs. 4 - 7, as viewed from the overhead sequence camera. The dark areas on the cones are regions of high localized heating due to shock wave impingement. Melt lines on the base plate are more difficult to identify.

Typical melt line tracings corresponding to the test conditions of Fig. 4 are shown in Figs. 8 and 9 for the overhead and side cameras, respectively. Study of the sequential photographs showed an orderly enlargement of the melt region on the cones with time; however, the base plate melt region appears to have formed during insertion of the model, since it is there in the first frame, with little or no further enlargement during the test. This point will be discussed further under Base Plate Data.

Post-test photographs of the models for runs 4 and 5 are given in Figs. 10 and 11. These photographs show regions of high heating on the interior base plate region (as well as on certain cones) which could not be seen by the cameras during the test.

The phase change paint data were evaluated as follows:

Cone data. The initial melt time, corresponding to maximum heating on each cone, was used to obtain the corresponding value of H/H_R, the ratio of measured local heat-transfer coefficient to theoretical model stagnation point heat-transfer

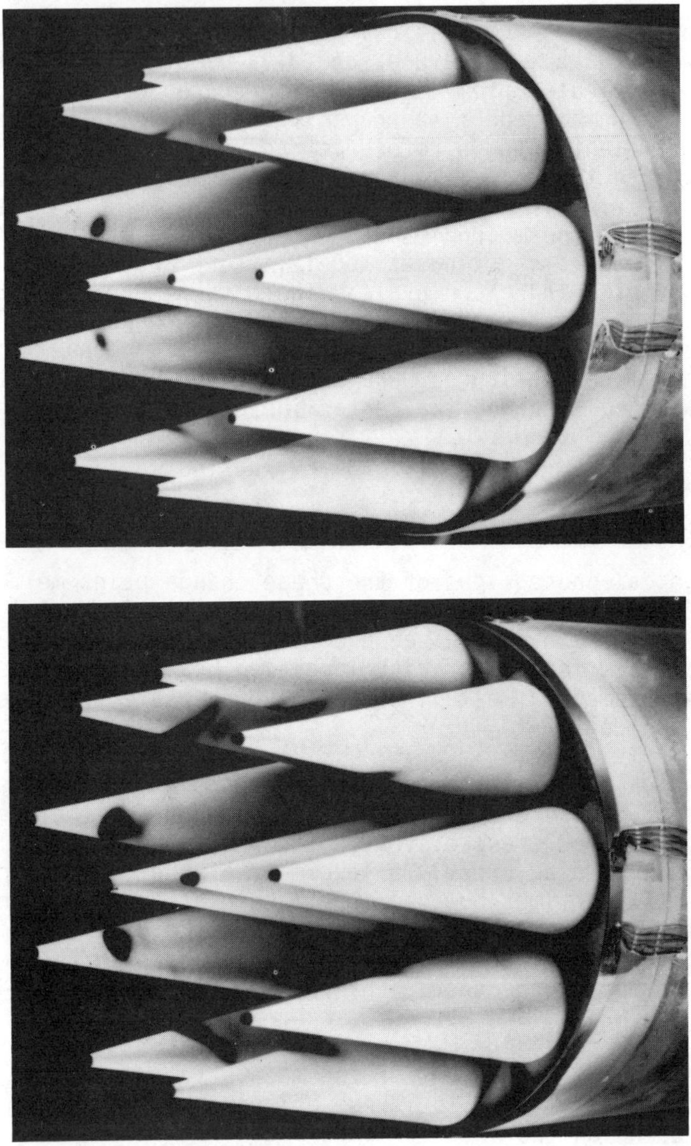

a)

b)

Fig. 4 Phase change paint data - run 4, $\alpha = 0$, $Re_{\infty}/ft = 1.0 \times 10^6$, $T_{PC} = 300^oF$: a) $\Delta t = 5.03$ s ; b) $\Delta t = 16.29$ s.

coefficient. Displays of these maximum heating ratios are given in Figs. 10 and 11.

In general, maximum values of H/H_R on the order of 1.2 were observed, except for a few cases in which melt occurred

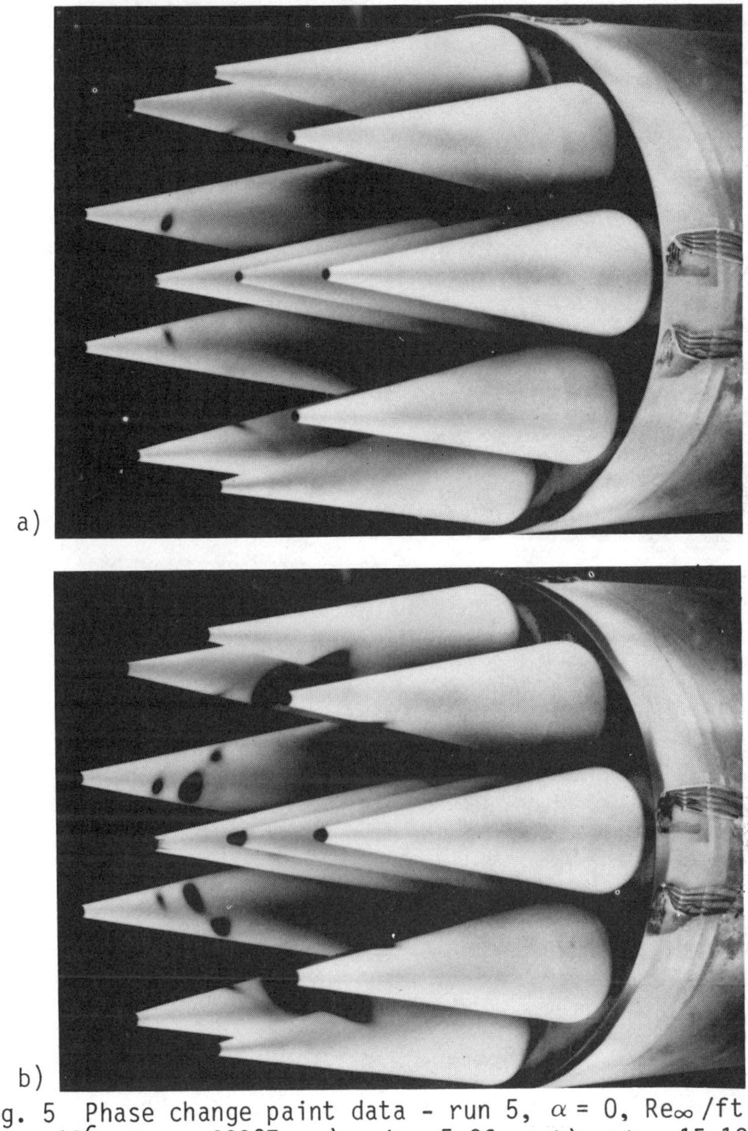

Fig. 5 Phase change paint data - run 5, $\alpha = 0$, $Re_\infty/ft = 1.0 \times 10^6$, $T_{PC} = 300^oF$: a) $\Delta t = 5.06$ s; b) $\Delta t = 15.18$ s.

very early in the run. For these cases, accuracy is questionable due to the finite frame viewing and also because of the larger uncertainty due to heating during model insertion. Maximum values of H/H_R for each test are given in Table 2, as observed from either the overhead or side camera. Variation in H/H_R for + or - 1 frame error in estimating melt onset are also included. It was noted that when the melt region

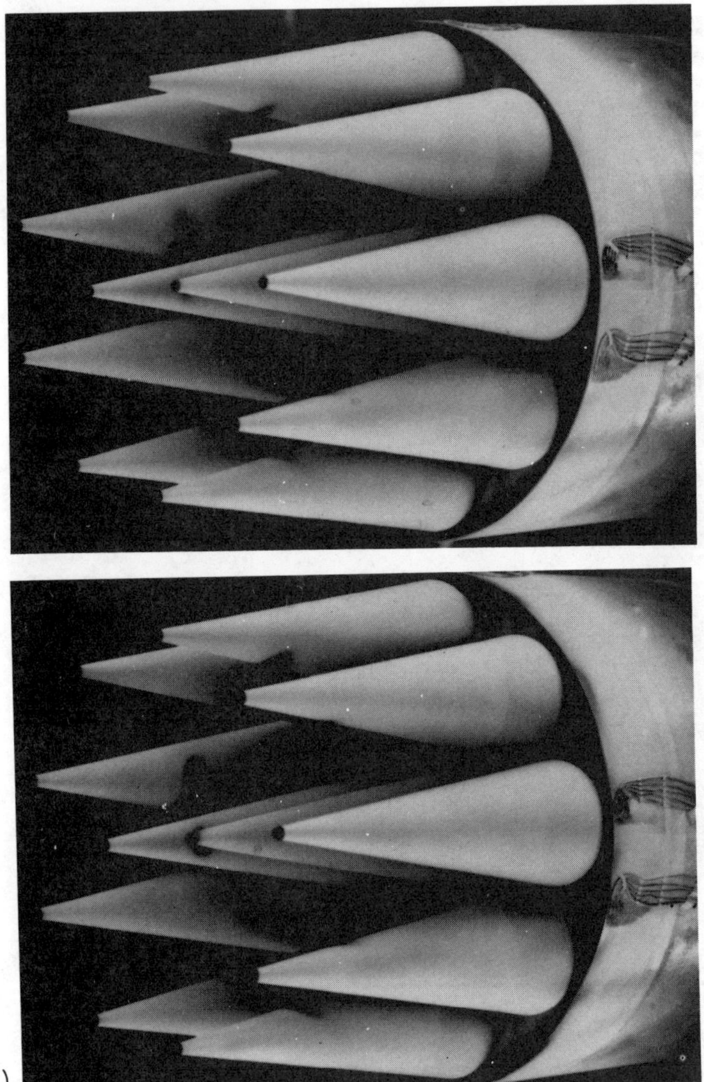

a)

b)

Fig. 6 Phase change paint data - run 7, $\alpha = 0$, $Re_\infty/ft = 1.0 \times 10^6$, $T_{PC} = 350^oF$: a) $\Delta t = 9.19$ s; b) $\Delta t = 16.22$ s.

for a given cone could be viewed equally well by either camera, melt onset was observed within one frame agreement between the two cameras.

It was also noted that for nominally identical test conditions, including paint melt temperature, different hot spot

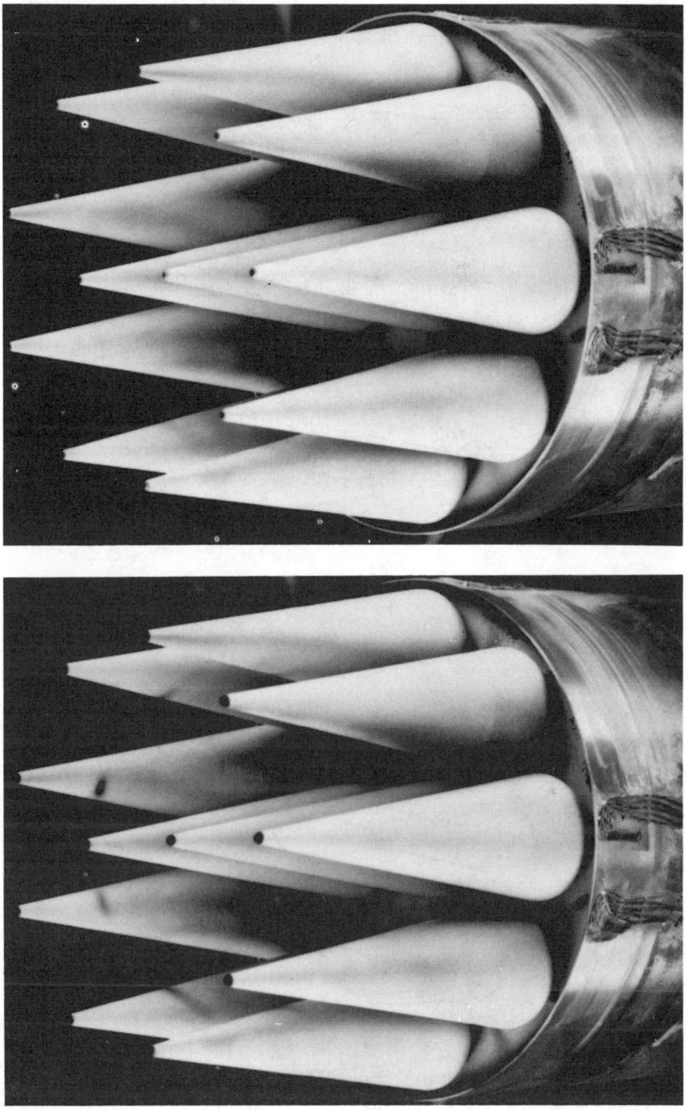

a)

b)

Fig. 7 Phase change paint data - run 13, $\alpha = 0$, $Re_\infty/ft = 0.5 \times 10^6$, $T_{PC} = 300^oF$: a) $\Delta t = 5.01$ s; b) $\Delta t = 15.11$ s.

locations were sometimes observed for a given cone. Thus, a nonrepeatability in results was evidenced, suggesting either unsteady and/or bistable shock formations. Comparison of Figs. 4 and 5 is particularly instructive; in Fig. 5, hot spots formed initially at the same location as in Fig. 4, but did not enlarge significantly during the run. Instead, addi-

MELT CONTOUR NO.	H/H_R
3	1.69
8	0.66
12	0.51
20	0.38

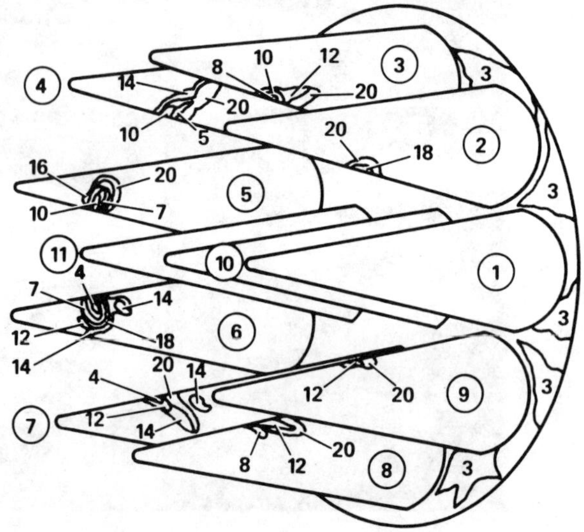

Fig. 8 Typical paint melt line progression (overhead camera) - run 4, $\alpha = 0$, $Re_\infty/ft = 1.0 \times 10^6$, $T_{PC} = 300^oF$.

MELT CONTOUR NO.	H/H_R
4	1.15
7	0.73
11	0.54
17	0.42
20	0.38

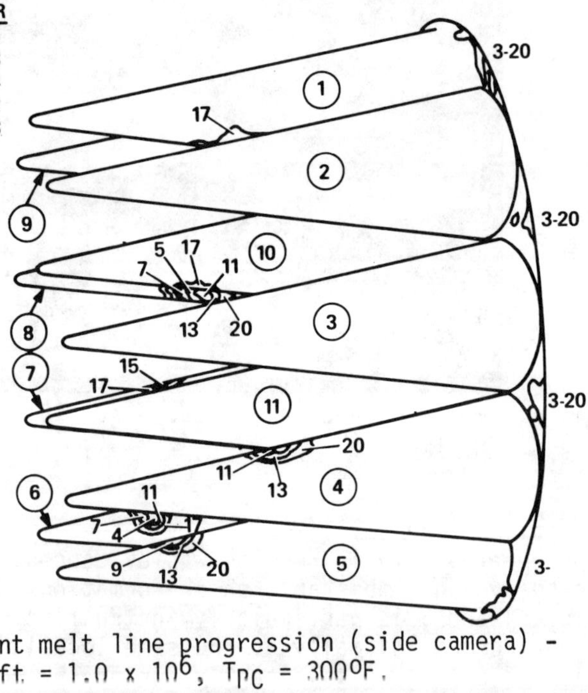

Fig. 9 Typical paint melt line progression (side camera) - run 4, $\alpha = 0$, $Re_\infty/ft = 1.0 \times 10^6$, $T_{PC} = 300^oF$.

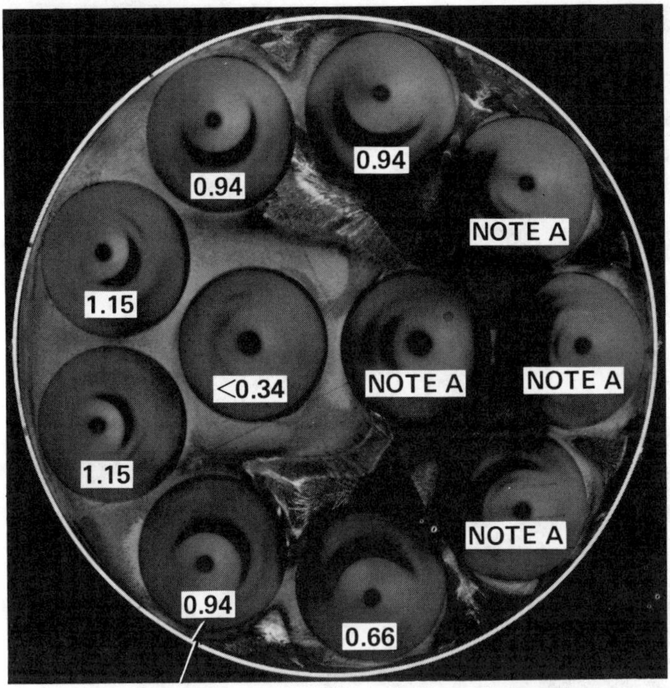

Maximum value of H/H_R observed during test, for each cone

Fig. 10 Post-test photograph - run 4, $\alpha = 0$, $Re_\infty/ft = 1.0 \times 10^6$, $T_{PC} = 300^oF$ (Note A: Maximum heating location could not be viewed by either camera).

Table 2. Maximum values of H/H_R for cones

Run no.	Nominal Re_∞/ft	α, deg	$(H/H_R)_{max}$	$(H/H_R)_{max}$ for ± 1 frame variation
3	1.0×10^6	0	1.18	1.05, 1.37
4	↓	↓	1.15	0.94, 1.69
5			1.64	1.15, —
7			2.43	1.66, —
8	0.8×10^6	0	2.87	1.99, —
9	↓	↓	1.17	1.01, 1.45
10	0.5×10^6	4	0.82	0.85, 0.79
11	↓	8	0.94	0.90, 1.00
12		12	<0.62	——
13		0	0.91	0.96, 0.87

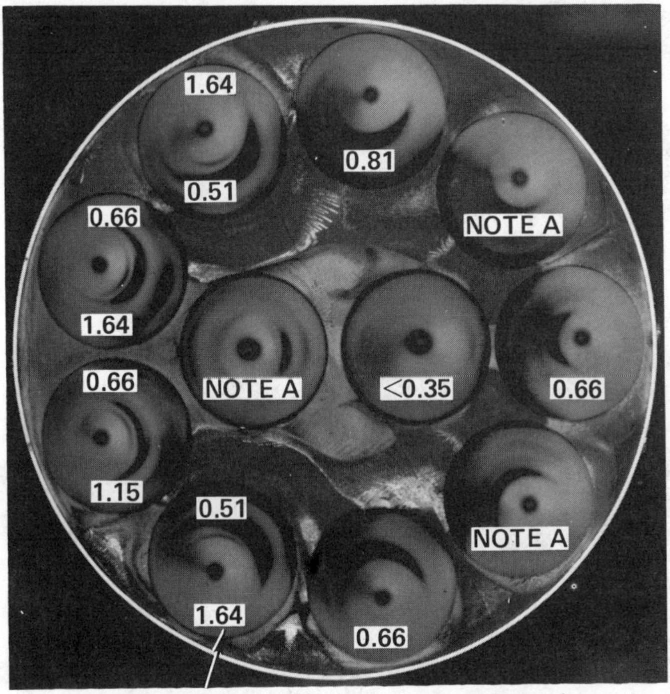

Maximum value of H/H_R observed during test, for each cone

Fig. 11 Post-test photograph - run 5, $\alpha = 0$, $Re_\infty/ft = 1.0 \times 10^6$, $T_{PC} = 300^oF$ (Note A: Maximum Heating Location could not be viewed by either camera).

tional hot spots appeared later in the run further downstream, and continued to enlarge. This comparison strongly suggests a bistable shock formation.

The use of higher melting point paints would have reduced the uncertainty in maximum H/H_R; however, there appeared to be a real variation in maximum heating level for otherwise identical test conditions. The still shadowgraphs cannot discriminate between a steady or unsteady shock pattern; hence it cannot be determined whether the maximum heating level of a given test is due to a spacially fixed shock pattern or is an average caused by an oscillatory shock pattern.

Reference to Table 2 shows that for $\alpha = 0$, the maximum heating ratio decreased moderately as Re_∞/ft decreased from 0.8 to 0.5×10^6. For $Re_\infty/ft = 0.5 \times 10^6$, the maximum heating ratio was independent of α for $0 < \alpha < 8$ deg, then de-

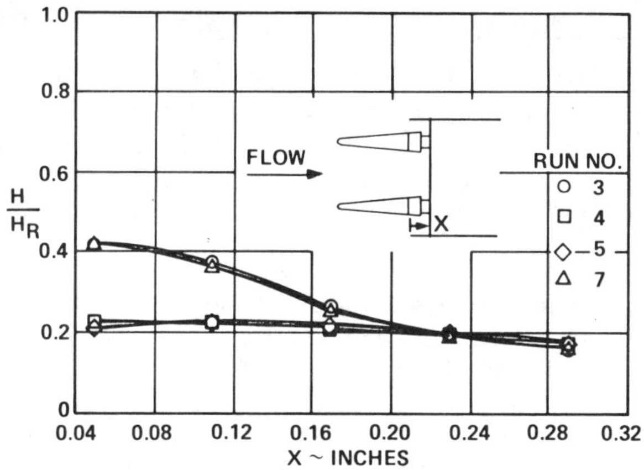

Fig. 12 Typical skirt thermocouple data - $\alpha = 0$, $Re_\infty/ft = 1.0 \times 10^6$, $\phi = 0$.

creased below the phase change paint melt threshold for $\alpha = 12$ deg.

Base plate data. As mentioned, the phase change paint technique did not yield completely satisfactory data on the base plate. The camera view angles did not permit the inner base plate region to be seen clearly because of view obstruction by the cones. The outer base plate region displayed paint removal in the first camera frame after the model reached its centerline position. In general, these regions did not enlarge during the test, suggesting that the initial paint removal was due to either a high transient heat pulse during model insertion, or high shear in the region of an attaching shear layer. This latter phenomenon has been reported by Keyes[3] in a shock impingement heat-transfer study. Keyes showed that the apparent heating rate obtained by interpreting mechanical removal of paint as melt was on the order of twice the heating rate obtained by more careful camera positioning to observe true melt onset.

Further qualitative information on base plate heating was obtained from the post-test headon photographs, such as Figs. 10 and 11. These photographs indicate paint removal in the inner region of the base plate, which could not be viewed clearly during the tests. Assuming that these regions are due to paint melt rather than mechanical removal, the lower-bound values of $(H/H_R)_{MAX}$ obtained for $\alpha = 0$ are given in Table 3.

Thermocouple Data

Typical heat-transfer data obtained from the thermocouple measurements on the base plate skirt are shown in Figs. 12 and 13 for the 0 and 180 deg meridians, for four tests at $\alpha = 0$ under nominally identical test conditions. Complete thermocouple heat-transfer data for all instrumented rays are given in Fig. 14 for run 4.

The heat-transfer distributions on the base plate skirt show the general character expected for reattachment heating on an annular ring due to radial outflow over the base plate (i.e., maximum at outer tip of skirt, decay inward towards base plate). However, complex heating patterns are evident, with highly nonuniform circumferential variations. Figs. 12 and 13 show two distinct heating levels along the same meridian for the same test condition, suggestive of the bistable flow pattern inferred earlier from the phase change paint results. The maximum observed value of H/H_R on the base plate ring was 1.25 on the 100 deg meridian at $\alpha = 4$ deg, compared with a maximum value of 0.72 at $\alpha = 0$.

Shadowgraph Data

A shadowgraph was obtained for each model run. High-speed motion picture coverage was not available due to paint camera coverage requirements. However, shadowgraphs for runs having identical test conditions do indicate the possible existence of an unsteady or bistable flow pattern.

Observation of the full RV configuration indicates that an interaction occurs between one of the two center cones and the outer cones. Fig. 15 shows the shadowgraph for run 4 while Fig. 16 is for run 7. Notice that in Fig. 15 the nose shock from cone 11 (third image from the bottom) has an angle of about 45 deg, while the shock from cone 10 (third image from top) has a smaller angle. This pattern is just reversed in Fig. 16.

Also note on the paint photographs that the high interaction heating for run 4 (Fig. 4) occurs close to the nose, consistent with the shock pattern of cone 11 with cones 6 and 7. This interaction heating location for run 7 (Fig. 6) is further aft on cone 7, again consistent with the shock pattern. Indeed, calculations made for the observed shock patterns show high localized pressures on the inboard side of the outer ring of cones and essentially subsonic flow on the inner cones. Examination of the photographs taken after each test show little, if any, interaction heating on the two inner cones.

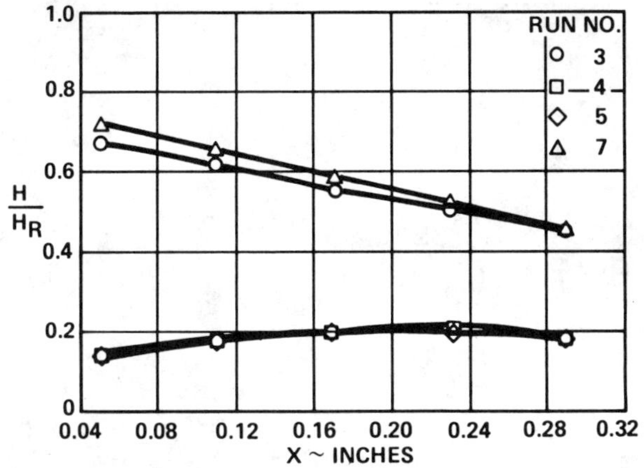

Fig. 13 Typical skirt thermocouple data - $\alpha = 0$, $Re_\infty/ft = 1.0 \times 10^6$, $\phi = 180$ deg.

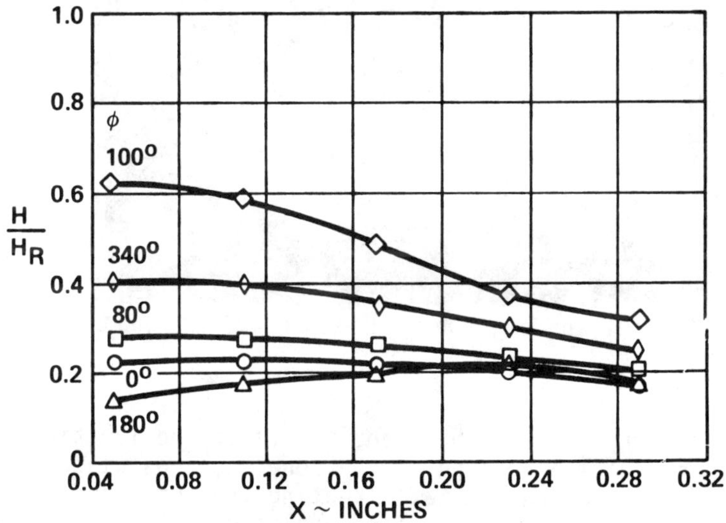

Fig. 14 Typical skirt thermocouple data - run 4, all instrumented rays $\alpha = 0$, $Re_\infty/ft = 1.0 \times 10^6$.

Table 3. Lower-bound values of $(H/H_R)_{MAX}$ for $\alpha = 0$

Run no.	Re_∞/ft	$(H/H_R)_{MAX}$
3, 7	1.0×10^6	>0.50
8	0.8×10^6	>0.58
13	0.5×10^6	>0.61

Fig. 15 Typical shadowgraph - run 4, $\alpha = 0$, $Re_{\infty}/ft = 1.0 \times 10^6$, $M_{\infty} = 8$.

A method for establishing the limits of the location of the high interaction heating on the cones was postulated, based on the shadowgraph data. Treating the cones and the base plate as separate free bodies analytically yields shock patterns shown in Fig. 17. The normal shock for the base plate and the coalescence of the undisturbed cone shocks occur at about the same axial location. This is assumed to be equivalent to a new effective flat-faced body which in turn causes a normal shock to form (dashed line). Examination of the test data suggests that the flow patterns range from undisturbed flow about the cone noses to a flow separation extending from one or the other of the two inner cone nosetips to the outer edge of the formed normal shock (Fig. 17). This tends to provide the limits of the location of the maximum interaction heating. Using the local flow conditions in regions

Fig. 16 Typical shadowgraph - run 7, $\alpha = 0$, $Re_{\infty}/ft = 1.0 \times 10^6$, $M_{\infty} = 8$.

I and II, iterations were made to solve for the flow conditions in regions III and IV such that the flow angles and pressures were the same. Thus, the shock patterns between regions I and III and II and IV were established and the flow conditions in regions V and VI were estimated. Pressure calculations for these patterns yield peak pressure ratios $P_V/P_I \approx 96$ for a freestream Mach number of 7.9.

Single Cone Results

The heating on the base plate for the single-cone configuration shows a wide range of values. Extremely high local heating rates are noted on the base plate immediately adjacent to the cone juncture. Fig. 18 shows melt onset in the third picture, corresponding to H/H_R of 1.23. Lower values of

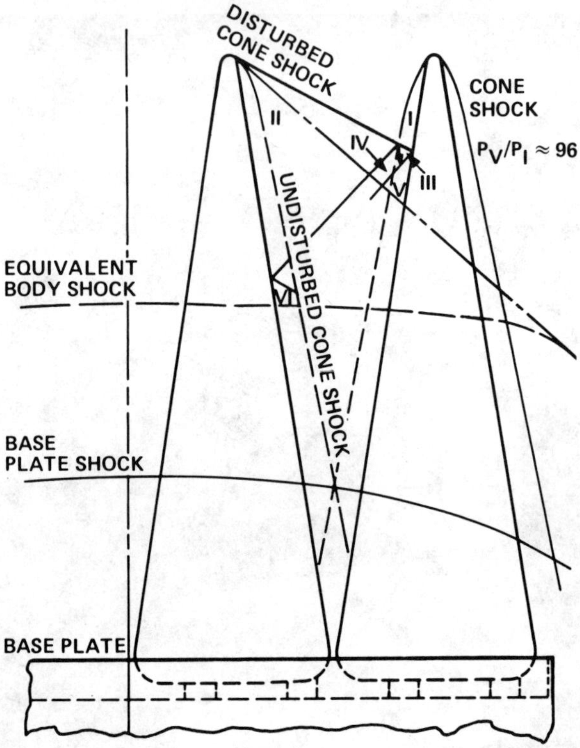

Fig. 17 Shock interaction patterns - $\alpha = 0$, Re_∞ /ft = 1.0×10^6, $M_\infty = 8$.

H/H_R are noted further out from the cone. The high local heating rates are probably caused by the multiple shock interaction of the cone oblique shock and the base plate normal shock when the flow is such that the shock about the cone is undisturbed (see Fig. 19), in the same manner as that incurred by recovery spoilers (Berman et al.[4]). The lower heating probably occurs when the unsteady nature of the flow causes a pattern having more of a normal shock about the nose of the cone (see Fig. 19), which diffuses the shock interaction phenomena. The shadowgraphs of Fig. 20 indicate an unsteady flow pattern, consistent with previous results for spiked noses (e.g., Wood[5]).

Prediction Techniques for Shock Interaction Heating

Base plate. At the outset of the study, it was expected that the base plate maximum heating could be approximated by modifying the method developed by Edney[6] for Type IV shock

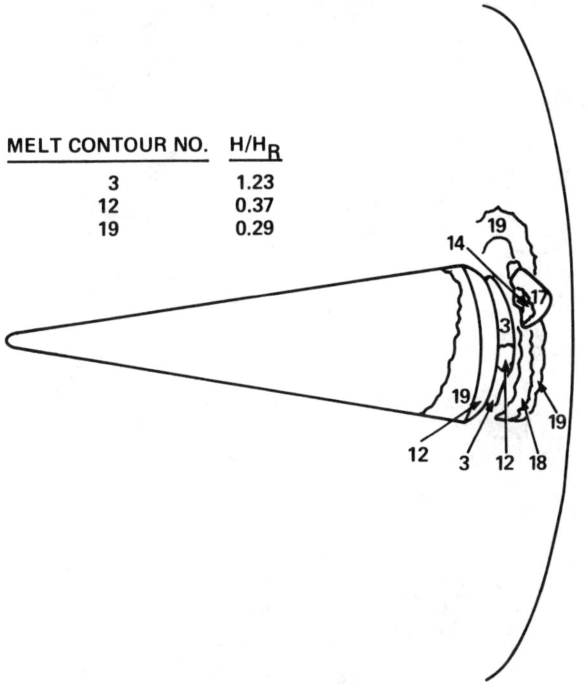

Fig. 18 Paint melt line progression - single cone at center of base plate, $\alpha = 0$, $Re_\infty/ft = 0.5 \times 10^6$, $T_{PC} = 200^oF$.

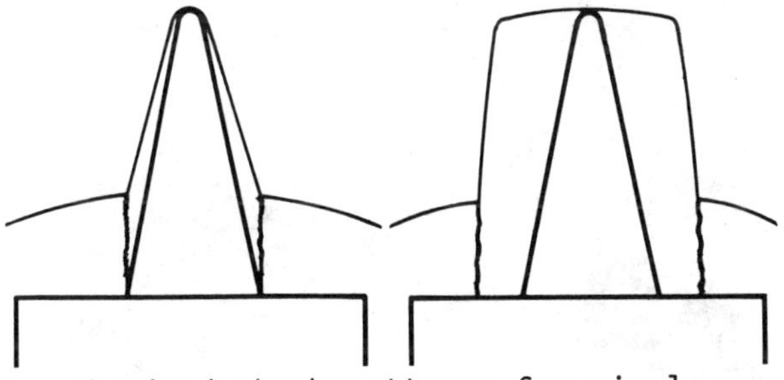

Fig. 19 Hypothesized shock patterns for single cone on base plate.

impingement on a hemisphere to apply to a flat-faced cylinder such as the base plate. In this method, the interference heating process is modeled by Edney as that due to an impinging supersonic jet generated by the intersection of a cone shock and the base plate normal shock. Referring to Fig. 21, Edney derived the following relation for shock impingement on

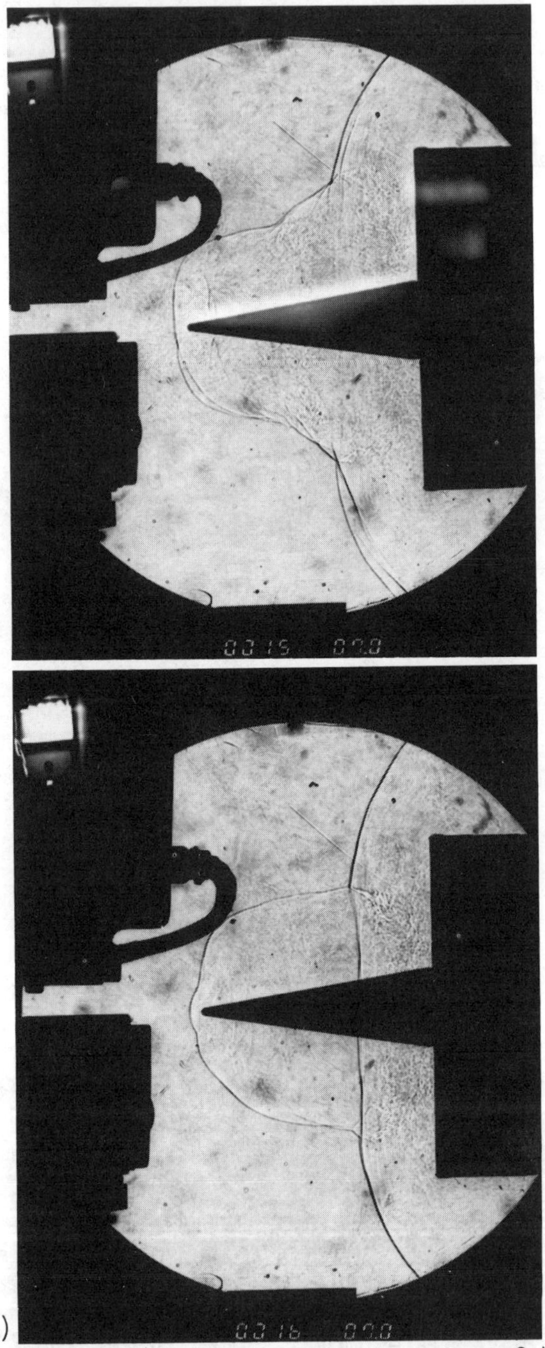

a)

b)

Fig. 20 Shadowgraphs - single cone at center of base plate - $\alpha = 0$, $Re_\infty/ft = 0.5 \times 10^6$, $M_\infty = 8$.

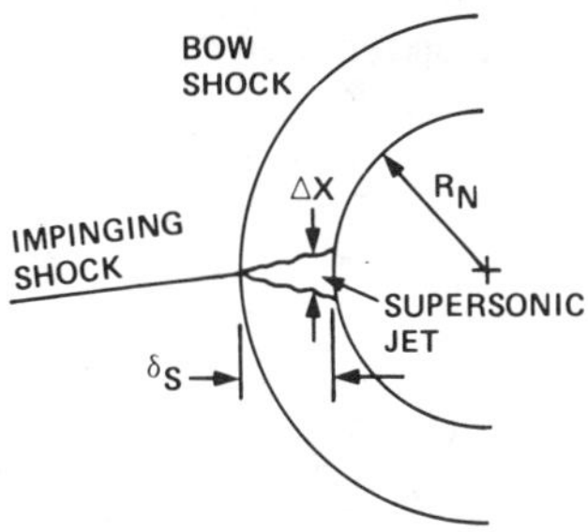

Fig. 21 Type IV shock interaction.

a hemisphere:

$$\dot{q}_{max}/\dot{q}_s \approx \left[\left(R_N/\Delta X\right)\left(P_{max}/P_s\right)\right]^{0.5} \tag{1}$$

in which subscripts max and S indicate maximum interference and normal stagnation point values, respectively.

To apply this relation to the flat base plate, the jet width ΔX was assumed to be proportional to the bow shock standoff distance δ_S. In this way, the value of $R_N/\Delta X$ derived by Edney from Schlieren photographs of a hemisphere at Mach 4.6 was generalized to a flat-face geometry at any Mach number.

From Serbin's relations (Ref. 7) for shock standoff distance for hemispheres and flat-faced cylinders

$$\frac{\left.\delta_S/R_B\right]_{\text{flat face}}}{\left.\delta_S/R_B\right]_{\text{hemisphere}}} = 1.54\left(\frac{\rho 2}{\rho 1} - 1\right)^{0.5} \tag{2}$$

where ρ_2/ρ_1 is the static density ratio across a normal shock. Assuming $\Delta X \sim \delta_S$,

$$\left.R_N/\Delta X\right]_{\text{flat face}} = \frac{\left.R_N/\Delta X\right]_{\text{hemi}}}{1.54\left(\rho_2/\rho_1 - 1\right)^{0.5}} \tag{3}$$

$$\approx 13/(\rho_2/\rho_1 - 1)^{0.5}$$

in which Edney's upper value of $\left.R_N/\Delta X\right]_{hemi} = 20$ has been used. Inserting Eq. (3) into Eq. (1) yields the final expression:

$$\dot{q}_{max}/\dot{q}_S = \left[3.6/(\rho_2/\rho_1 - 1)^{0.25}\right]\left[(P_{max}/P_S)^{0.5}\right] \tag{4}$$

The ratio P_{max}/P_S is a function of the Mach number and flow deflection angle produced by the shock intersection. Since the flow deflection angle is not known without detailed optical information, the maximum value of P_{max}/P_S was read from Fig. 22, as computed by Edney. Thus for $M_\infty = 8$ of the present tests, $P_{max}/P_S = 13$ and $\dot{q}_{max}/\dot{q}_S = 8.9$. To convert this heat flux ratio to the heat-transfer coefficient ratio H_{max}/H_R requires division by the ratio $(R_{baseplate}/R_{nose})^{0.5} = (5/0.936)^{0.5} = 7.31$. Hence, $H_{max}/H_R = 1.21$ is predicted by the modified Edney method, which is compatible with the experimental value of $H_{max}/H_R > 0.6$ deduced previously. Coincidentally, this predicted value of H_{max}/H_R on the base plate equals the maximum value of H/H_R on the cones in the shock interaction region.

Cones. For shock interference heating on adjacent cones, the following empirical relation is suggested:

$$H_{max}/H_U = (P_{max}/P_U)^n \tag{5}$$

in which subscript U indicates undisturbed conditions. Hung[8] has shown that the exponent n depends on the type of body boundary layer. Hung derived the following values of n for different types of interactions: all laminar, 0.7; all turbulent, 0.8; laminar boundary layer tripped to turbulent by impinging shock, 0.8.

Analysis of the shock-wave structure shown by the shadowgraphs yielded a ratio of $P_{max}/P_U = 96$ (see Shadowgraph Data). From the phase change paint evaluation, a ratio of $H_{max}/H_U = 30$ was selected as a typical nominal value (see Re-entry Vehicles). Then $n \approx 0.75$ from Eq. (5), in reasonable agreement with the range of Hung's exponents.

Application of Data To Flight

Re-entry Vehicles

It is customary to correlate shock-wave interference heating by defining an interaction heating factor H/H_U, where subscript U indicates the undisturbed heat-transfer coefficient in the absence of interaction (e.g., Hung,[8] Holden,[9] etc.). If the interaction is all laminar, a laminar value of H_U is used. If the interaction is transitional (such as a laminar undisturbed flow being tripped by the shock-wave impingement) or all turbulent, a turbulent value of H_U

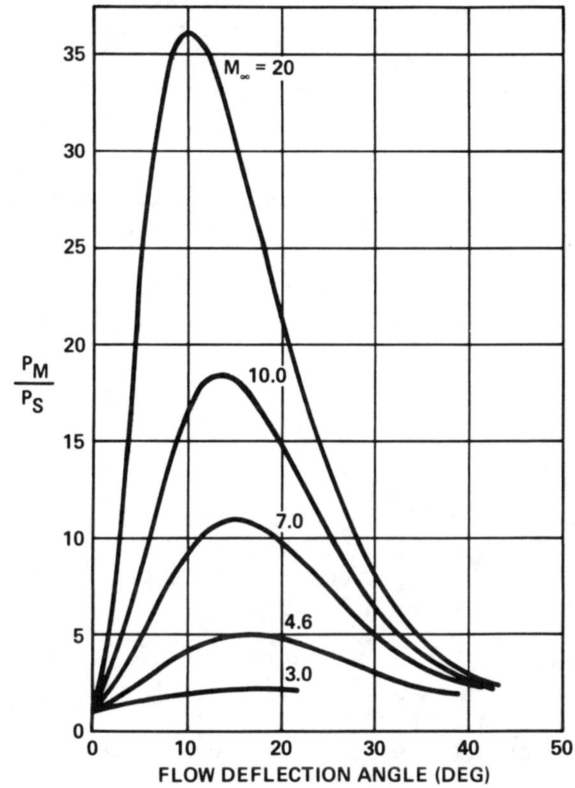

Fig. 22 Maximum impact pressure from Type IV shock interaction (Edney[6]).

is used. In the present case, values of H_U were determined theoretically using the GE Three-Dimensional Viscous Code,[10] at the approximate location of maximum interference heating. The resulting values of H/H_U for $\alpha = 0$ are shown in Table 4.

It is noted that the interaction heating factor based on a turbulent value of H_U is on the same order as the factor based on a laminar value of H_U. This situation exists because of the low local Reynolds numbers ($Re_S \approx 20{,}000 - 30{,}000$) at the location of maximum heating.

The interaction heating factors developed in this section can be applied to the undisturbed heating rates computed for a single re-entry vehicle to estimate the heating rates due to shock impingement for the multiple re-entry vehicle configuration. These increased heating rates can then be used in a

Table 4. Value of H/H_U for $\alpha = 0$

Re_∞/ft	$(H/H_U)_{laminar}$	$(H/H_U)_{turbulent}$
0.5×10^6	21	24
0.8×10^6	30	28
1.0×10^6	30	27

transient thermal response analysis to predict the surface temperature rise during ascent heating following shroud removal.

Base Plate

Since the absolute magnitude of the base plate maximum heating rates were not obtained, reliance must be placed on the theoretical method developed in Prediction Techniques for Shock Interaction Heating. Application of Eq. (4) for maximum base plate heating to flight conditions requires determination of P_{max}/P_S from Fig. 22 as a function of M_∞, calculation of ρ_2/ρ_1 as a function of M_∞ and altitude, and calculation of $\dot{q}_S$ to a hemisphere whose diameter is equal to the base plate diameter. This is a relatively simple procedure which should yield reasonable predictions of maximum base plate heating for ascent heating configurations of the type shown in Figs. 1 and 2.

Summary and Conclusions

The primary observations and conclusions derived from evaluating the test data are summarized below:

1) The wind-tunnel tests provided interaction heating data on a multiple re-entry vehicle/deployment module base plate configuration. Test conditions simulated flight Mach number, Reynolds number, and altitude.

2) The tests indicated complex flow patterns with indication of unsteady or bistable flow.

3) The test data validated the empirical relation for maximum shock interference heating on the re-entry vehicles in terms of shock-induced pressure rise.

4) The test data for the base plate, while not clearly defining maximum heating, are consistent with the analytical maximum heating predictions.

5) While the interaction flow patterns are too complex to provide detailed heating and pressure analysis along the cones and base plates, predictions of maximum values appear possible.

Acknowledgment

This work was sponsored by the General Electric Company Independent Research and Development Program.

The authors wish to acknowledge the cooperation of the staff of the Von Kármán Gas Dynamics Facility, Tunnel B, Arnold Engineering Development Center in conducting the test program described herein, particularly D.E. Boylan for supervising the tests and data reduction. The efforts of E.M. Storer of GE-RSD in reducing the paint melt data are also appreciated.

This effort was conducted under the direction and support of V. Corbo, IR&D Program Director for GE-RSD.

References

[1]Fay, J. A. and Riddell, F. R., "Theory of Stagnation Point Heat Transfer in Dissociated Air," Journal of Aero Sciences, Vol. 25, No. 2, Feb. 1958, pp. 73-85.

[2]Jones, R. A. and Hunt, J. L., "Use of Fusible Temperature Indicators for Obtaining Quantitative Aerodynamic Heat-Transfer Data," NASA TR-R-230, Feb. 1966.

[3]Keyes, J. Wayne, "Shock Interference Pear Heating Measurements Using Phase Change Coatings," Journal of Spacecraft and Rockets, Vol. 13, Jan. 1976, pp. 61-63

[4]Berman, R., Hecht, A., Metzger, J., and Nestler, D., "Separated Hypersonic Turbulent Flow Measurements on Spoilers," Journal of Spacecraft and Rockets, Vol. 14, No. 3, March 1977, pp. 163-169.

[5]Wood, C. J., "Hypersonic Flow Over Spiked Cones," Journal of Fluid Mechanics, Vol. 12, Part 4, pp. 614-624, 1962.

[6]Edney, Barry, "Anomalous Heat Transfer and Pressure Distributions on Blunt Bodies at Hypersonic Speeds in the Presence of an Impinging Shock," FFA Report 115, Stockholm, 1968.

[7]Serbin, H., "Supersonic Flow Around Blunt Bodies," Journal of Aero Sciences, Jan. 1958, pp. 58-59.

[8]Hung, F. T., "Three-Dimensional Shock Wave Interference Heating Prediction," AIAA Paper 77-756, June 1977, Albuquerque, NM.

[9]Holden, M. S., "Shock-Wave - Turbulent Boundary Layer Interaction in Hypersonic Flow," AIAA Paper 72-74, Jan. 1972, San Diego, CA.

[10]Hecht, A. M. and Nestler, D. E., "A Three-Dimensional Boundary Layer Computer Program for Sphere-Cone Type reentry Vehicles," GE-RSD Document No. 78SDR2255-1, July 1978. (AFFDL-TR-78-67, Vol. I.); also AIAA paper No. 79-0306, Jan, 1979, New Orleans, LA.

RE-ENTRY VEHICLE SOFT-RECOVERY TECHNIQUES

Donald J. Rigali* and Marlyn W. Sterk
Sandia National Laboratories, Albuquerque, N.Mex.

and

Lt. Col. Juri Randmaa‡
USAF Ballistic Missile Office, Norton Air Force Base, Calif.

Abstract

A technique to soft recover high ballistic coefficient re-entry vehicles from ICBM re-entry conditions has been developed and demonstrated. To date, two different types of vehicles have been soft recovered, utilizing the mass jettison, parachute recovery technique described herein. The fabrication and assembly of two additional RVs of different designs are presently underway in preparation for flight test. A technique to allow an increase in the severity of the environment from which an RV can be recovered is presently being analyzed and ground tested with plans to flight test it within two years. Descriptions of all of these vehicles and a summary of the flight test results are presented.

Introduction

Many re-entry vehicle (RV) designers and aerothermodynamicists have long considered an intact, soft recovered RV the ultimate verification of nosetip, heatshield, antenna window, and recession instrumentation performance, as well as the final validation of re-entry analytical predictive capabilities. Of recent major interest is the recovery evaluation of possible dispersion contributors (such as asymmetric shaping, roll torque generators, and differential ablation) in contrast to the general RV survivability aspects determined by the design integrity/margins of the nosetip, heatshield, and antenna windows. As the RV ballistic coef-

---

Presented as paper 80-1455 at the AIAA 15th Thermophysics Conference, Snowmass, Colorado, July 14-16, 1980.

*Department Manager, Advanced Systems Development Department.

+Division Supervisor, Advanced Systems Division 1.

‡Manager, MX Weapons Systems, Advanced Planning Requirements.

ficient, β, is increased, the severity of the corresponding re-entry environment increases, producing the phenomena of interest mentioned above. However, without an active recovery system, the corresponding impact velocity also increases severely with β, negating any possibility of obtaining meaningful performance information from recovered RV debris. The range of β considered in these vehicle designs is 9800 to over 19,000 kg/m^2.

A successful RV soft recovery technique must slow the vehicle down to a survivable impact velocity, but not change the normal flight conditions until after the vehicle has experienced peak environment or the selected environment of interest. This constraint means that the recovery process cannot be initiated until the RV is at or below the altitude of interest. Many soft recovery concepts have been investigated to date. However, the ones of primary interest are those that require minor or no changes to the outer heat protective surfaces, i.e., elimination of the concepts which would possibly destroy or modify any prerecovery initiated phenomena. For this reason, a recovery concept predicated on jettisoning a large portion of the RV mass out the base of the vehicle was chosen for the first stage of the RV slowdown sequence in the vehicles described here. Additionally, mass jettison reduces the load experienced by both the RV and any later deployed drag devices by the ratio of the vehicle mass before and after mass jettison.

The desires of the system analyst and the phenomenologist usually differ regarding soft recovery philosophy. The former is interested in the RV condition just prior to impact on a normal trajectory, whereas the latter is interested in obtaining a "snapshot" of the RV condition at various points in the RV trajectory by essentially "freezing" the RV condition at recovery initiation. This type of freezing recovery of several similar vehicles at different points on similar trajectories would permit this analyst to validate the various inputs that coalesce into the total heat protective system performance. However, it is clear that no real system can provide such an instantaneous slowdown and there will always be a significant amount of ablation after the recovery system initiation, even if only because the extremely hot RV surface is still exposed to the atmosphere. However, if a recovery system can provide a β decrease of about 10, the frozen nosetip case can be approached. Likewise, if a β decrease of 2 to 3 is accomplished just after occurrence of the maximum aerothermal environment, the resulting nosetip shape will approach that of the impact shape in a nonperturbed trajectory. This is due to the somewhat lowered ablation rates being compensated for by the corresponding longer time resulting from the recovery process.

The vehicle soft recovery systems described in this paper provide a β decrease due to mass jettison of about 2 to 3 and a β decrease of 7 to 10 for the mass jettison/flap deployment system utilized in the Advanced Recovery System (ARS) vehicle described later. In addition, all of the vehicles described have a single- or dual-stage parachute retardation system to further slow them prior to Earth impact.

Flight Test Programs

The first soft recovery program conducted by Sandia National Laboratories was the Nosetip Recovery Vehicle (NRV) program initiated in late CY 1973 as a subtask of the Ad-

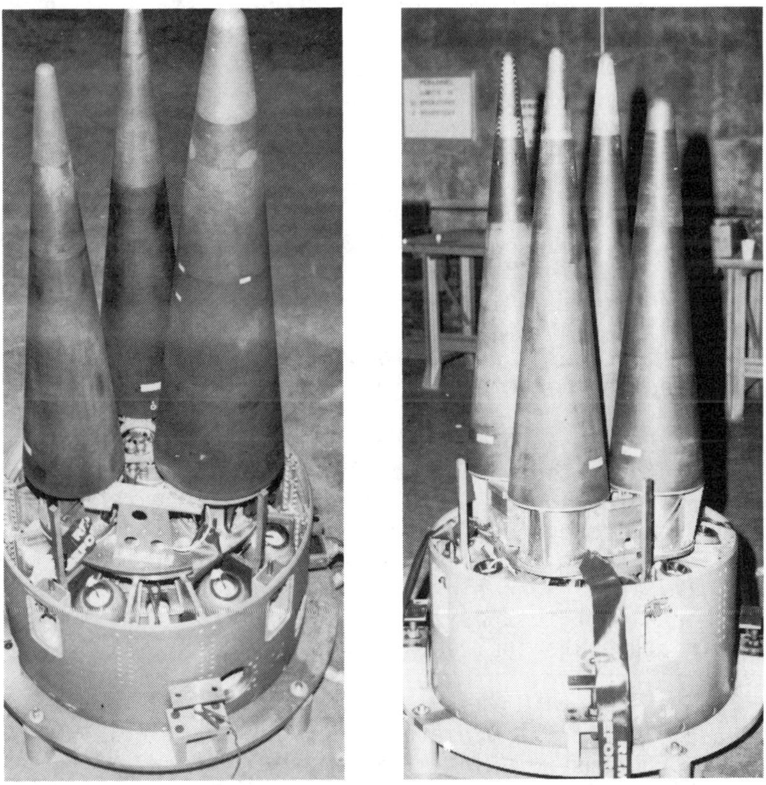

Fig. 1 ANT-II and ANT-IIIA payloads.

vanced Nosetip Test (ANT) program. The ANT and Interim Recovery System (IRS) programs make up the Sandia small re-entry vehicle flight test programs sponsored by the Advanced Ballistic Re-entry Systems (ABRES) Directorate of the Air Force Ballistic Missile Office (BMO). These programs were originated to simultaneously test three or four RVs on a single Minuteman I booster system, and to develop a very cost effective nosetip test bed for obtaining and evaluating re-entry technology data. The test programs were designed to permit the use of multiple test beds flown under exactly the same test conditions which provided the following advantages. First, it helped to eliminate the problems associated with correlating the results of many designs and materials over a period of many years. Another advantage was that it reduced the cost of obtaining re-entry flight technology data, since about half of the costs involved in flight testing RVs are directly related to the booster and range support. Additional economies are also realized since small RVs can be fabricated less expensively than large RVs.

A series of seven small vehicle flight tests is planned, consisting of four ANT flights and three IRS flights (both with three or four RVs per flight). To date four ANT flight tests have been completed. Photographs of two of the ANT payloads are shown in Fig. 1 and illustrate the following: a three and a four RV payload, the range of nose radii tested (from 0.019 to 0.036 m), and two different base diameters (0.254 and 0.318 m). These vehicles ranged in weight from 36 to 113 kg and all had telemetry systems of 100 to 150 channels. Figure 2 illustrates the general design layout of the standard ANT vehicles and summarizes the type of instrumentation that is transmitted throughout re-entry. Three ANT flight tests were conducted in clear air, and one was conducted in characterized weather.

The first IRS payload of three RVs (one recovery and two ANT RVs) is to be tested in February, 1981 in a clear air environment. The second flight of four RVs (one recovery RV) is being rescheduled for test in characterized weather. The projected third IRS flight will be a test of three recovery RVs in a clear air environment. Table 1 is a summary of the flight test payloads and re-entry conditions for the 25 miniature RVs included in these small RV flight test programs.

In addition to the miniature recovery vehicles, two further test programs were initiated in the past two years to recover full-scale RVs: the Large Ballistic Recovery Vehi-

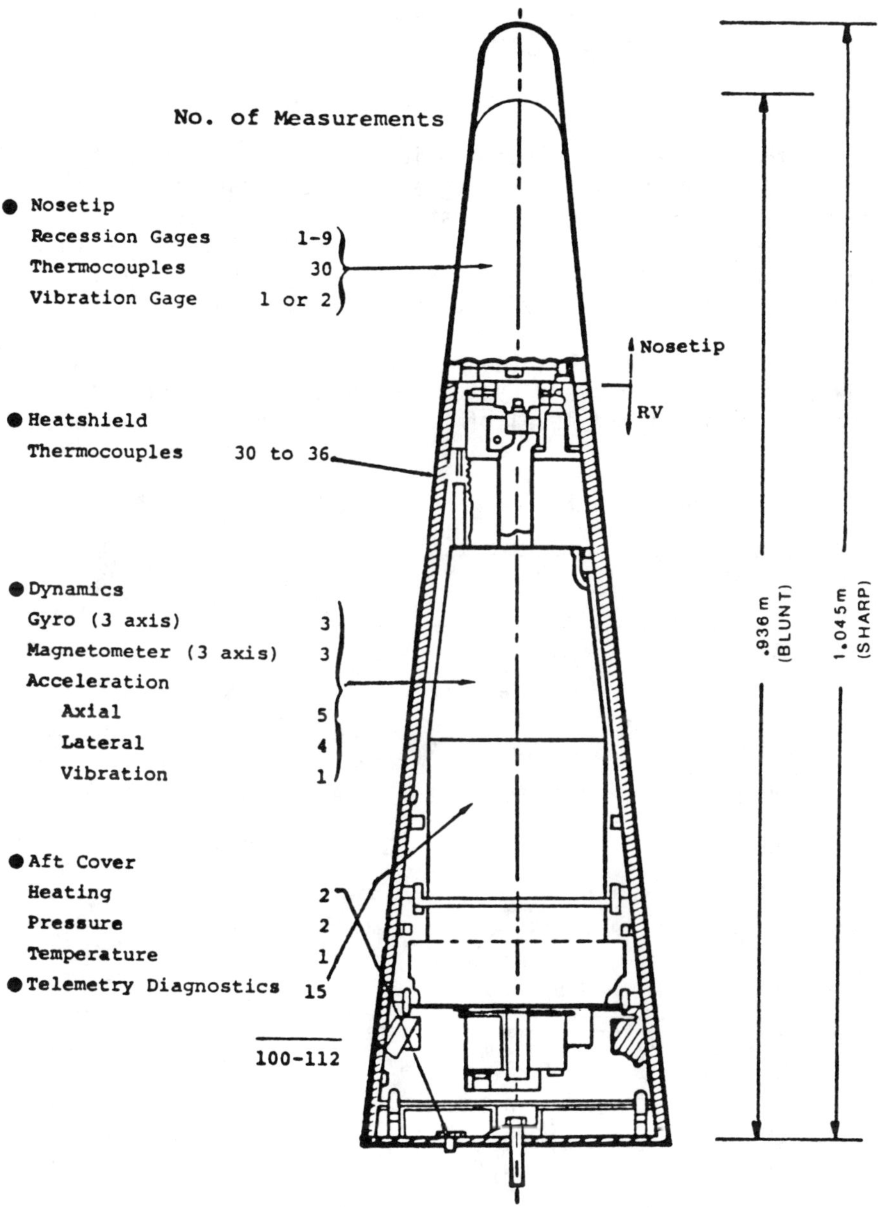

Fig. 2 ANT standard RV.

cle (LBRV) and the Systems Technology Vehicle (STV) program. Both of these vehicles utilize almost identical recovery systems which will be described later in the discussion of the LBRV design.

General Recovery System Design

There are presently three weight classes of RVs that have had recovery systems designed and fabricated: 50, 150, and 300 kg. These recovery system designs are capable of recovering, after peak aerothermal environments, a RV with a ballistic coefficient as high as 13,000 kg/m from a re-entry trajectory with a peak dynamic pressure as high as 6000 kPa.

Figure 3 depicts how the maximum dynamic pressure capability of these recovery systems at mass jettison has increased from the NRV level of 3300 kPa to the IRS-1 level of

Table 1 ANT and IRS flight tests

Flight	Date	Payload	Velocity m/s	Angle deg
ANT I	2/72/76	4 A[a]	6800	-30
ANT III	6/16/78	4 A	6800	.-30
ANT II[f]	7/4/78	4 A	6800	-30
ANT IIIA	11/27/78	1 H[b], 2 A	6800	-30
IRS I	2/81	1 I[c], 2 A	6810	-26
IRS II[f]	TBD	1 I, 2 A 1 M[d]	6810	-26
IRS III	TBD	3 13[e]	6830	-22

A[a] - ANT RVs, D_B = 0.254 or 0.267 m, Wt = 40 kg, R_N = 0.019 or 0.032 m.

H[b] - Helfire RV, D_B = 0.318 m, Wt = 113 kg, R_N = .036 m.

I[c] - IRS Recovery RV, D_B = 0.318 m, Wt = 66 kg, R_N = 0.036 m.

M[d] - MINT RV, D_B = 0.203 m, Wt = 34 kg, R_N = 0.036 m

13[e] - IRV-3, D_B = 0.330 m, Wt = 75 kg, R_N = 0.019 m.

11[f] Characterized weather (not clear air).

5800 kPa. This was accomplished by adding a mass fragmentation technique to the jettison mass, refining the overall vehicle design, improving the recovery sequence fuzing system, and increasing the capacity of the parachute system. All of these subsystems were fine tuned and verified through an extensive series of ground tests, including high speed rocket sled tests, before the final vehicle design was completed.

The mass fragmentation system was added to assure separation of the ejected mass from the recovery vehicle. Since the ejected mass is heavier and smaller in diameter than the recovery vehicle, and it is in the wake of the vehicle, its drag is insufficient to assure a clean separation of the two bodies, even at a differential velocity of 100 m/s at mass ejection. Therefore, the concept of mass fragmentation, i.e., the fragmentation of a high density ballast material

Table 2 Recovery comparison.

	NRV[1]	LBRV	IRS-1	IRS-2	IRS-3	ARS
Recovery Conditions						
Altitude, km	8.6	5.6	6.3	7.6	6.6	3.7
Velocity, km/s	3.83	3.78	4.20	4.97	4.57	4.27
Peak Aero/ Thermo Conditions						
β_N, KG/M^2	9,800	13,200	13,200	14,700	13,700	19,500
Dynamic Pressure, kPa	3400	5200	5800	6900	6400	7700
Stagnation Pressure, atm	64	98	108	125	120	143
Afterbody Heat Rate, MW/m	6.8	37	14	15	39	17

[1]NRV re-entry conditions were 5, 7 km/s and -25 deg. All others are ICBM velocities with re-entry angles between -22 and -27 deg.

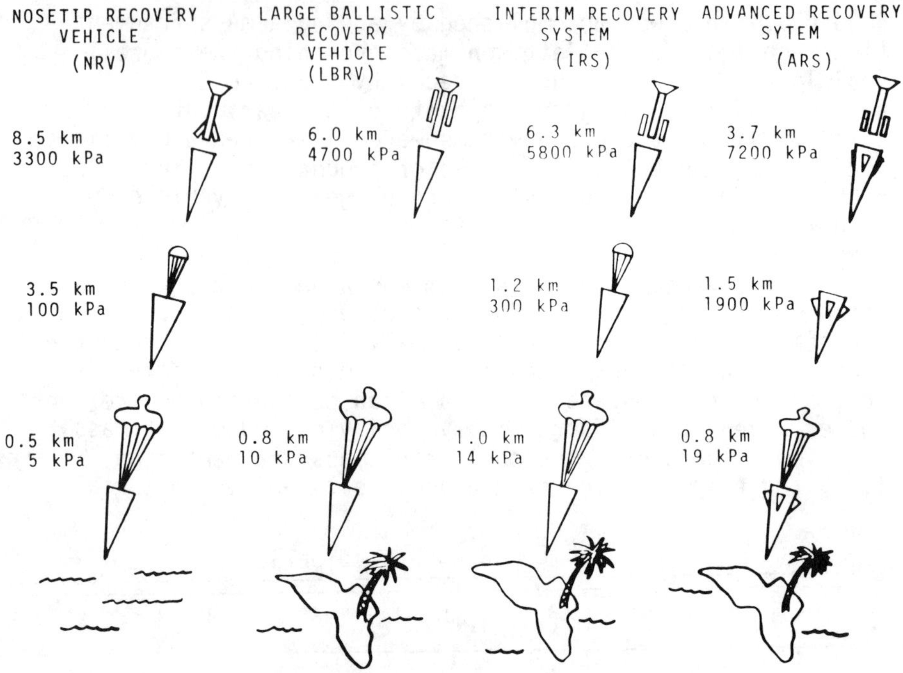

Fig. 3 Recovery concept evolution.

from a lightweight, high drag center section was evolved to assure separation. A relatively high lateral velocity is imparted to the ballast fragments to blow them out of the wake into the high dynamic pressure environment where they quickly drag-separate from the RV. The weight of the ejected center section is low enough, and its drag is high enough, that it also quickly separates from the RV after the high density ballast has been removed.

The first-stage ribbon parachute on IRS is similar to that on NRV, except that the IRS chute is made entirely of Kevlar-29 and is more than four times stronger. The final recovery stage on all of the vehicles is a guide surface parachute of various size and strength with an integral ram air-filled flotation bag.

A comparison of the recovery conditions and the peak aerothermal conditions for all of the recovery flight tests discussed herein is presented in Table 2. Note the increase in peak dynamic pressure (3400 kPa) and stagnation pressure (63 atm) recorded on NRV to that planned for ARS, 7700 kPa and 143 atm, respectively. Also, by changing vehicle cone angle and/or nose radius, a severe nosetip environment or afterbody heating environment can be obtained, as planned on the ARS and IRS-3 programs, respectively.

A re-entry trajectory for a typical IRS RV recovery is plotted in Fig. 4 as RV velocity vs altitude. The RV impact velocity without recovery would be 2400 m/s and with the addition of mass jettison, the impact velocity drops to 100 m/s. Similarly, the first-stage chute and second-stage chute slow the vehicle to about 60 m/s and 25 m/s, respectively, at impact.

NRV Design

The original NRV subtask in the ANT program was prompted by the desire to demonstrate the feasibility of a high β, ICBM recovery, as well as a strong interest in recovering a portion of the ANT experiments. Therefore, maximum compatibility with the ANT design and original schedule was desired. ANT requirements dictated a sphere cone re-entry vehicle with a ballistic coefficient of at least 9800 kg/m and a weight of approximately 50 kg.

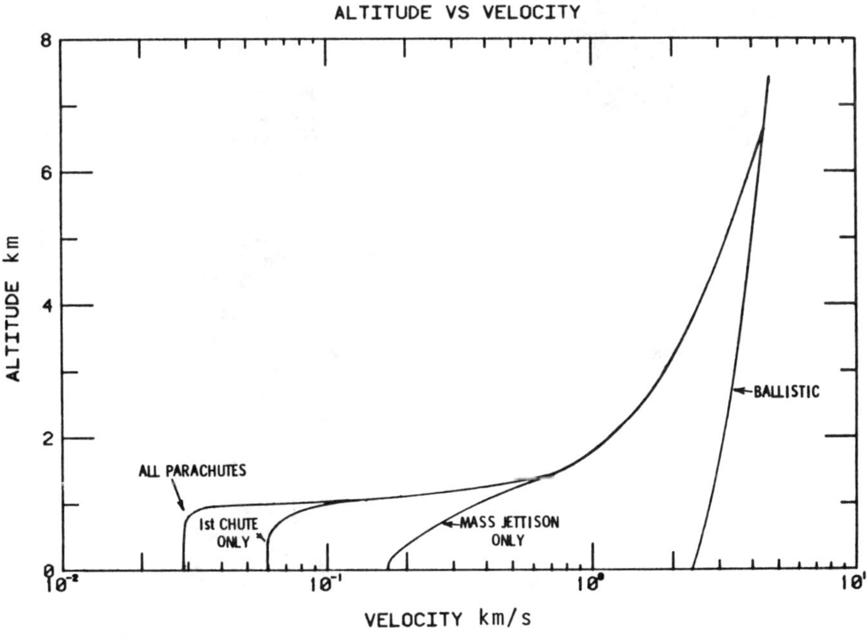

Fig. 4 Typical recovery RV trajectory profiles.

Fig. 5 Recovery and mass jettison configuration.

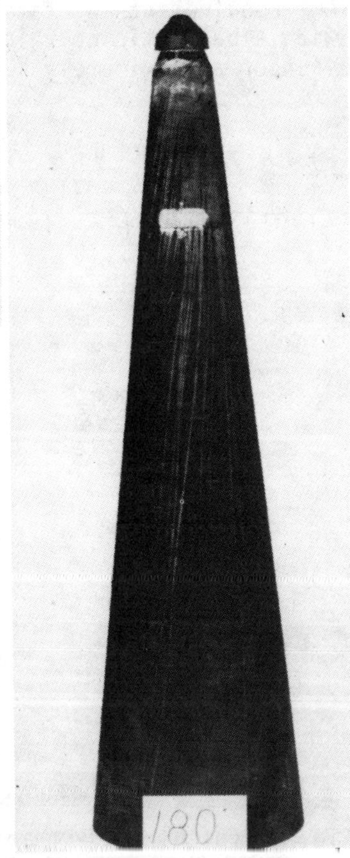

Fig. 6 Recovered NRV.

For the NRV program, the baseline recovery initiation altitude was chosen to be 7.6 km. This ensured that peak aerothermal environments would be experienced before the normal ballistic trajectory was interrupted and that the total integrated heating would be in excess of 90% of that experienced if ballistic flight were continued to impact. If recovery were not initiated, the vehicle would have impacted at a velocity in excess of 1200 m/s. Mass jettison was chosen as the method of obtaining the additional deceleration needed to slow to parachute operating velocities corresponding to a dynamic pressure of 96 kPa (1974 state-of-the-art technology). This was accomplished by ejecting 60% of the mass of the vehicle at an altitude of about 7.6 km, thereby increasing the deceleration by a factor of 2.5. Approximately the last 0.15 m of the aft heatshield and about half of the 31 channel telemetry/instrumentation system were ejected from the recovery portion of the NRV to obtain a mass jettison fraction as high as possible. Figure 5 shows the overall configuration of the recovery and mass jettison sections with the high density ballast section split to form aerodynamic drag flaps. The 0.15 m portion of the RV afterbody and the ballast drag flaps were added to the mass jettison section to provide sufficient drag in the RV wake to assure contained separation of the two bodies. In the NRV design, as well as all of the other recovery RV designs, the parachute container forms a piston and the ejected mass forms a cylinder for the piston-cylinder principle that provides the energy to drive the two sections apart when a hot gas generator mounted in the end of the cylinder pressurizes this ejection system. This system imparts a relative separation velocity of about 75 m/s to provide sufficient separation distance so that the mass moves far enough aft in the RV wake to develop sufficient aerodynamic drag to continue the separation.

Prior to an ICBM flight test on a Minuteman booster, a meaningful demonstration test of the NRV system on a re-entry flight with reasonably severe conditions was desired. This test was conducted using the Sandia STRYPI booster system. This flight test was successfully flown on Sept. 18, 1975 from Wallops Island, Virginia, on a STRYPI VII booster, which placed it at re-entry conditions of $V = 5.7$ km/s and $\gamma = -25$ deg at an altitude of 91.4 km. Normal ballistic flight was continued until shortly after peak heating rate and peak deceleration had been experienced. The recovery process was then initiated at an altitude of approximately 8.6 km. The recovery portion of the vehicle

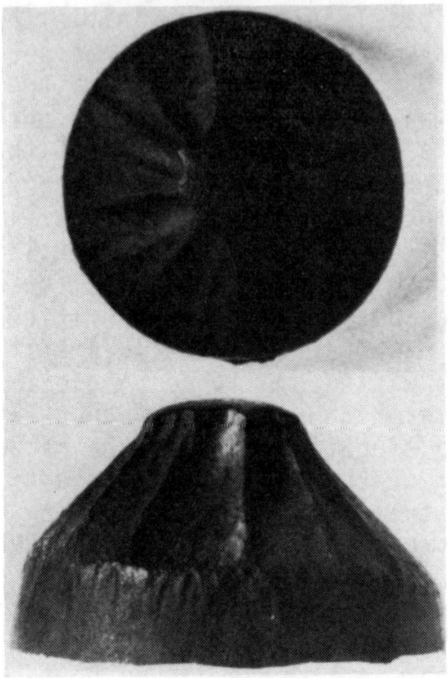

Fig. 7 NRV nosetip.

was "soft landed" as planned and removed from the water about 2 h later by the Coast Guard Cutter Cherokee. This event marked the first time that a high-performance re-entry vehicle had been soft landed and recovered. Figure 6 is a photograph of the recovered vehicle, and Fig. 7 shows two closeup views of the graphite nosetip.

IRS Design

The IRS-1 payload (see Fig. 8) has one recovery vehicle and two non-recovery (ANT) vehicles. The program objectives are: 1) to recover a high-performance re-entry vehicle from an ICBM trajectory after experiencing peak re-entry environments; 2) to establish relationships between recovered hardware and the end points of the telemetered data; 3) to evaluate, by inspection, the material performance of the nosetip, antennas and the heatshield of the recovered RV; and 4) to fly auxilliary nonrecovery payloads to obtain performance data on the state-of-the-art nosetip heatshield and antenna window materials.

The remaining discussion will cover only the recovery RV. Figure 9 shows a cutaway of the RV with identifiers showing the significant RV parts. The 70 kg class RV consists of a sphere cone body having a carbon-carbon nosetip with a radius of 0.032 m and a carbon-phenolic heatshield with a

Fig. 8 IRS-I payload.

base radius of 0.159 m. The forward RV internal volume is dedicated to the telementry/instrumentation and the aft volume to the recovery systems and some instrumentation.

The 110 channel telemetry/instrumentation system measures parameters like nosetip recession [using the Backscatter Radiation Ablation Gage (BRAG) and the acoustic gage] heatshield performance (using thermocouples) and RV dynamics (using accelerometers, magnetometers, and rate gyros). Only about 10% of the RV instrumentation is located on the mass jettison assembly and is ejected at the mass jettison event.

The performance information monitored during re-entry will be telemetered out the antennas to a ground based telemetry receiving station. These results will then be compared to the recovered article after the test. But even the inflight measurements do not always reveal localized phenomenology (e.g., gouging) occurring on the vehicle, whereas postflight inspections of the recovered RV should show important and interesting details.

The scenario of events taking place during the recovery sequence is shown in Fig. 10 along with the corresponding aero environment at each stage. At a preselected altitude (6.6 km), 63% of the RV mass is explosively ejected from the vehicle to initially slow the vehicle. Then, 0.080 s later,

80% of the ejected mass is explosively fragmented into smaller pieces in order to keep the ejected mass from overtaking the RV and colliding with it.

The lightweight RV continues on its trajectory for about 5-1/2 s more. A preset timer then fires a gas generator which forces out a parachute container lid, which pulls the first-stage Kevlar ribbon parachute out into the flowstream. This slows the RV down to a low enough velocity that the final guide surface parachute will function properly (this event is also timed). The second-stage parachute contains a ram-air flotation bag which will buoy the RV after it impacts the water. These events identify all the critical functions that must occur for the RV to soft impact at about 30 m/s, instead of the normal 2400 m/s.

LBRV Design

The LBRV program was initiated at an Air Force request because of the need to recover a full-scale 150 kg class RV for evaluation of the nosetip and heatshield materials. The complete program from go-ahead to actual recovery, took less than ten months to complete because of the existing capability developed under the IRS program. The IRS technology was readily adapted for this flight, except the mass fragmentation system, which was developed for this program and then incorporated into the IRS program. The recovery scenario was identical to that outlined for the IRS program, except for the mass jettison altitude of approximately 5 km.

The LBRV was successfully flight-tested on a Minuteman I booster from Vandenberg AFB to the Kwajalein Missile Range Lagoon in July 1979. This flight test was the first recovery demonstration of a high-performance RV from an ICBM trajectory after experiencing peak environments. Figure 11 is a series of radot camera frames (every 0.050 s) showing the recovery sequence in operation on the LBRV program. Frame A shows the RV just before mass ejection; Frame B shows the first indication of the mass jettison event (venting of the hot gas); Frame C shows the initiation of mass fragmentation; Frames D, E and F show the separation of the bodies and the optical traces of the fragments.

The RV continued on its trajectory until impact, with an impact velocity less than 120 m/s. The parachute system which would have further slowed the RV to about 30 m/s did not function on this flight because an electrical connector

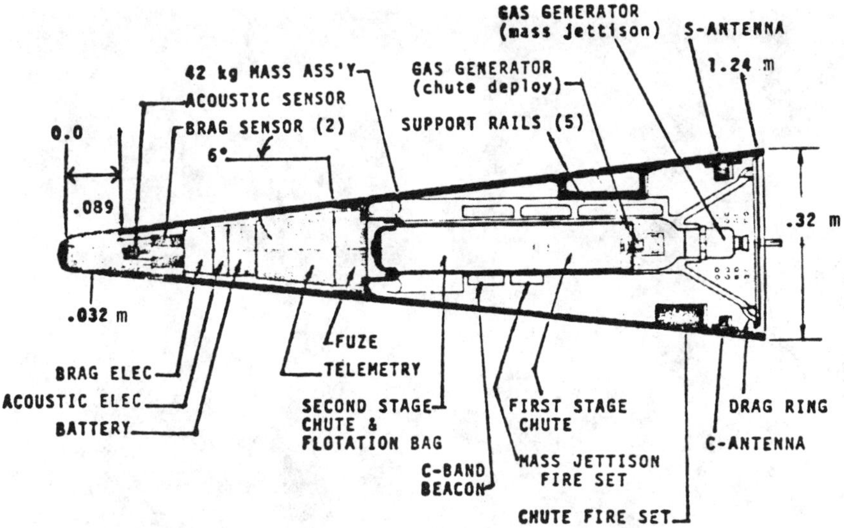

Fig. 9 Interim recovery system vehicle.

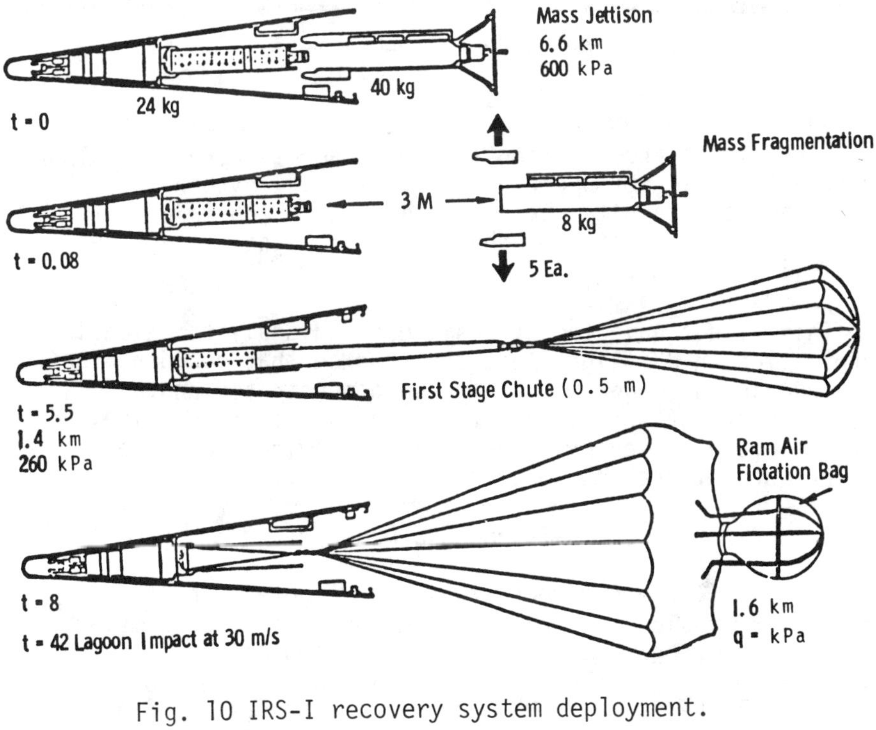

Fig. 10 IRS-I recovery system deployment.

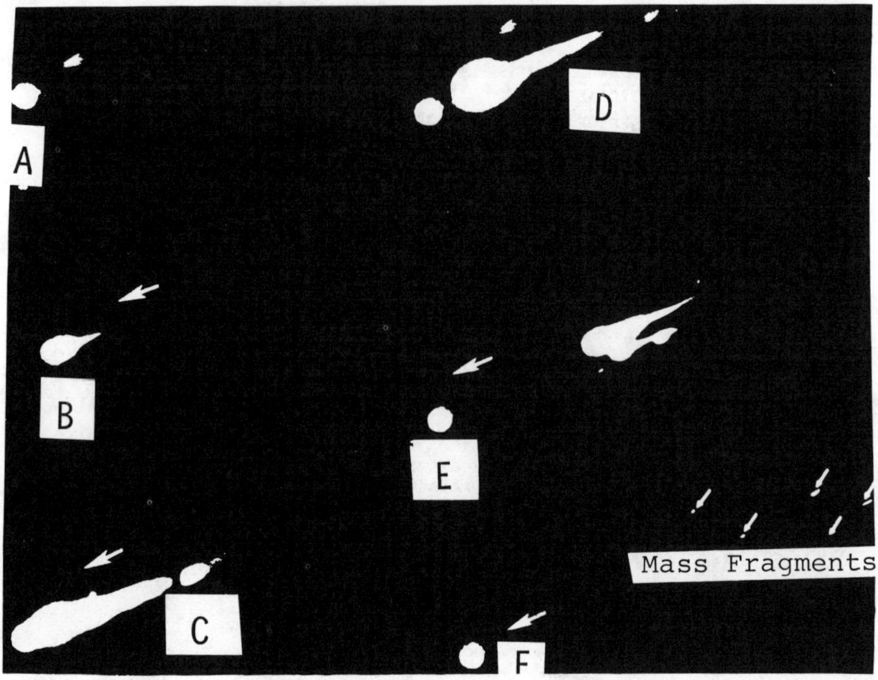

Fig. 11 Radot photographs of LBRV mass jettison.

opened due to the 600 g mass jettison shock and prevented the fire signal from getting to the parachute fire set. The fact that it did not operate did little, if anything, to alter the test results; because the RV was recovered from soft sand on the bottom of the Kwajalein Lagoon with very little impact damage.

This flight gave the RV designers and analysts valuable information on RV flight test performance that could not be obtained in any other way. It also demonstrated another advantage of soft recovery, i.e., the ability to examine soft recovered re-entry hardware to determine the cause of failure of internal components.

IRS-II and ARS Design

The IRS-II payload consists of one recovery and three nonrecovery vehicles. This program has objectives similar to the IRS-I program, except for the following: 1) the test is to be conducted in characterized weather instead of clear air; and 2) the recovery RV includes a retardation flap experiment to evaluate the ARS concept on an ICBM trajectory.

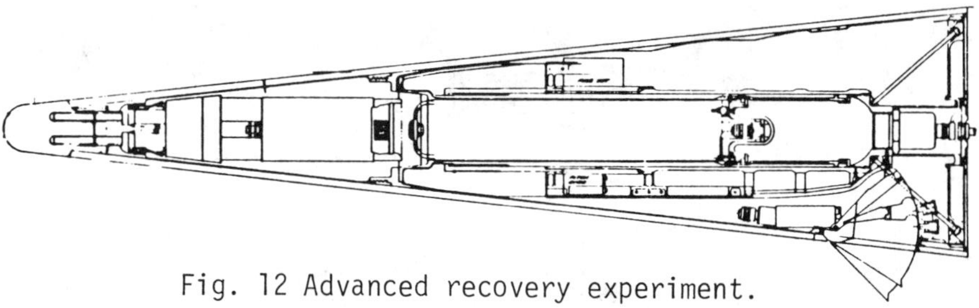

Fig. 12 Advanced recovery experiment.

This experiment is designed primarily to collect flight test data on the flap performance, rather than act as a necessary, required, drag-producing device. That is, the experiment has been configured so that even if the flaps produce only a fraction (as little as 20%) of their expected drag, the mass jettison/parachute system performance is sufficient to soft recover the RV.

Figure 12 shows a cutaway of this RV which is very similar to the IRS-I RV. The significant difference is the addition of six retardation flaps and the corresponding larger length and base diameter (0.337 m). In order to better understand the significance of these flaps, the recovery scenario will be discussed first. Figure 13 shows the IRS-II deployment sequence which begins at about 7.6 km. At this altitude, approximately half of the mass is ejected and the six flaps are simultaneously deployed to the first position (23 deg). Then, about 2 s later, the flaps are forced out to their final position (45 deg). The two-stage flap deployment is necessary in order to limit the deceleration (to about 200-220 g) as well as to limit the aerothermal flap environment. In addition to the deceleration limitations, previous flight test and ballistic range data have indicated that flaps exposed to severe aerothermal environments suffer extremely high ablation rates at angles much over 30 to 40 deg, and the two-stage flap concept was designed around these limitations. Referring to Figs. 12 and 13, note that the flaps are automatically extended to their initial position by the mass ejection guide rails as the mass moves rearward. Spring loaded pins lock the flaps in place at the 23 deg position until six gas generator driven pistons push the flaps out to the 45 deg position about 2 s later.

After the RV has been slowed on its trajectory by the mass change and the two-stage drag flaps, the first-and second-stage parachutes are deployed 4-1/2 and 7 s later, respectively. This completes the recovery sequence and the vehicle should soft impact at about 30 m/s.

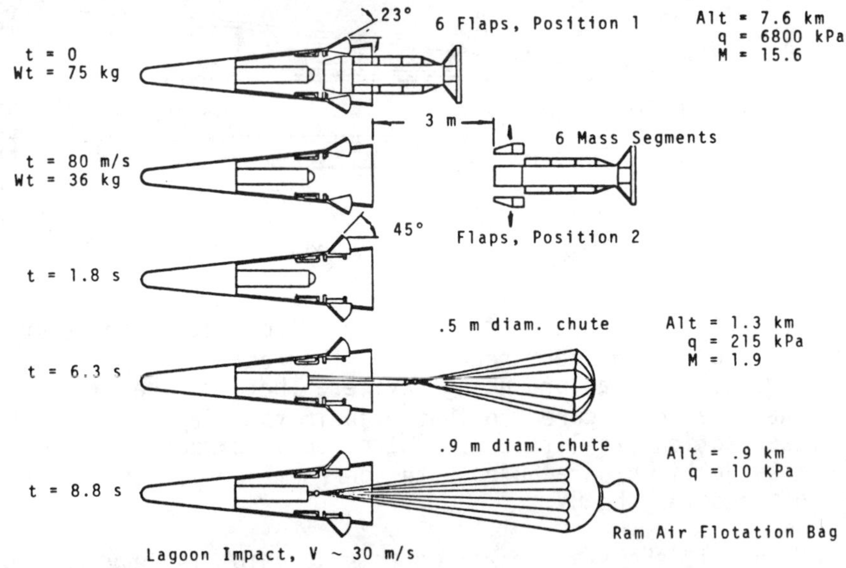

Fig. 13 IRS-II Recovery system deployment.

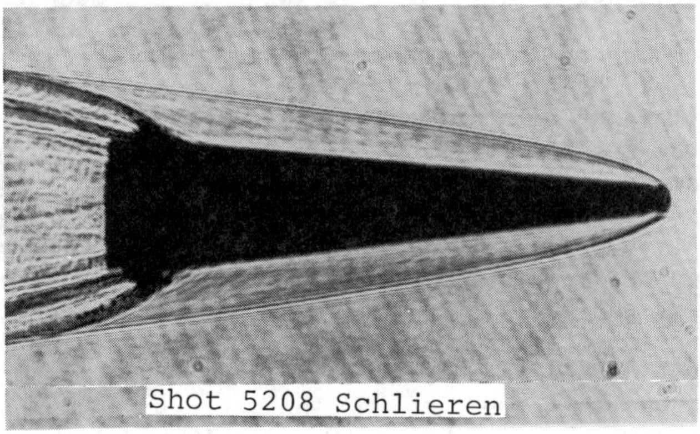

Fig. 14 Schlieren of IRS-II RV at M = 13.

The flap drag data collected to date (from the AEDC Ballistic Range G) are summarized in Table 3. The preflight conditions for IRS-II and ARS are also listed for comparison. Figure 14 is a Schlieren photograph of an IRS-II ballistic range model with the flaps at the 25 deg position at a Mach number of 13. To verify the estimated performance of these flaps at lower Mach numbers, a series of wind tests of 0.8 scale models of these vehicles has been completed.

Table 3 Flap drag data summary

Mach	Re/ft	Drag Coefficient		
		No flaps	25 deg flaps	45 deg flaps
8	13×10^6	--	0.157	0.418
10	16×10^6	0.066	0.141	--
13	15×10^6	0.065	0.133	--

IRS-II Flight Conditions:

Flaps to 23 deg (MJ); M = 15, Re/ft = 50×10^6

Flaps to 45 deg; M = 7, Re/ft = 30×10^6

ARS Flight Conditions:

Flaps to 23 deg (MJ); M = 13, Re/ft = 70×10^6

Flaps to 45 deg; M = 8, Re/ft = 50×10^6

The factor of two increase in drag listed in Table 3, coupled with a mass jettison of approximately half of the RV weight, produces a β reduction factor of four for ARS. When the flaps are extended to the second position about 1 to 2 s later, the drag increases by another factor of more than 2.5 for an accumulative β decrease of over 10. This now allows one to consider the recovery of very high β RVs and/or the recovery of moderately high β vehicles with a very low recovery initiation altitude. With this performance capability, the experiment "freezing" concept discussed earlier can be approached.

Conclusions

The recovery of moderately high ballistic coefficient re-entry vehicles has been successfully demonstrated for a 50 kg class RV from a MRBM trajectory and a 150 kg class RV from an ICBM trajectory utilizing the mass jettison/parachute recovery technique. Similar design concepts have been applied to another 50 kg class RV and a 300 kg class RV for ICBM flight test. These vehicles (except IRS-II and III) have been tested on ICBM MMI missions.

Any existing RV known to the authors can be readily modified to accommodate this soft recovery concept without significant changes to the outer heat protective shell. In addition, a drag flap system has been designed and is being ground tested to verify its performance for application to relatively high ballistic coefficient RVs and/or soft recovery from more severe re-entry environments.

Acknowledgments

This work was jointly sponsored by the ABRES Directorate of the Ballistic Missile Office of the United States Air Force and the United States Department of Energy under contract DE-AC094-76DP00789.

Chapter III. Planetary Entry

THERMAL PROTECTION SYSTEM FOR THE GALILEO MISSION ATMOSPHERIC ENTRY PROBE

R. A. Brewer* and D. N. Brant+

General Electric Company, Philadelphia, Pa.

Abstract

Carbon phenolic has been selected as the forebody heatshield material for the Galileo probe mission. The probe is currently designed to enter the Jupiter atmosphere at an entry velocity of 48.2 km/s with an entry-path angle corridor of -7 to -10.2 deg, resulting in several centimeters of the carbonaceous heatshield being consumed by sublimation. The base thermal environment is also very severe, more than an order of magnitude greater than that of the Pioneer Venus probes, and phenolic nylon has been selected as the base thermal protection system. Unique design techniques have been developed to determine the probe's heat protection requirements. The predicted thermal environment, material response, and safety margin philosophy are discussed in this paper.

Nomenclature

A = area
B' = blowing coefficient
C_H = film coefficient
C_{Ho} = zero mass-addition film coefficient
D_B = base diameter
H_o = stagnation enthalpy
h = enthalpy

Presented as Paper 80-0358 at the AIAA 18th Aerospace Sciences Meeting, Pasadena, Calif., Jan. 14-16, 1980.

*Supervisor, Thermodynamics Analysis, Re-entry Systems Division.

+Manager, Aerothermodynamics Engineering, Re-entry Systems Division.

$\dot{m}$ = mass-loss rate
N_S = shock standoff distance
P = pressure
$\dot{q}$ = heat-transfer rate
R = radius
R_N = nose radius
S = wetted length
V_E = entry velocity
Z = axial length
β = ballistic coefficient
γ_E = entry-path angle
θ_c = cone semiangle
θ_s = shock angle
θ_b = base angle, body angle
ψ = heat-transfer ratio

Subscripts

B = base
C = convective
Cyl = cylinder
i = initial
L = laminar
LB = last body point
R = relative, radiative
S = stagnation
T = turbulent
W = wall

Introduction

The project Galileo mission will be launched in 1984 for an encounter with the planet Jupiter after a 2-1/2 yr transit to the planet. This paper discusses the design and analysis performed to define the heat protection requirements for the atmospheric entry probe. The primary objective of the probe is to determine the planet's atmospheric composition, atmospheric structure, and radiative energy balance. A single probe will enter the planet's hydrogen/helium atmosphere at an entry-path angle of approximately -8.6 deg with a relative entry velocity of approximately 48 km/s. Entering the atmosphere at such a high velocity will result in shock-layer heating capable of consuming by sublimation several centimeters of the best carbonaceous ablative thermal protection materials available. Based on extensive flight experience, including a successful Pioneer Venus probe mission, carbon phenolic has been selected as the forebody heatshield material.

The mass loss associated with this ablation is a major fraction of the total mass of the probe and is not confined principally to the stagnation region. The probe drag coefficient is altered because of shape change and the probe base area is reduced. The combined effects result in a ballistic coefficient that decreases significantly during entry. Reductions in ballistic coefficient result in less severe thermal environments and lower mass loss; thus, the problem is fully coupled in that instantaneous probe mass and shape must be known to describe accurately the velocity. Since the entire probe forebody undergoes significant ablation, the need arises for a shape-change code that treats the forebody with a built-in trajectory sensitivity for variable ballistic coefficient. Such a code has been developed and has been used in assessing the heatshield requirements for the Galileo probe mission.

The heat protection system for the Jupiter probe has to protect the probe's scientific payload from the combined convective/radiative heating during entry generated by the most severe combination of entry conditions and possible atmospheric composition. Also, the heatshield performance must be predictable so that the changing probe mass and shape are known, allowing the probe ballistic coefficient to be determined accurately during entry, thus permitting accurate reconstruction of the upper atmosphere density/altitude profile.

A study has been made to evaluate the impact of off-nominal entry conditions including the effect of encountering an off-nominal atmospheric composition. Atmospheric composition has a significant effect on the radiative heat transfer and, hence, on the heatshield ablation performance. This analysis is used together with estimates of uncertainties in the prime parameters that affect the heatshield design to determine the required heatshield thickness distribution necessary to ensure a high probability of mission success.

The severity of the thermal environment predicted for the Jupiter probe also extends to the base region. To compute the base convective and radiative heat transfer, an engineering model of the base flow has been constructed. In this model, the predicted heat-transfer rates are much greater than those experienced by the Pioneer Venus probes. Heatshield material tradeoff studies resulted in the selection of phenolic nylon as the heatshield material for the base region, which has been qualified extensively with flight data on military space programs.

The verification of the entire probe heatshield design is accomplished through use of NASA Ames Research Center and GE-RSD test facilities, including radiative heating simulation using a CO_2 gas dynamic laser and combined radiative/convective heating tests using the Giant Planet hyperthermal arc facility (GPF).

A discussion of the analytical methods employed, heatshield materials used, and the design approach for the Galileo probe heatshield subsystem is included in this paper.

Heatshield Subsystem Description

Carbon phenolic and phenolic nylon heatshield materials were selected for the Jupiter mission because both are backed by extensive ground and flight test experience. The selected manufacturing processes are fully developed; carbon phenolic on the Pioneer Venus program and phenolic nylon on military space programs. High-quality heatshield sections with known material properties and developed quality-control procedures are therefore ensured without the need for costly high-risk materials development programs. Sufficient test data are available on the selected materials and processes from the Pioneer Venus and military space programs to ensure that design performance is not affected by exposure of the probe to the storage, prelaunch, STS launch, interplanetary cruise, and approach-phase environments.

The carbon phenolic heatshield is fabricated using the techniques and processes developed for the Pioneer Venus probes. The heatshield thickness varies from 9.47 cm (3.73 in.) at the stagnation point to 4.72 cm (1.86 in.) at the end of the cone. The design incorporates a chopped-molded carbon phenolic (CMCP) nosecap and a tape-wrapped carbon phenolic (TWCP) frustum. Both heatshield sections are fabricated using FM 5055G, a flight-proved carbon phenolic prepreg. The FM 5055G prepreg incorporates a carbon-opacified resin system to prevent in-depth radiation heating penetration during Jupiter entry.

The nosecap is fabricated using a highly staged, chopped (nominal 1.25 cm square) prepreg molded at high pressure [$\sim 34.5 \times 10^6$ Pa ($\sim$5000 psi)] into a solid billet. The billet is then machined to the required configuration. The Jupiter probe nosecap billet is nearly identical in size to that fabricated for the Pioneer Venus Large Probe, thus eliminating the need for process development. This CMCP laminate construction has been tested successfully both in flight (Pioneer Venus probes) and in a rocket-exhaust ablative environment at surface pressures in excess of those expected for the Jupiter mission.

The forebody frustum is fabricated using an aft-facing tape and broadgoods. The 30 deg tape angle reduces shield fabrication complexity by decreasing the tape width requirements 50% compared to a 20 deg tape angle construction.

The CMCP and TWCP sections are mated before assembly to the probe aeroshell structure using a carbon-filled (35%) Dow Corning silicone bond as the adhesive at the contoured interface joint, resulting in an integral forebody shield design. The nosecap frustum joint reduces the frustum maximum tape width requirements, increases the bonded surface at the joint, and prevents formation of a direct light path to the aeroshell structure during entry heating.

The forebody heatshield is bonded to the aeroshell structure using a Dow Corning bond DC-93 500 by nesting the aeroshell structure into the supported heatshield assembly. The heatshield is first partially filled with catalyzed adhesive and the structure assembly is then lowered into the shield section, displacing the adhesive and filling the heatshield gap. This technique was successful in producing void-free bonded assemblies for the Pioneer Venus program.

The phenolic nylon afterbody heatshield sections, including the aft cover, mortar cover, and aft ring, are fabricated from a two-dimensional woven prepreg as a parallel to surface laminate. The heatshield thickness varies from 0.83 cm (0.326 in.) at the probe afterbody centerline to 0.46 cm (0.18 in.) at the base of the aft cover and aft ring sections.

The phenolic nylon fabrication procedure uses precut, shaped, preimpregnated broadgoods made from 91-LD phenolic resin and Wellington Sears SN-19 nylon cloth, hand applied to a male fabrication mandrel. This technique has been used extensively for forebody heatshield applications, and several hundred successful Earth re-entry flight tests have been conducted using shields fabricated by this process.

Forebody Design Techniques

The predicted surface recession of the probe on entry into Jupiter is expected to be significant, resulting in large amounts of the carbonaceous heatshield being consumed in ablation. About 25% of the probe mass is predicted to be lost during entry, resulting in significant probe shape change. Since the radiative heat-transfer rate is sensitive to probe velocity and changing body shape (e.g., blunting in the stagnation region), the prediction of the thermal environment and heatshield response must be coupled to include effects of changing probe mass, area, and body shape. To accommodate these important, implicitly related design considerations, a steady-state shape-change code was developed for the design of the carbonaceous heatshield systems for the outer planets. This code is the Thermodynamics Outer Planets Insulation Code (TOPIC), the details of which are presented in Ref. 1.

Some of the more important aspects of the design technique include: 1) shock shape and pressure distribution, 2) graphite thermochemistry for H_2/He mixtures, 3) convective and radiative heat-transfer predictions, 4) heat-transfer blockage correlations, and 5) ballistic-coefficient variations and trajectory computations. A primary emphasis has been placed on rapidly incorporating the theoretical techniques and the results from the NASA centers that have been developing the technical base for the design of the Jupiter probe. Since the original TOPIC code was developed in 1976, a number of advancements in outer planetary technology have been made. Among these are the development of an approximate inviscid flowfield code that includes approximate radiation formulations. This code is the Maslen-Olstad Shock Technique (MOST) code, Ref. 2. At the inception of the TOPIC code, only laminar boundary-layer solutions were available from the viscous shock-layer solutions of HYVIS (NASA-LRC), Refs. 3 and 4. More recently, turbulent boundary-layer solutions have become available from the HYVIS code, Refs. 5 through 7. The incorporation of the MOST code and the latest HYVIS solutions into the TOPIC code eliminates several shortcomings of the original technique, and these are described in this section.

The TOPIC code represents an engineering solution for the probe heatshield response to the intense thermal environment of entry into Jupiter. The current version of the code is similar to that described in Ref. 1; however, there are several significant improvements. These include: 1) the incorporation of the Maslen-Olstad Shock Techniques (MOST) code into TOPIC, 2) inclusion of turbulent blockage correlations for both convective and radiative heat transfer, and 3) a technique to determine the peripheral shape change of the probe.

The flowfield formulation of the TOPIC code consisted of a set of correlations that related shock shape to local S/R_N and pressure to local body angle. These relationships compared quite favorably to exact solutions obtained from the GE 3D Flow Field code, Ref. 8, when applied to sphere cones. However, there are at least two factors that have been reconsidered: ablated shapes and nonadiabatic effects (shock cooling). A simplified flowfield technique developed by Maslen (Ref. 9) and modified by Olstad (Ref. 10) to account for radiation and large shock cooling has been incorporated into the TOPIC code. This method is the Maslen/Olstad Shock Technique (MOST) and it provides the shock-shape definition and pressure distribution over blunt sphere cones in addition to the nonadiabatic radiative heat-transfer rates. The shock layer formed around a smooth, symmetric body at hypersonic speeds is quite thin and, downstream of the stagnation point, the flow direction in the shock layer is nearly parallel to the shock. Maslen developed a simplified system of equations that adequately describes the flow and thermodynamic properties in the shock layer for an inviscid, nonreacting, nonradiating gas for a given shock shape. The approach is inverse in that the shock geometry is initially specified and the body geometry computed.

This approach was employed and modified by Olstad (Ref. 10) to include radiation effects. Tangent slab approximations are used based on Nicolet's (Ref. 11) RAD/EQUIL code, and a frequency step-function model for the absorption coefficient was also incorporated (Ref. 2). Thermochemical equilibrium is assumed.

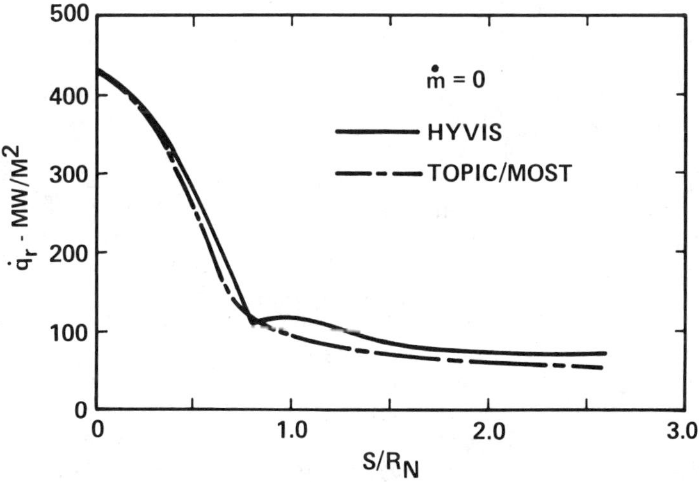

Fig. 1 Comparison of radiative heating rates at peak heating (t = 111.3 s).

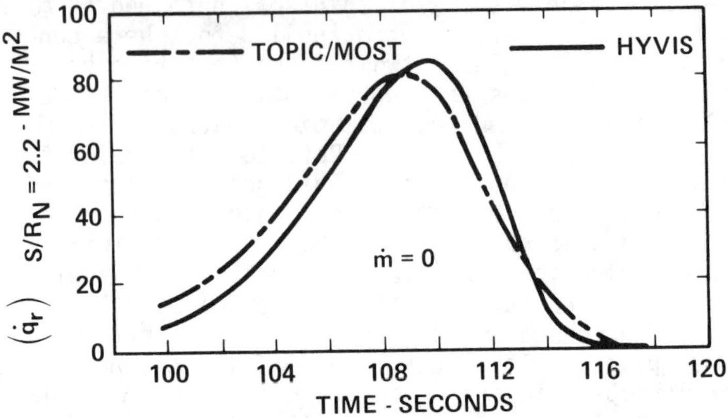

Fig. 2 Comparison of radiative heating rates at $S/R_N = 2.2$.

A comparison of the pressure distribution over a sphere cone for a nominal entry condition has been made, and the HYVIS results (obtained from the study reported in Ref. 6) and the pressure distribution determined using the MOST/TOPIC approach compare very well. The nonadiabatic radiative heat-transfer rate distribution over the body at peak heating (for $\dot{m} = 0$) is shown in Fig. 1 compared with the "exact" solution of HYVIS (Ref. 6). The time-history comparison at the end of cone location is shown in Fig. 2 where a slight time-phasing error is evident, the TOPIC/MOST solution peaking before the HYVIS solution. The inviscid properties developed in the MOST technique are used for the boundary-layer edge properties in the convective heat-transfer expressions described in Ref. 1. A comparison of the convective heat-transfer rates predicted using this approach with the more exact solutions of HYVIS (Ref. 6) are shown in Fig. 3 for an end-of-cone location, with no mass addition.

Calculations of the radiation heat transfer show that significant amounts of energy may be absorbed by the ablating carbonaceous heatshield. Reliable estimates of the radiation blockage must be made to ensure that a realistic heatshield design evolves. Recent studies by Moss (Refs. 5 to 7) have evaluated the radiation blockage for a 45 deg sphere cone. The studies, for several Jovian atmospheres, encompass an entire trajectory using the coupled solutions of the viscous shock-layer code, HYVIS. These solutions include the effects of turbulent boundary layers on the convective and radiative heat transfer. Previous correlations of the mass-transfer blockage reported in Ref. 1 were based solely on laminar boundary-layer solutions. These correlations are still valid for laminar

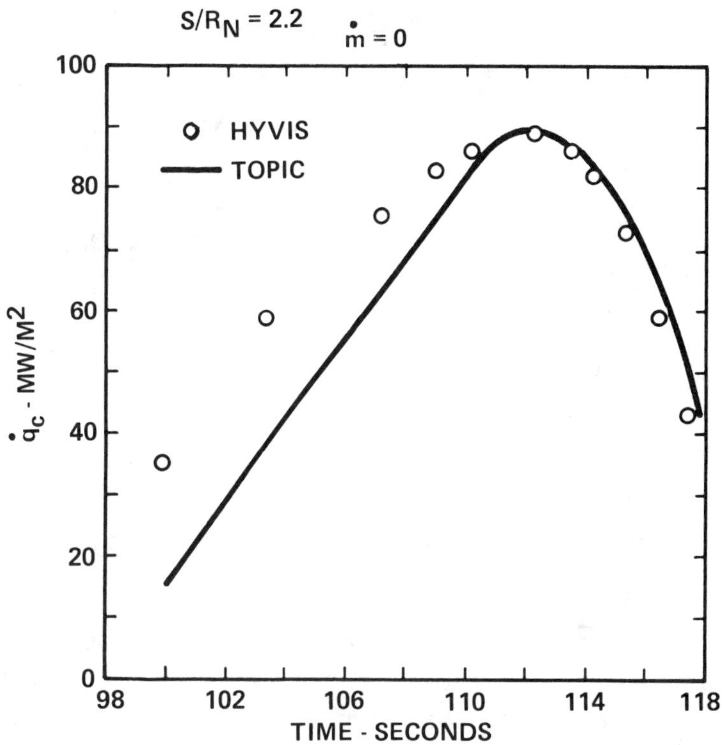

Fig. 3 Comparisons of convective heating calculations.

flow; however, new correlations have been developed for turbulent-flow conditions. The convective blockage correlation is shown in Fig. 4, where C_H/C_{HO} (ψ_{CT}) is presented as a function of B'. Correlations of the radiative blockage for two model atmospheres of Jupiter are shown in Figs. 5 and 6, for volumetric gas mixtures of 89% H_2 - 11% H_e (nominal) and 78% H_2 - 22% H_e (heavy), respectively. Similar correlations exist for the warm atmosphere. These correlations are expressed as a function of local body position (S/R_N), shock standoff distance (N_S), and an average mass-addition rate $\frac{\int \dot{m}\, dA}{\int dA}$. The turbulent-flow solutions of HYVIS show less blocking effectiveness compared with laminar-flow solutions for both convective and radiative heat transfer. Figures 7 and 8 show the good agreement between TOPIC and HYVIS mass-loss rates for two entry trajectories evaluated. These trajectories assume a constant shape and constant probe mass during entry. Since neither of these assumptions is correct for a probe entry into Jupiter, the TOPIC code is used to determine

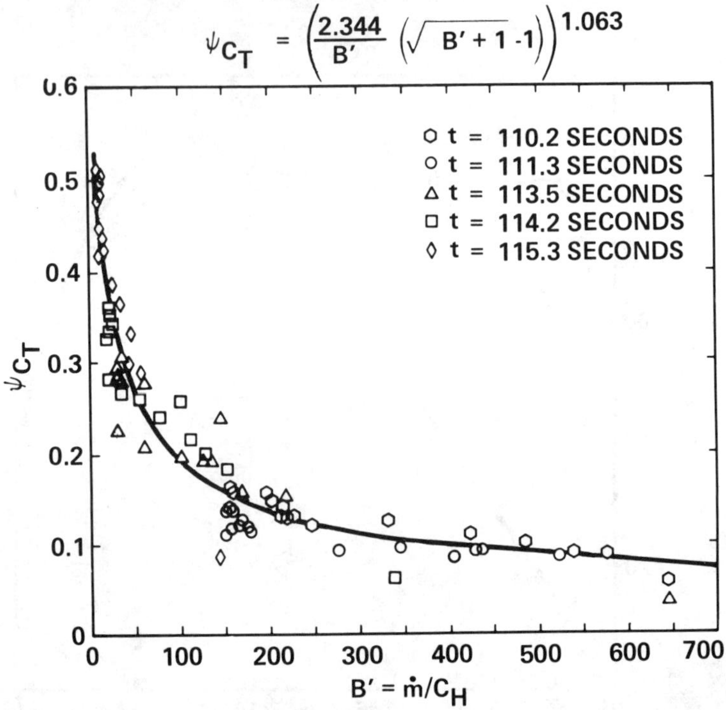

Fig. 4 Convective blockage for turbulent boundary layers.

the heatshield performance in the design studies, accounting for both of these effects.

The definition of the shape change at the periphery of the forebody heatshield is important because it impacts the probe area change and, hence, the variation of the ballistic coefficient. Method-of-characteristics solutions have been obtained to develop pressure-gradient correlations for the periphery region, and these are used in conjunction with the momentum integral equations described in Ref. 1 to calculate the convective heating. The radiative heating is computed by tracking the difference between the shock angle and the local body angle. At the last body point where $\theta_S - \theta_B$ is less than 10 deg, the radiative heat-transfer rate is determined and stored. The radiative heating behind that location on the cylinder is then defined as

$$\dot{q}_{R_{Cyl}} = \dot{q}_{R_{LB}} \cos(\theta_S - \theta_B)$$

where $\dot{q}_{R_{LB}}$ is the radiative heating at the last body point and $(\theta_S - \theta_B) < 10$ deg.

$$\psi R_{T_{PREDICTED}} = \frac{4.289 \times S/RN^{0.3033}}{\left(\frac{\int \dot{m}\, dA}{\int dA}\right)^{0.3871} \times N_s^{0.5382}}$$

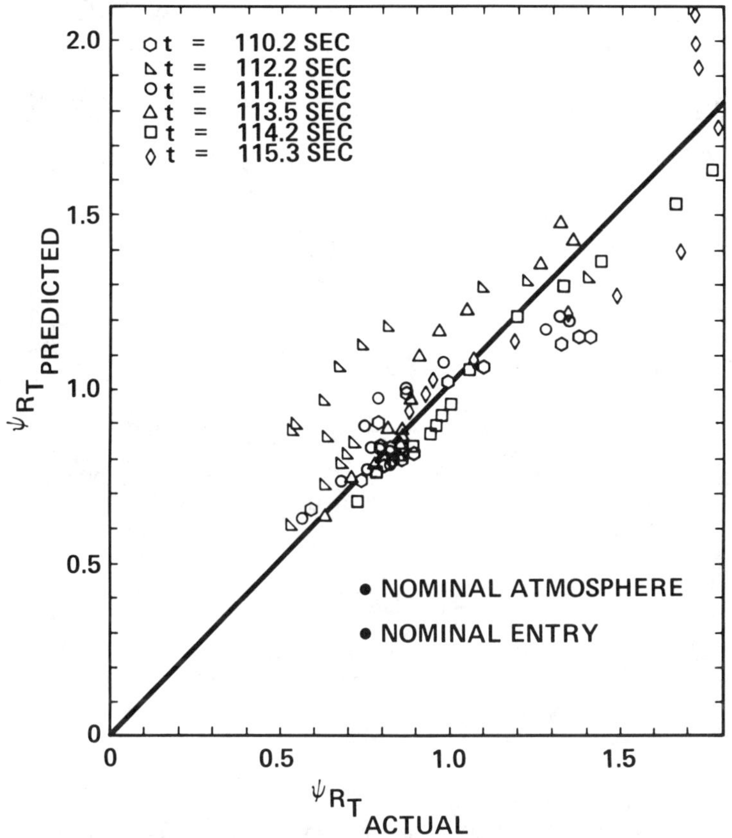

Fig. 5 Radiation blockage.

This reduced thermal environment on the cylindrical section of the forebody results in a smooth shape change.

Forebody Heatshield Design

The primary design criterion for the forebody heatshield is to protect the shield/structure bondline from exceeding a maximum allowable temperature limit of 644 K (700°F) before deceleration module/descent module separation. In addition, the structure maximum allowable temperature is 589 K (600°F).

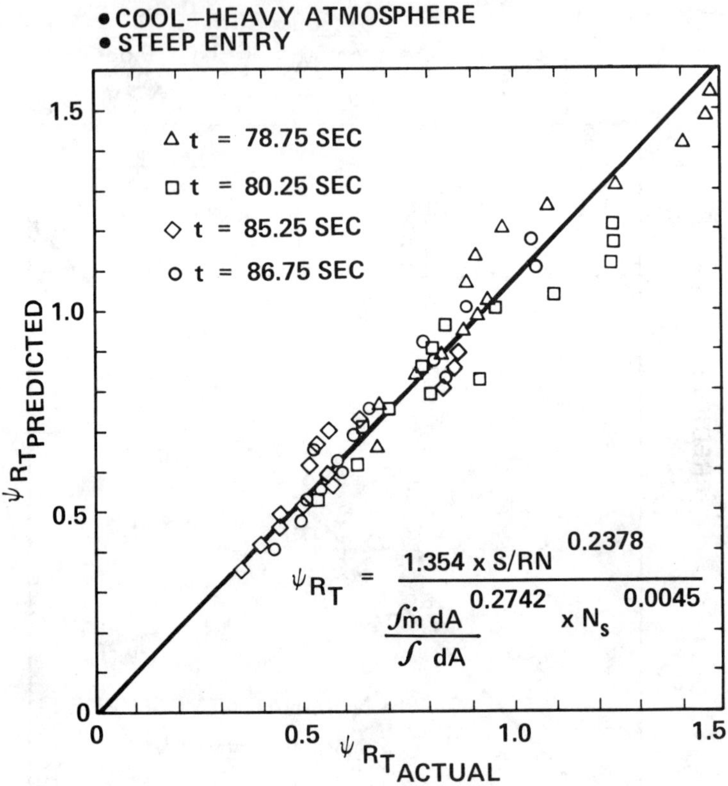

Fig. 6 Radiation blockage.

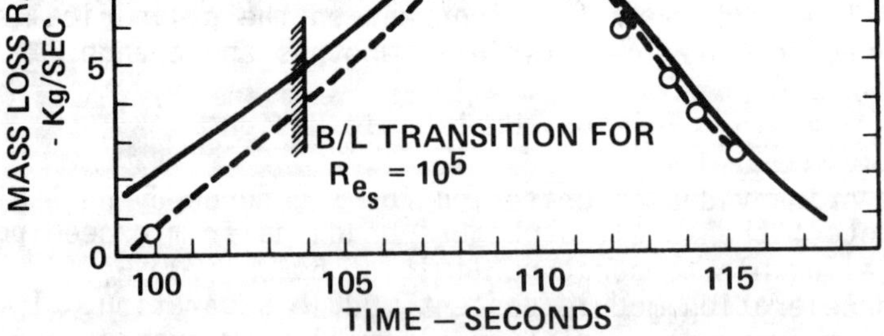

Fig. 7 Comparison of mass loss rates for nominal entry case.

Table 1 Entry conditions and configuration

Entry Conditions

- Altitude = 450 km
- V_{E_R} = 48.2 km/s
- γ_{E_R} = -8.6 (±1.6 deg)
- Entry Latitude = -5.5 deg
- Azimuth Angle = 82 deg
- Atmosphere = 89% H_2 - 11% H_e (by volume)
  (78% H_2 → 100% H_2 (for off-nominal)

Configuration

- θ_c = 44.25 deg
- R_N/R_B = 0.543
- R_N = 13.42 in. (0.341m)
- D_B = 49.4 in. (1.2255m)
- Probe Mass = 290 kg

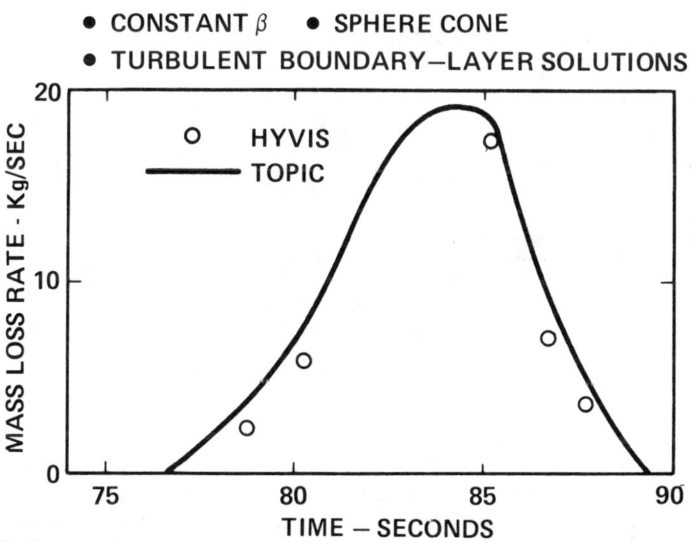

Fig. 8 Comparison of mass-loss rates for cool-heavy steep trajectory.

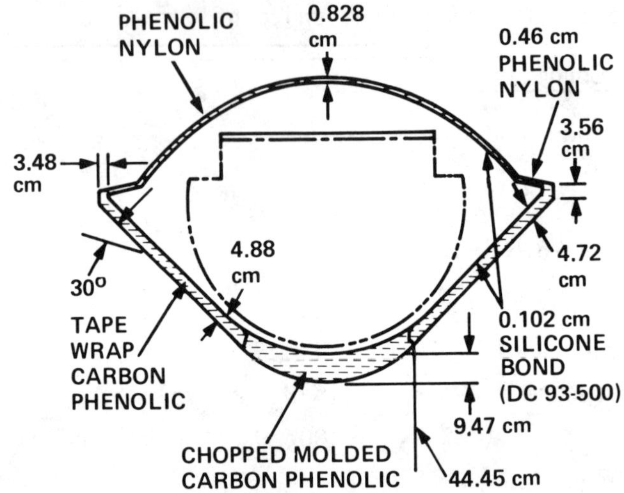

Fig. 9 Heatshield subsystem.

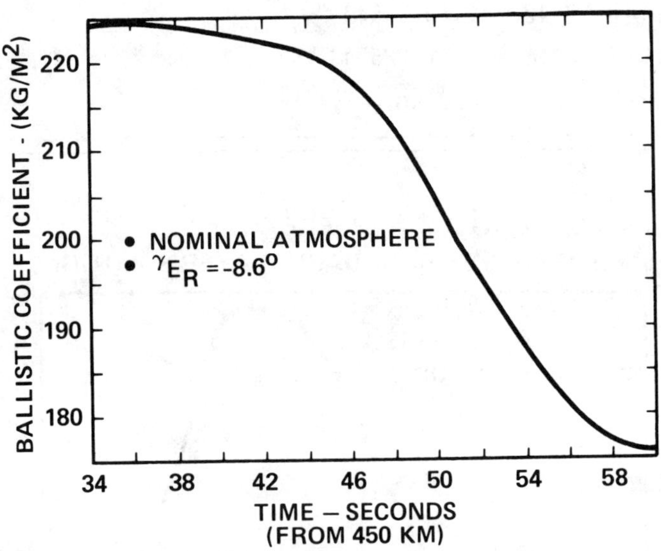

Fig. 10 Ballistic-coefficient history.

The generalized procedure used to define the shield requirements is to define the thermal environment during entry, determine the heatshield recession over the probe forebody, and determine the insulation requirements to maintain the bondline temperature below the allowable limit. Finally, a safety margin is applied to account for uncertainties in environment and material response. Heatshield temperature gradients throughout the probe trajectory are generated for thermostructural analysis of the shield/structure/bond system. The design conditions used in determining forebody heatshield requirements are shown in Table 1 and the Jupiter atmosphere definition is presented in Ref. 14. The heatshield is sized for a 290 kg nominal probe pre-entry mass, and the heatshield subsystem design is shown in Fig. 9.

A conservative approach is used for boundary-layer transition, where a wetted-length Reynolds number of 100,000 is assumed for the transition criterion. This value was selected after reviewing the available blunt-body flight data, as well as relevant wind tunnel data. Specific data for application to this mission do not exist, and the conservative approach used in this design is assumed to be more realistic than that used in the benchmark solutions (i.e., boundary-layer transition occurring very close to the stagnation point at all times during entry). This relatively low-transition Reynolds number yields early transition over the entire frustum. The predicted trajectory for the nominal atmosphere (γ_{ER} = -8.6 deg) accounts for the variable ballistic coefficient shown in Fig. 10. The stagnation pressure and recovery enthalpy histories have been evaluated for this nominal trajectory, and the peak stagnation pressure is about 9 atm.

Stagnation-point heat-transfer rates predicted for the nominal entry conditions are shown in Fig. 11. The radiative heat-transfer rate for zero mass addition peaks at approximately 450 MW/m^2, with the main radiation heat pulse occurring over a time period of about 20 seconds. Accounting for mass addition reduces the radiation to about 150 MW/M^2 at peak heating. This reduction in the radiative heating has a major impact on the probe heatshield design. The convective heating for zero mass addition is also shown in Fig. 11, peaking at a value of about 150 MW/m^2. Massive blowing occurs at the stagnation point, reducing the convective heat transfer to an insignificant level. Thus, at the stagnation point the carbon phenolic heatshield performance is dominated by radiation heat transfer. On the frustum of the probe, the carbon phenolic is subjected to a severe thermal environment that includes radiative and turbulent convective heating, Fig. 12. The turbulent convective heating is approximately one-third of the total incident heat load. The results shown in Fig. 12 indicate that ablation injection does not reduce significantly the radiative heat load on the frustum and, in fact, causes a slight increase in the radiant heat tansfer near peak heating.

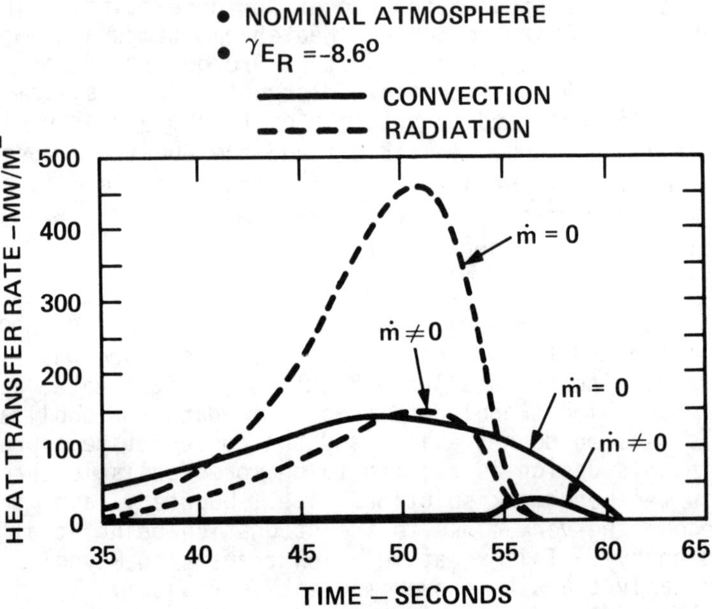

Fig. 11 Stagnation-point heat transfer.

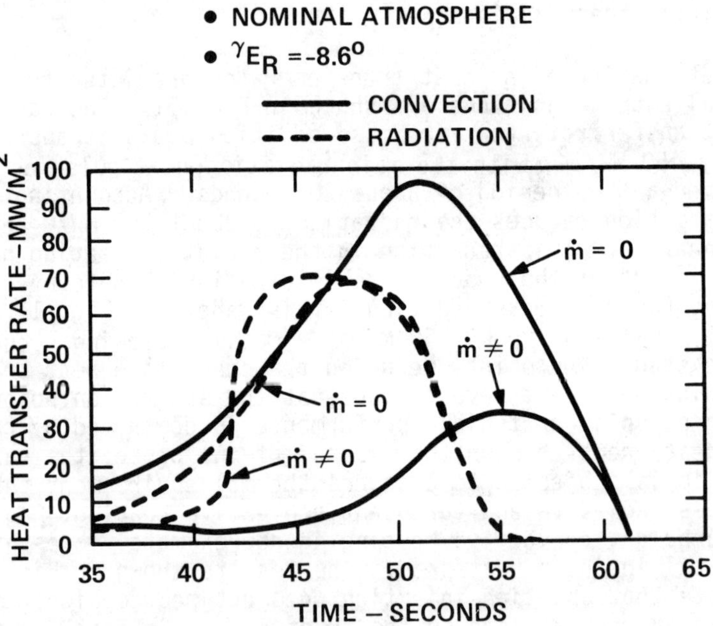

Fig. 12 Frustum heat transfer.

This is because of the turbulence significantly altering the extent and composition of the ablation layer, as discussed by Moss in Ref. 5. The most significant effect is that turbulence brings the high-temperature shock layer gases closer to the surface, thereby causing the highly absorbing molecules to dissociate.

The carbon phenolic heatshield recession profile for the nominal entry is predicted to be a smooth process, and the predicted shape change at the forebody heatshield periphery is shown in Fig. 13, using the techniques described previously.

The insulation requirements have been determined using a one-dimensional heat-conduction code, Reaction Kinetic Ablation Program (REKAP) (Ref. 15) that has been modified to permit shield sizing for the Jupiter probe to be accomplished. Approximately 0.51 - 0.76 cm (0.2 - 0.3 in.) of carbon phenolic is required to satisfy the bondline temperature criterion previously described.

Atmospheric composition and entry-path angle affect the entry heat-transfer rates for this mission. The entry-path angle effects are relatively small and are included in the

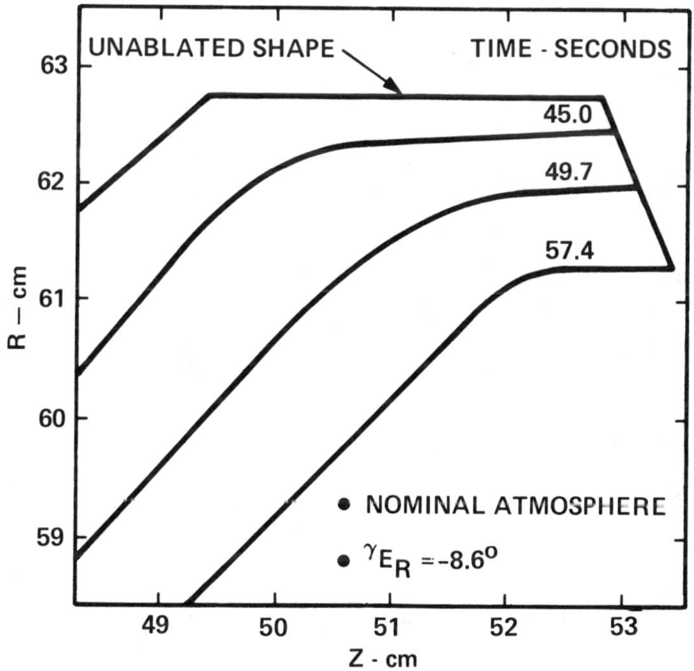

Fig. 13 Predicted peripheral shape change.

shield safety margin, but the effect of atmospheric composition (H_2 - H_e) on the heatshield requirements is much more significant. Figure 14 shows the large increase in the frustum radiative heat-transfer rates as the percentage of hydrogen in the atmosphere is reduced. The temperature in the flowfield increases with increased amounts of helium, since helium has fewer internal degrees of freedom to accommodate the energy compared with hydrogen. The frustum radiative heating is a factor of two greater for the heavy atmosphere (78% H_2) than for the nominal atmosphere (89% H_2), whereas the light atmosphere (100% H_2) yields less than one-half of the radiative heating compared to nominal. The nominal predicted requirements for the forebody heatshield for a nominal entry condition (nominal atmosphere, γ_E = -8.6 deg) have been determined, and safety margins are applied to obtain the flight heatshield thicknesses. The margins were determined by assessing the uncertainties of the primary contributors to the

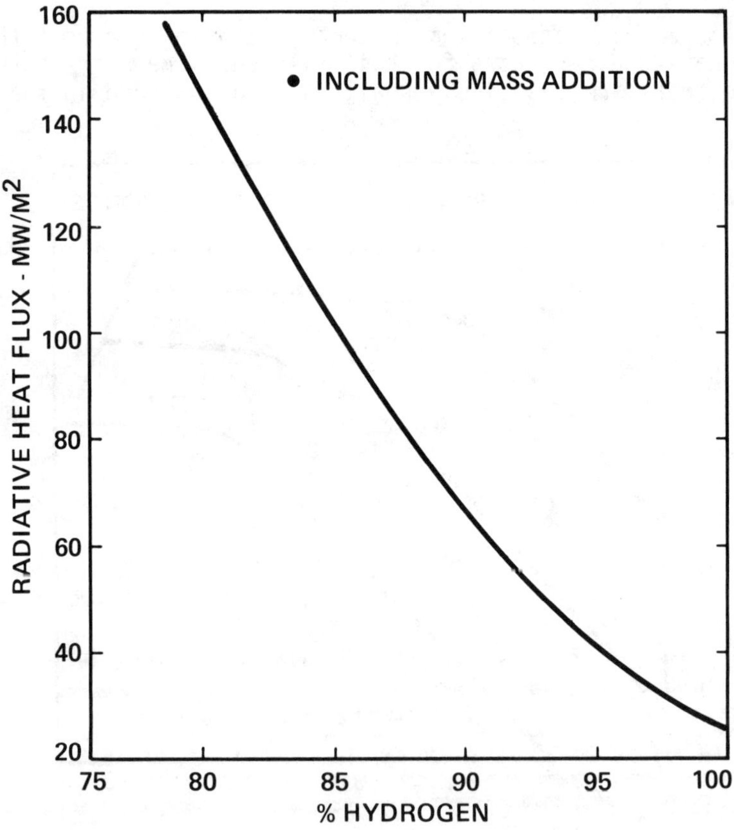

Fig. 14 Frustum radiative heat flux.

heatshield thermal performance (including spallation) and root-sum-squaring those effects to yield the required safety factors. The safety margins used are 1.34 on the frustum and 1.50 at the stagnation point, where the uncertainty of the CMCP heatshield is greater than for the 30 deg TWCP shield used on the frustum.

An assessment has been made of the heatshield design for an encounter with the worst-case thermal environment of the cool-heavy atmosphere, for a steep entry-path angle, $\gamma_E =$ -10.2 deg. Figure 15 shows that for a nominally predicted environment and ablation performance, the heatshield designed for a nominal atmosphere entry with safety margin will survive the worst-case environment with almost zero margin, since the maximum bond and structure temperatures are predicted to be very close to the maximum allowables. Thus, the probe has a 50% probabiity of surviving a cool-heavy steep-entry encounter.

Base Heatshield Design

Base heating to the probe has a significant contribution to the total probe thermal protection requirements and probe mass loss. Before the Jupiter probe program, limited investigations had been conducted to determine the base heating to blunt sphere-cone configurations. This work dealt primarily with the convective heating distribution and has not addressed

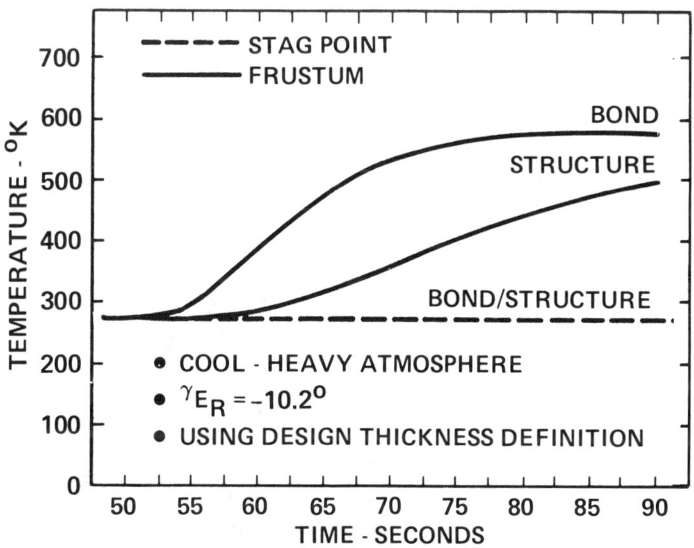

Fig. 15 Bond and structure temperature history.

the contribution to the total heating from the radiative component. To determine the afterbody thermal protection requirements for the Jupiter probe, it was necessary to develop analytical techniques to model the afterbody environment. These techniques have been described in Ref. 16, and more recent modifications have been made to include the effects of the entrainment of the forebody ablation species in the base flow. These techniques are to be reported in the near future.

The primary design criterion for the base thermal protection system is to protect the aft cover/heatshield bondline from exceeding a maximum allowable temperature of 644 K (700 K) before the aft cover extraction by the pilot parachute. The maximum allowable structure temperature is 589 K (600°F).

Predictions have been made for the entry conditions and probe configuration described in Table 1 and Fig. 9. Both de-

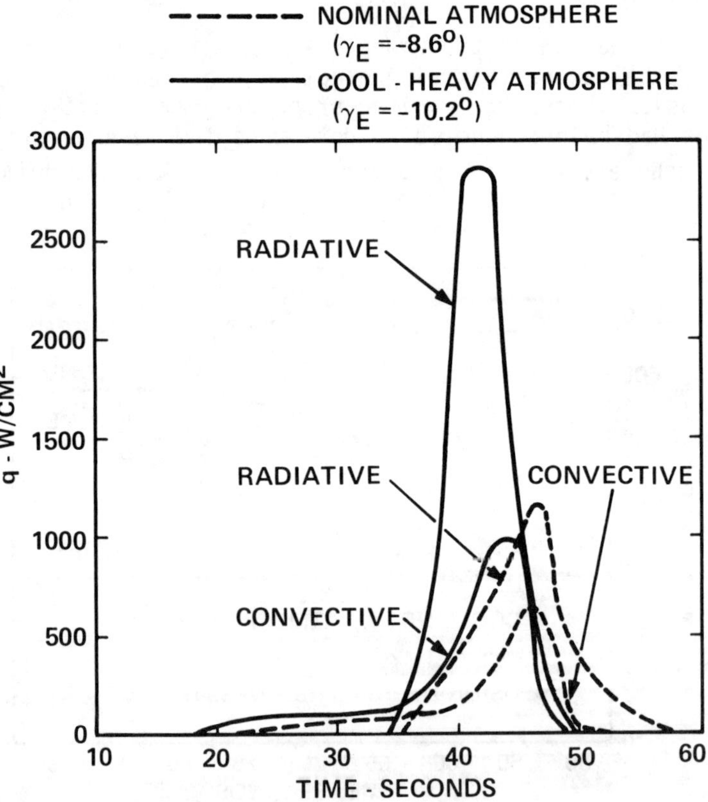

Fig. 16 Base heat-transfer histories ℄ heating.

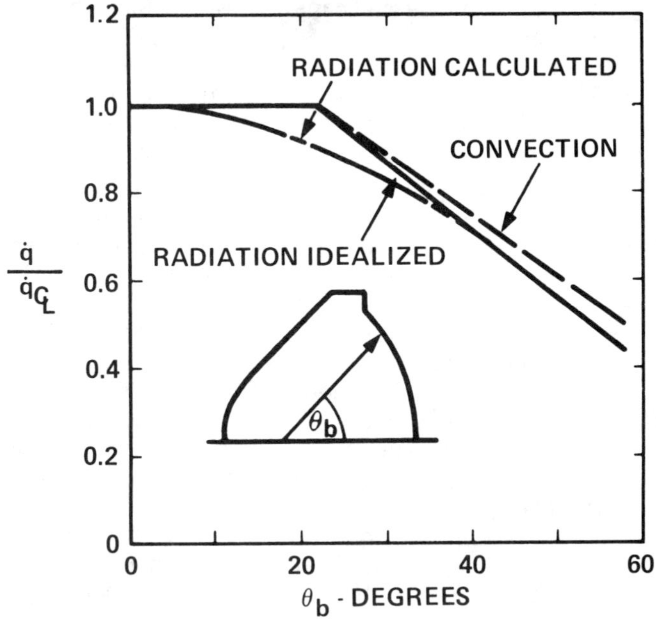

Fig. 17 Heat-transfer distribution at peak heating.

sign conditions, nominal atmosphere, nominal entry-path angle (-8.6 deg) and cool-heavy atmosphere, steep entry-path angle (-10.2 deg) were evaluated for the 290 kg probe.

The convective and radiative heat-transfer rates predicted for the base centerline are shown in Fig. 16 for the nominal atmosphere (γ_E = -8.6 deg), and cool-heavy atmosphere (γ_E = -10.2 deg). The radiative heating is greater than the convective heating for both conditions; however, the heat-transfer rates for the cool-heavy, steep-entry condition are much higher than for the nominal atmosphere entry case. The recommended distribution of the heating around the aft cover is shown in Fig. 17 as a function of body angle, θ_b. It is recommended that both convective and radiative heating be maintained constant to θ_b = 22 deg, with a linear distribution to θ_b = 58 deg, as shown in Fig. 17. This distribution was recommended by Nestler (GE-RSD) based on theoretical calculations of the laminar convective heating distributions over a hemispherical aft cover. The radiative heating distribution was determined as shown in Fig. 17, then simplified to a two-stage linear curve as approximated for the convective heating.

Using the predicted heat-transfer rates previously described, the heatshield thickness distribution over the aft cover has been determined with the Reaction Kinetics Ablation Program (REKAP), Ref. 15. Previous studies resulted in the selection of phenolic nylon as the most suitable material for the aft cover thermal protection, and the thermal model described in Ref. 17 has been used to determine the heatshield thickness, temperature gradients, and recession histories. The one-dimensional analyses were made for the phenolic nylon (PN) bonded to the structure with DC93-500.

A summary of the nominal predictions of the heatshield requirements and recession are shown in Table 2. For the nominal atmosphere (γ_E = -8.6 deg), the PN thickness requirements vary from 0.47 cm (0.187 in.) at the base centerline to 0.27 cm (0.106 in.) at the base edge. The recession varies from 0.37 cm (0.145 in.) to 0.13 cm (0.050 in.). Owing to higher heating rates, the PN requirements for the cool-heavy atmosphere (γ_E = -10.2 deg) entry are greater, varying from 0.78 cm (0.306 in.) at the base centerline to 0.41 cm (0.160 in.) at the base edge. Note that of the 0.78 cm PN required at the base centerline, 0.7 cm is recession, and only 0.08 cm of PN is required to limit the bond and structure temperatures to their maximum allowable values. Current design philosophy is that a safety factor of 1.65 is applied to the nominal predicted recession, with no margin on the insulation requirements. Using this design philosophy yields total PN shield thickness for the nominal atmosphere (γ_E = -8.6 deg) case, which is less than the nominal thickness required for a cool-heavy atmosphere (γ_E = -10.2 deg) entry (with no margin). Thus, this latter entry condition defines the aft cover heatshield thicknesses.

Table 2 Summary of nominal H/S requirements for base

Atmosphere	Path Angle (deg)	Location	Thickness (cm)	Recession (cm)
Nominal	-8.6	℄	0.475	0.368
Nominal	-8.6	Edge	0.269	0.127
Cool-heavy	-10.2	℄	0.777	0.701
Cool-heavy	-10.2	Edge	0.406	0.315

Design criteria — bond: 644 K max; structure: 589 K max.

Figure 18 shows the temperature histories for the PN/bondline interface and the bondline/structure interface for the phenolic nylon nominal design thickness. It can be seen from this curve that the bondline temperature peaks much sooner than the structure temperature. In fact, it occurs soon after peak heating, which is approximately 40-45 s after entry. The structure temperature peaks near parachute deployment for this entry case, 113.5 s for the nominal atmosphere (γ_E = -8.6 deg). Initial heatshield temperature at entry was assumed to be 18°F (-7°C).

Heatshield Design Verification

Verification of the heatshield subsystem design is accomplished through extensive complex analyses, supported and substantiated by comprehensive thermal and mechanical tests. For this reason, a primary emphasis is placed on refining prediction techniques and verifying their accuracy by test wherever possible. The heatshield design validation program is conducted by: 1) refinement of the entry environment and heatshield material-response analytical techniques, and 2) conduction of an ablation test program to provide material-response design data for both heatshield materials and local perturbation areas. Ablation testing has been conducted in the NASA-ARC Giant Planet Arc, NASA-ARC Gas Dynamic Laser, and in the GE-RSD laser facility. The Giant Planet Arc is the only facility in the country with the potential of providing a

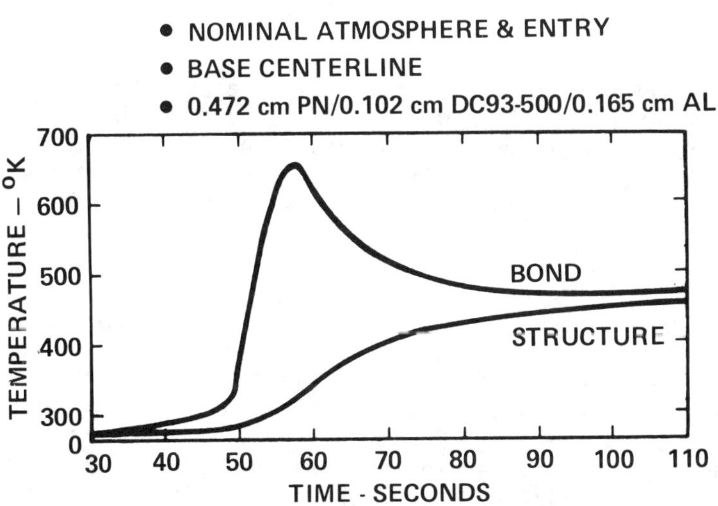

Fig. 18 Bond and structure temperature history.

combined convective/radiative thermal environment in an H_2/H_e atmosphere that can even approach the extremely hostile environment predicted for Jupiter entry. Because of the difficulty in achieving even reasonable simulation of these conditions, a flexible program of heatshield design has been undertaken that can accommodate lack of data or unexpected results.

Analysis of the heatshield material response to the intense environment of entry will involve use of data and techniques evolved by the entire technical community. This will include coupling of codes already available at GE-RSD and at NASA/ARC and use of updated analytical information and test data emanating from NASA/ARC. The analytical techniques will be calibrated continuously against results from the test program. As additional data and/or techniques are developed, analyses will assess the impact on heatshield design margins and/or risks. It is anticipated that a continuing update on thickness and/or margins will occur as the program progresses.

A base heatshield test program is also being conducted to verify the phenolic nylon ablation model in the H_2/H_e environment as well as in a laser radiation environment. These tests are conducted in NASA/ARC and GE-RSD facilities, and the tests are performed to verify the heatshield performance and qualify the thermal protection design in attchment, access, and shield penetration regions. In addition to technique development and ablation test programs, both heatshields are structurally qualified for the mission by complete destructive testing of a prime heatshield (process approval), and systems-level maximum-loads testing when attached to their respective structural subsystems.

By following the approach outlined above, the flight model heatshield will be qualified and verified by the best analytical and experimental techniques available. It will contain design margins sufficient to account for all remaining unknowns, therefore permitting successful mission completion.

Conclusions

The thermal protection systems for the forebody and aft cover of the Galileo probe are carbon phenolic and nylon phenolic, respectively. These materials satisfy the requirements of the mission and are validated by extensive ground and flight test experience. Unique design techniques have been developed to ensure that the thermal environment of the probe is well characterized and that the sensitivity of the heatshields to the potential variations in atmosphere compositon and entry conditions is evaluated. Adequate thermal margins have been applied to the nominal heatshield requirements to ensure a high probability of mission success for an encounter with the nominal atmosphere. The current design described in this paper has zero margin for an encounter with the cool-heavy atmosphere, steep-entry (-10.2 deg) condition.

References

[1]Brewer, R. A., Brant, D. N., and Fogaroli, R. P., "Development of a Steady-State Shape Change Ablation Code for the Design of Outer Planet Probes," AIAA Paper 77-95, Jan. 1977.

[2]Zoby, E. V., Sutton, K., Olstad, W. B., and Moss, J. N., "An Approximate Inviscid Radiating Flow-Field Analysis for Outer Planet Entry Probes," AIAA Paper 78-189, Jan. 1978.

[3]Moss, J. N., Anderson, E. C., and Bolz, C. W., Jr., "Viscous-Shock Layer Solutions with Radiation and Ablation Injection for Jovian Entry," AIAA Paper 75-671, May 1975.

[4]Moss, J. N., Anderson, E. C., and Bolz, C. W., Jr., "Aerothermal Environment for Jovian Entry Probes," AIAA Paper 76-469, July 1976.

[5]Moss, J. N., Simmonds, A. L., and Anderson, E. C., "The Impact of Turbulence on a Radiating Shock Layer with Coupled Ablation Injection," AIAA Paper 78-1186, July 10-12, 1978.

[6]Moss, J. N., "A Study of the Aerothermal Entry Environment for the Galileo Probe," AIAA Paper 79-1081, AIAA 14th Thermophysics Conference, June 4-6, 1979.

[7]Moss, J. N., "Flow Field Analysis for Future Planetary Entry," AIAA Paper 79-0918, Hampton, Va., May 8-11, 1979.

[8]Daywitt, J., Brant, D. N., and Bosworth, F., "Computational Technique for Three-Dimensional Inviscid Flow Fields About Re-entry Vehicles," Vol. I - Numerical Analysis, Report SAMSO TR-79-5, April 1978.

[9]Maslen, S. H., "Inviscid Hypersonic Flow Past Smooth Symmetric Bodies," AIAA Journal, Vol. 2, No. 6, June 1964, pp. 1055-1061.

[10]Olstad, W. B., "Nongray Radiating Flow About Smooth Symmetric Bodies," AIAA Journal, Vol. 9, No. 1, Jan. 1971, pp. 122-130.

[11]Nicolet, W. E., "User's Manual for the Generalized Radiation Transfer Code (RAD/EQUIL)," CR-116353, NASA, 1969.

[12]Fogaroli, R. P., and Brant, D. N., "Re-evaluation of Graphite Thermochemical Ablation," GE TFM-9151-060, Oct. 1968.

[13]JANAF Thermochemical Tables - Second Edition, NSRDS - NB537, U.S. Government Printing Office, Washington, D. C., June 1971.

[14]Jupiter Atmosphere Thermodynamic State Properties Models, NASA Document JP-590.04, Jan. 1978.

[15]Gordon, P., "Analysis of One-Dimensional Heat Conduction Computer Program," GE-RSD TIS-R66S D10, Mar. 1966.

[16]Brant, D. N., and Nestler, D. E., "Development of an Afterbody Radiative and Convective Heating Code for Outer Planet Probes," AIAA Paper 78-882, 2nd AIAA/ASME Thermophysics and Heat Transfer Conference, Palo Alto, Calif., May 1978.

[17]Kottock, S., "The Development of a Phenolic Nylon Model for a Reaction Kinetics Ablation Program," GE-RSD TIS-655D-320, Sept. 1965.

VISCOSITY OF MULTICOMPONENT PARTIALLY IONIZED GAS MIXTURES ASSOCIATED WITH JOVIAN ENTRY

Bassem F. Armaly*
University of Missouri-Rolla, Rolla, Mo.

and

Kenneth Sutton†
NASA Langley Research Center, Hampton, Va.

Abstract

An approximate method is proposed for predicting the viscosity of partially ionized gas mixtures. This technique expresses the viscosity of a mixture in terms of the viscosities of the individual pure components, is simple in form, and does not require large computer run times nor storage. Thus, the technique is suitable for use with complex flowfields and heat-transfer calculations. Results for gas mixtures which are representative of the atmospheres of Jupiter, Earth, and Venus are presented and it is shown that the results compare favorably with detailed kinetic theory analysis.

Introduction

The flowfield and the heat transfer to bodies re-entering the Earth's atmosphere and entering the atmospheres of the planets are predicted by detailed and complex computer codes.[1-4] The high temperatures associated with these missions ionize and dissociate the surrounding gases, causing a large number of different molecular species to exist in the boundary layer and in the viscous shock layer that surrounds the probe. For

---

Presented as Paper 80-1495 at the AIAA 15th Thermophysics Conference, Snowmass, Colo., July 14-16, 1980.

*Professor, Thermal Radiative Transfer Group, Department of Mechanical and Aerospace Engineering.

†Research Leader, Aerothermodynamics Branch, Space Systems Division.

example, during the entry phase of the Jupiter probe, temperatures of 20,000 K can be experienced and when ablation products are considered, over 20 different molecular species can exist in the boundary layer of the probe[5] (H_2, H, H^+, He, He^+, C, C_2, C_3, C^+, C_2H, C_3H, C_4H, C_2H_2, O_2, O, O^+, CO_2, CO, N_2, N, N^+, CN, HCN, e^-, etc.). The viscosity of such a mixture is needed in order to predict the flowfield and the heat transferred to the probe. A recent study[5] indicated a 25% to 40% difference in convective heating rates due to different models of predicting the transport properties of the mixture.

Several investigators have examined the viscosity of partially ionized gas mixtures. Biolsi[6] reported on the viscosity of H_2-He gas mixtures (the atmosphere of Jupiter), Yos[7] reported on the viscosity of air (the Earth atmosphere) and Freeman and Oliver[8] reported the viscosity of CO_2-N_2 gas mixtures (the atmosphere of Venus). They utilized the Chapman-Enskog[9] kinetic theory formulas and provided appropriate collision integrals for the various interactions to predict the viscosities of these mixtures at high temperatures. These detailed studies require large computer run times and storage, and thus it is not practical to include them directly in the existing detailed and complex flowfield codes. They serve, however, as a reference to evaluate proposed approximate methods for predicting these properties.

Several approximate methods have been proposed in the literature[10-15] to evaluate the viscosity of gas mixtures. The most widely used approximation is the one by Wilke.[12] This approximation, however, has been shown to fail drastically at high temperatures or specifically when ionization occurs in the mixture.[16]

The purpose of this paper is to present a new approximate method for predicting the viscosities of gas mixtures which is similar to Wilke's approximation in that it requires only the viscosities of the pure components and is easy to use. However, the present method is more accurate at conditions where ionization occurs in the gas mixture.

Development

The viscosity of a multicomponent, n components, gas mixture is given[9] in terms of the ratio of two determinants as follows:

$$\eta_m = (-1)\frac{\begin{vmatrix} H_{11} & H_{12} & H_{13} & \cdots & H_{1n} & x_1 \\ H_{12} & H_{22} & H_{23} & \cdots & H_{2n} & x_2 \\ H_{13} & H_{23} & H_{33} & \cdots & H_{3n} & x_3 \\ \cdot & \cdot & \cdot & & \cdot & \cdot \\ \cdot & \cdot & \cdot & & \cdot & \cdot \\ \cdot & \cdot & \cdot & & \cdot & \cdot \\ H_{1n} & H_{2n} & H_{3n} & \cdots & H_{nn} & x_n \\ x_1 & x_2 & x_3 & \cdots & x_n & 0 \end{vmatrix}}{\begin{vmatrix} H_{11} & H_{12} & H_{13} & \cdots & H_{1n} \\ H_{12} & H_{22} & H_{23} & \cdots & H_{2n} \\ H_{13} & H_{23} & H_{33} & \cdots & H_{3n} \\ \cdot & \cdot & \cdot & & \cdot \\ \cdot & \cdot & \cdot & & \cdot \\ \cdot & \cdot & \cdot & & \cdot \\ H_{1n} & H_{2n} & H_{3n} & \cdots & H_{nn} \end{vmatrix}} \qquad (1)$$

where

$$H_{ii} = \frac{x_i^2}{\eta_i} + \sum_{\substack{k=1 \\ k \neq i}}^{n} \frac{2x_i x_k}{\eta_{ik}} \frac{M_i M_k}{(M_i + M_k)^2} \left[\frac{5}{3A^*_{ik}} + \frac{M_k}{M_i}\right] \qquad (2)$$

$$H_{ij} = -\frac{2x_i x_j}{\eta_{ij}} \frac{M_i M_j}{(M_i + M_j)^2} \left[\frac{5}{3A^*_{ij}} - 1\right] \quad i \neq j \qquad (3)$$

$$\eta_i = C\sqrt{MT}/\sigma_{ii}^2 \Omega_{ii}^{(2,2)*} \qquad (4)$$

$$A^*_{ij} = \Omega^{(2,2)*}_{ij}/\Omega^{(1,1)*}_{ij} \tag{5}$$

$$\eta_{ij} = C\ \sqrt{2M_iM_jT/(M_i + M_j)}/\sigma^2_{ij}\Omega^{(2,2)*}_{ij} \tag{6}$$

and C is a constant that depends on the units used; x_i is the mole fraction of the ith component; M_i is the molecular weight of the ith component; T is the temperature; η_i is the viscosity of the pure i^{th} component; A^*_{ij} is the quantity defined in Eq. (5); η_m is the viscosity of the mixture; H_{ii},H_{ij} are quantities defined in Eqs. (2) and (3); and σ_{ij} is the collision diameter associated with interaction ij.

The functions $\Omega^{(1,1)*}_{ij}$ and $\Omega^{(2,2)*}_{ij}$ are the dimensionless collision integral which describes the force field and the interaction between two particles in a potential field (in the case of pure gas, the two particles are identical to each other). The numerical evaluation of the collision integral is difficult, but possible, if the potential function is specified. The exact form of the potential function may not be known and then an approximate form is used to model the interaction[6-8]. Each state is described by a different potential energy function and the collision integrals for the interaction are obtained by averaging over the collision integrals for each state.

The difficult task in utilizing the above relations for calculating the viscosity of a mixture is the evaluation of collision integrals for the various interactions between similar and dissimilar particles. Each interaction requires a potential function that describes accurately the force field resulting from the relative motion between the two particles. For example, a 3 component gas mixture requires 6 interactions and a 27 component mixture (simulating ablation products) requires 378 interactions. The complexities and the large number of interaction potentials which are required for calculating the viscosity make the use of the above detailed formulation impractical for use in a complex flowfield calculation. Only recently, Biolsi[6] applied the above formulation to evaluate the viscosity of H_2-He gas mixtures at high temperatures. This mixture generates 21 different interactions and it is the only exact calculation of a mixture viscosity with that many interactions.

The expression for the viscosity of a mixture, Eq. (1), can be replaced by an infinite series equivalent to[10]

$$\eta_m = \sum_{i=1}^{n} \frac{x_i^2}{H_{ii}} - \sum_{i=1}^{n} \sum_{\substack{j=1 \\ j\neq i}}^{n} \frac{x_i x_j H_{ij}}{H_{ii} H_{jj}} + \sum_{i=1}^{n} \sum_{\substack{j=1 \\ j\neq i}}^{n} \sum_{\substack{k=1 \\ k\neq i}}^{n} \frac{x_j x_k H_{ij} H_{ik}}{H_{ii} H_{jj} H_{kk}} - \cdots \tag{7}$$

From detailed calculations,[17] the off-diagonal elements of the matrix H_{ij} is seen to be much smaller than the diagonal element H_{ii}. In fact, if one assumes that the magnitude of $A^*_{ij} = 5/3$,[11] then H_{ij} becomes identically equal to zero. This assumption is very poor for some ion-atom interactions,[17] but it is reasonable for neutral atoms and molecules interactions. Many existing approximations utilize this assumption, $A^*_{ij} = 5/3$, to simplify the expression for the mixture's viscosity from an infinite series to only one sum which is given by

$$\eta_m = \sum_{i=1}^{n} \frac{x_i^2}{H_{ii}} \tag{8}$$

In order to utilize Eq. (8), values for A^*_{ij} and $\sigma^2_{ij}\Omega^{(2,2)*}_{ij}$ should be made available for all the needed interactions. These values, however, are not normally available in the literature. Biolsi[17] reported the behavior of these functions for the case of H_2-He gas mixtures in the temperature range between 1000 K to 25,000 K and accounted for both dissociation and ionization by examining the 21 different types of interactions that occur in this mixture. This study and the one by Yun and Mason[18] provide the only detailed data of this type in the literature that are associated with interactions of a specific mixture. The following observations can be made:

1) The parameter A^*_{ij} can be approximated as being independent of temperature, but its magnitude differs for different types of interactions and it is often different than the value of 5/3 assumed in some approximations.

b) The parameter $\sigma_{ij}^2\Omega_{ij}^{(2,2)*}$ is a fairly strong function of temperature and its magnitude differs for different types of interactions.

The last feature is undesirable because it requires a different functional relation for each interaction. To remedy this behavior, it is proposed to express σ_{ij}^2 in terms of the pure gas collision diameter σ_i, σ_j, and the pure gas viscosity, η_i, η_j, as follows:

$$\sigma_{ij}^2 = \frac{(\sigma_i + \sigma_j)^2}{4} = \frac{CT^{1/2}}{4}\left[\left(\frac{M_i^{1/2}}{\eta_i\Omega_{ii}^{(2,2)*}}\right)^{1/2} + \left(\frac{M_j^{1/2}}{\eta_j\Omega_{jj}^{(2,2)*}}\right)^{1/2}\right]^2 \tag{9}$$

The advantage of this form is the fact that the parameter σ_{ij} is expressed in terms of individual pure gas properties, which are more readily available than the ones associated with interactions between two different species. By using the above relation in the expression for η_{ij}, Eq. (6) becomes

$$\eta_{ij} = [32M_iM_j/M_i + M_j]^{1/2} \Big/ [(M_i^{1/4}/\eta_i^{1/2})F_{ij} + (M_j^{1/4}/\eta_j^{1/2})B_{ij}]^2 \tag{10}$$

where

$$F_{ij} = \left\{\Omega_{ij}^{(2,2)*}/\Omega_{ii}^{(2,2)*}\right\}^{1/2} \qquad B_{ij} = \left\{\Omega_{ij}^{(2,2)*}/\Omega_{jj}^{(2,2)*}\right\}^{1/2}$$

The advantage of this form over Eq. (6) is the fact that the coefficients F_{ij}, B_{ij}, are not strong functions of temperature and, for most cases, they can be assumed to be independent of temperature.

The Wilke approximation[12] can be related to the above development if the following values are assumed for the magnitudes of the three constants:

$$A_{ij}^* = 5/3$$

$$F_{ij} = 1.0$$

$$B_{ij} = 1.0$$

If these values are substituted in Eqs. (2) and (10), then Eq. (8) reduces to the well-known Wilke form which is given by

$$\eta_m = \sum_{i=1}^{n} \frac{x_i M_i^{1/2}}{\sum_{j=1}^{n} \frac{x_j M_j^{1/2}}{[8(M_i + M_j)]^{1/2}} \left[\frac{M_i^{1/4}}{\eta_i^{1/2}} + \frac{M_j^{1/4}}{\eta_j^{1/2}}\right]^2} \qquad (11)$$

Additional and less realistic assumptions are, in some cases[10,11] imposed on the detailed development to generate simpler expressions for calculating the viscosity of mixtures. These expressions could be quite adequate over a limited temperature and composition range, but should not be extended to regions where exact data are not available for comparisons and evaluations.

Proposed Approximation

Equations (2), (8), and (10) are the governing relations for the proposed approximation. The viscosities of the pure components are assumed to be known and the magnitude of A^*_{ij}, F_{ij}, B_{ij}, are needed in order to utilize these relations. The last two parameters are not normally tabulated in the literature, and since the objective of this study was to develop a simple approximation for use in computing the viscosity of mixtures in a complex flowfield, the final expression must remain simple. Various methods were examined to approximate the behavior of these parameters and the following is the one which was finally adopted for this study. The exact results of Biolsi[17] (collision integrals in the temperature range between 1000 K to 25,000 K) for the various interactions in a hydrogen-helium gas mixture were used to calculate the magnitude of η_{ij} as a function of temperature from Eq. (6). A value of unity was imposed on the magnitude of F_{ij} in Eq. (10). Equation (10) was then used to determine the magnitude of B_{ij} at the various temperatures. Surprisingly, the magnitude of B_{ij} was found to be a weak function of temperature, and in fact, could be considered constant for most of the interactions. Another important feature which appeared in this approximation is the fact that B_{ij} has the same magnitude for similar types of interactions, i.e., atom-atom, ion-atom, etc.

The off-diagonal member in Eq. (1), H_{ij}, was assumed to be negligible without forcing the magnitude of A^*_{ij} to be equal to 5/3. Exact results[17] indicate that indeed H_{ij} is significantly smaller than H_{ii} and that the magnitude of A^*_{ij}

could be significantly different than 5/3, especially for some ion-atom interactions. The exact results[17] were used to estimate a magnitude of this constant A^*_{ij} and it was assumed to be independent of temperature.

The above procedure generated the following guidelines for the values of these constants:

1) $A^*_{ij} = A^*_{ji} = 1.25$ for all interactions except atom interactions with its own ion. For that case, the following values are recommended:

$$A^*_{H-H^+} = 0.18$$

$$A^*_{He-He^+} = 0.025$$

$$A^*_{C-C^+} = 1.10$$

For other interactions of atoms with their own ions, the value can be assumed to be equal to that of $C - C^+$ interactions.

2) $F_{ij} = F_{ji} = 1.0$ for all interactions.

3) $B_{ij} = B_{ji}$ = as outlined below:

= 0.2 for atoms and molecules interactions with electrons

= 0.15 for atoms and molecules interactions with ions

= 0.78 for atoms and molecules interactions with atoms and molecules

= 1.0 for all other interactions

The above results were generated from the limited exact collision integrals data of Biolsi[17] and of Yun and Mason.[18] They are assumed to be suitable for other types of interactions where exact data are not available. They should be considered somewhat empirical in nature and their suitability should be confirmed by comparison with exact results as is done in the Results and Discussion section. For some interactions, the magnitude of B_{ij} might vary slightly with temperature, and if exact collision integrals data are available (as is the case for hydrogen-helium interactions), this functional form could be determined and can be used in the calculations without complicating the evaluation procedure.

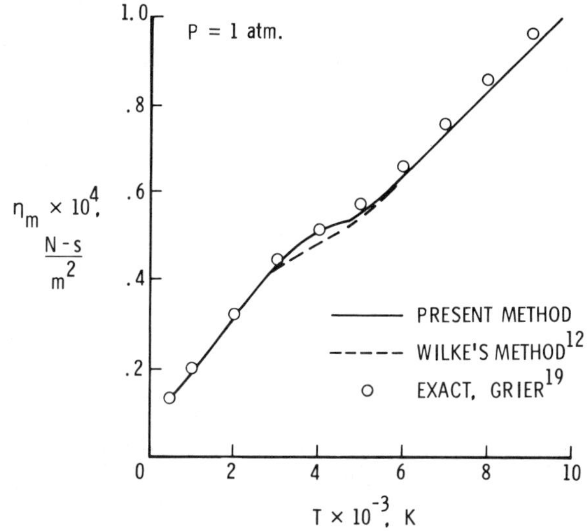

Fig. 1 Viscosity of dissociating hydrogen (H_2, H).

The proposed approximate method consists of utilizing the stated values for the three constants (A^*_{ij}, F_{ij}, B_{ij}) and Eqs. (2), (8), and (10) to evaluate the viscosity of the mixture. This method appears to offer significant improvement over the Wilke's approximation.[12]

Results and Discussion

The proposed approximate method was used to predict the viscosity of various partially ionized and dissociated gas mixtures. The results were compared with Wilke's approximation[12] and with exact kinetic theory analyses.[6-8,19,20]

The viscosities of dissociated hydrogen and of ionized atomic hydrogen are presented in Figs. 1 and 2, respectively, as a function of temperature. The results are for an equilibrium composition at a pressure of 1 atm. The first mixture involves the interactions of two species, and the second mixture involves the interactions of three species. It is clear from these figures that Wilke's approximation[12] does predict with good accuracy the viscosity of dissociated hydrogen, but it fails for ionized hydrogen. The good agreement between the results by the proposed method and the exact values is due to the fact that all the associated collision integrals for these mixtures have been published by Biolsi[17] and the role of each interaction could be approximated with

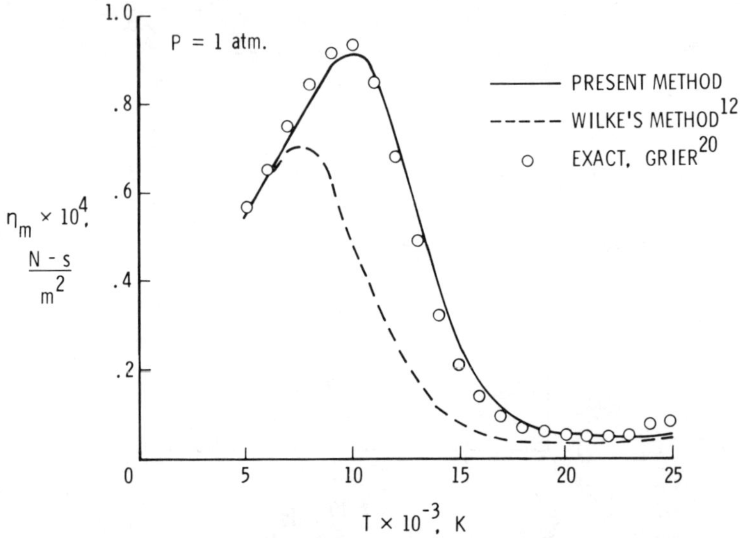

Fig. 2 Viscosity of ionized atomic hydrogen (H, H^+, e^-).

good accuracy. For example, the parameter B_{ij} for the hydrogen atom-hydrogen ion interaction was found to vary slightly with temperature as given by the following approximation:

$$B_{H-H^+} = 0.01 + (7 \times 10^{-6})T$$

where T is the temperature in degrees Kelvin. This relation was used in this study to describe this interaction in place of the proposed constant value. That is the only parameter and the only interaction requiring variation with temperature. The viscosities of all the individual species were calculated by using Eq. (4) and the published collision integrals.[17]

The viscosity of an 0.89 H_2 - 0.11 He mixture, by volume, is presented as a function of temperature in Fig. 3. The results are for equilibrium compositions at the various pressures. The Wilke approximation[12] deviates from the exact values when ionization begins to occur in the mixture. The proposed approximation provides excellent agreement with the exact values. Here again, this good agreement is due to the fact that all the associated collision integrals for this mixture have been published[17] and, as such, they were used to establish the appropriate values for the constants in the proposed approximation. The viscosities of the individual species were calculated by using Eq. (4) and the published[17] col lision integrals for these interactions. This mixture received significant attention recently in the open literature because it represents the atmospheric composition of Jupiter.

The proposed approximation was applied to gas mixtures representative of the Earth atmosphere (air) and a Venus at-

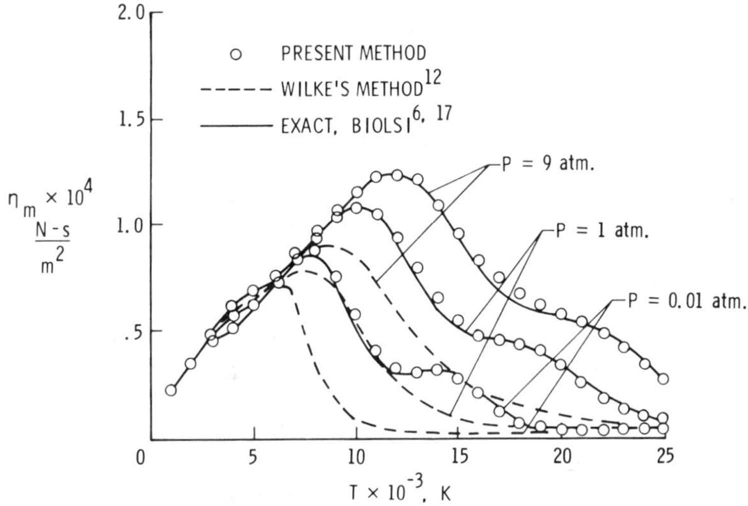

Fig. 3 Viscosity of a hydrogen-helium mixture, 0.89 H_2 - 0.11 He by volume (H_2, H, H^+, He, He^+, e^-).

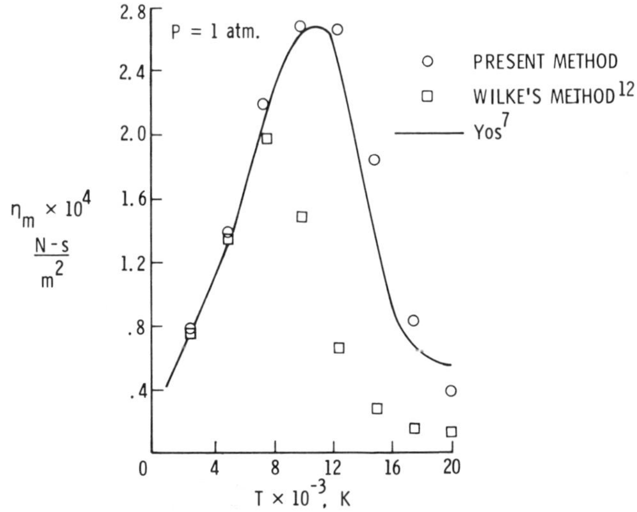

Fig. 4 Viscosity of air (N_2, O_2, NO, N, O, NO^+, N^+, O^+, e^-).

mosphere model (0.50 CO_2 - 0.50 N_2 by volume). The first mixture involves the interactions of nine different species and the second involves the interactions of 15 different species. These mixtures were selected for comparisons because detailed (but not exact) kinetic theory analyses have been performed by Yos[7] and by Freeman and Oliver,[8] For these mixtures, many of the associated collision integrals are not available in the literature. The following assumptions were employed to evaluate the viscosities of the individual components which are needed in the proposed approximation:

1) The collision integrals for all the ion-ion interactions were assumed to be equivalent to the available H^+ - H^+ interactions.[17]

2) Lennard-Jones potentials were used on some atomic and molecular species for which exact collision integral data or viscosities are not available.

The results are presented as a function of temperature for an equilibrium composition at a pressure of one atmosphere in Figs. 4 and 5. They clearly demonstrate again the weakness of the Wilke approximation and the accuracy of the proposed approximation. Wilke's approximation deviates from the detailed results when ionization occurs in the mixture.

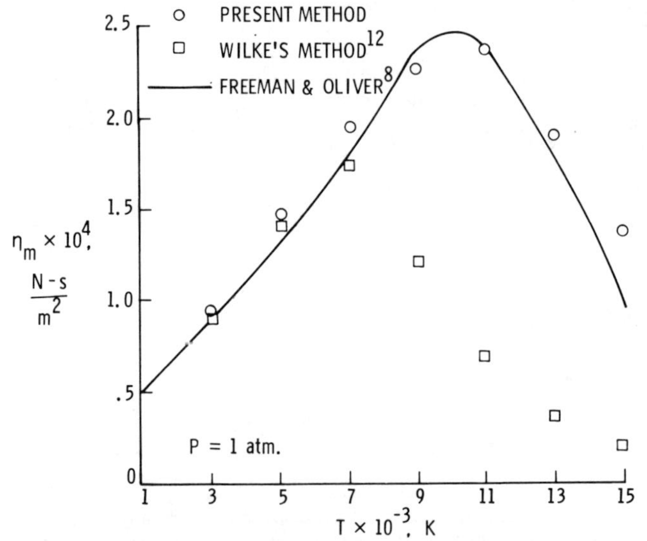

Fig. 5 Viscosity of a carbon dioxide-nitrogen mixture, 0.5 CO_2 - 0.5 N_2 by volume (CO_2, CO, CN, C_2, N_2, O_2, NO, C, N, O, NO^+, C^+, N^+, O^+, e^-).

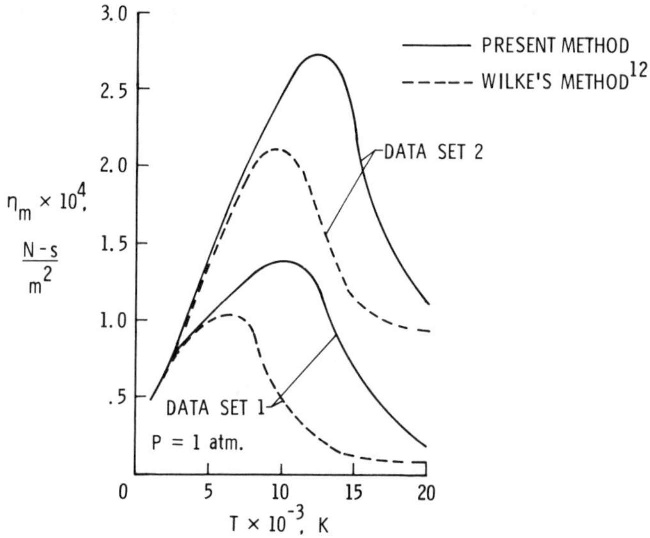

Fig. 6 Viscosity of an ablation products mixture, 0.92 C - 0.06 O - 0.02 H by elemental mass fractions (C_2H, C_2H_2, C_3H, C_4H, O_2, CO, CO_2, C_2, C_3, H_2, O, C, H, O^+, C^+, H^+, e^-).

The present method was also applied to predict the viscosities of a mixture consisting of 0.92 carbon, 0.06 oxygen, and 0.02 hydrogen by elemental mass fractions (the composition of a carbonphenolic heat shield such as proposed for the Jupiter probe).[21] The results are presented as a function of temperature for an equilibrium composition at a pressure of one atmosphere in Fig. 6. The figure presents a comparison between the present method and the Wilke method using two sets of data to calculate the viscosities of the individual species. This mixture generates 17 different species and for many of these species, the exact collision integrals and viscosities are not available in the literature. The assumptions which were used for the previous two mixtures were also employed here to calculate the viscosities of the individual components and the data are referred to as "Data Set 1." The Wilke method and the individual viscosity data referred to as "Data Set 2" represent that used by Moss,[21] et. al. in detailed calculations of viscous shock-layer flow around the Jupiter probe. The viscosities of the individual species, especially ions and electrons, appear to be very poorly predicted in the existing computer code[21] and the data for the ions and electrons are responsible for the difference between the results for the two sets of data. The figure demonstrates, as expected, that the choice of the mixing model and the choice of the viscosities for the individual species affect significantly the predicted mixture's viscosity. Exact results for this mixture are not available for comparisons.

Conclusion

A new approximate method is proposed for evaluating the viscosity of partially ionized gas mixtures. Results calculated with the method compare reasonably well with results calculated by exact and detailed kinetic theory analyses. For ionized gas mixtures, the present approximate method provides a significant improvement over the widely used Wilke approximation without significantly increasing the complexity of the calculations. It is recommended that the present method be considered for use in flowfield calculations where ionization occurs in the gas mixture.

References

[1]Moss, James N., "Stagnation and Downstream Viscous Shock Layer Solution With Radiation and Coupled Ablation Injection," AIAA Paper 74-73, Jan. 1974.

[2]Kumar, Ajay, Tiwari, S. N., and Graves, Randolph A., Jr., "Effect of Small Angle of Attack on the Radiating Viscous Shock Layer Solutions for Jovian Entry," Outer Planet Entry Heating and Thermal Protection, Progress in Astronautics and Aeronautics, Vol. 64, edited by R. Viskanta, AIAA, New York, 1979, pp. 147-164.

[3]Sutton, Kenneth, "Characteristics of Coupled Nongray Radiating Gas Flow With Ablation Product Effects About Blunt Bodies During Planetary Entries," Ph.D. Dissertation, North Carolina State University, Raleigh, 1973.

[4]Moss, James N., Anderson, E. C., and Bolz, C. W., Jr., "Aerothermal Environment for Jovian Probes," AIAA Paper 76-469, July 1976.

[5]Anderson, E. C., Moss, James N., and Sutton, Kenneth, "Turbulent Viscous Shock Layer Solution With Strong Vorticity Interaction," AIAA Paper 76-120, Jan. 1976.

[6]Biolsi, L., "Transport Properties in the Jovian Atmosphere," Journal of Geophysical Research, Vol. 83, No. A3, 1978, p. 1125.

[7]Yos, J. M., "Transport Properties of Nitrogen, Hydrogen, Oxygen, and Air to 30,000^{o}K," Technical Memorandum, RAD-TM-63-7, AVCO Corp., Wilmington, Mass., 1963.

[8]Freeman, G. N. and Oliver, C. C., "High Temperature Thermodynamic and Transport Properties of Planetary CO_2 - N_2 Atmosphere," AIAA Journal, Vol. 8, Sept. 1970, p. 1687.

[9]Hirschfelder, J. O., Curtis, C. F., and Bird, R. B., Molecular Theory of Gases and Liquids, John Wiley and Sons, New York, 1954.

[10]Herning, F. and Zipperer, L., "Calculation of the Viscosity of Technical Gas Mixtures from the Viscosity of the Individual Gases," Gas- u. Wasserfach 79, 49-54, 69-73, 1936.

[11]Reichenberg, D., "New Simplified Methods for the Estimation of the Viscosities of Gas Mixtures at Moderate Pressures," NPL Report Chem. No. 53, May 1977.

[12]Wilke, C. R., "A Viscosity Equation for Gas Mixtures," Journal of Chemical Physics, Vol. 18, No. 4, April 1950, p. 517.

[13]Saxena, S. C. and Tanzman, A., "A Note on the Calculation of Viscosities for Multicomponent Gas Mixtures," High Temperature Science, Vol. 6, No. 3, 1974, p. 215.

[14]Svehla, R. A., "Estimated Viscosities and Thermal Conductivities of Gases at High Temperatures," NASA TR R-132, 1962.

[15]Brokaw, R. A., "Approximate Formulas for the Viscosity and Thermal Conductivity of Gas Mixtures," Journal of Chemical Physics, Vol. 29, No. 2, 1958, p. 391.

[16]Blake, L. H., "Approximate Transport Calculations for High-Temperature Air," AIAA Journal, Vol. 8, Sept. 1970, p. 1698.

[17]Biolsi, L., "Transport Properties in the Atmosphere of Jupiter," NASA Progress Report, Jan. 1977 - Dec. 1978, CR-158094, 1978.

[18]Yun, K. S. and Mason, E. A., "Collision Integral for the Transport Properties of Dissociating Air at High Temperatures," Physics of Fluids, Vol. 5, No. 4, 1962, p. 380.

[19] Grier, N. T., "Calculation of Transport Properties and Heat Transfer of Dissociating Hydrogen," NASA TN D-1406, 1962.

[20] Grier, N. T., "Calculation of Transport Properties of Ionized Atomic Hydrogen," NASA TN D-3186, 1966.

[21] Moss, J. N., Simmonds, A. L., and Anderson, E. C., "The Impact of Turbulence on a Radiating Shock Layer With Coupled Ablation Injection," AIAA Paper 78-1186, 1978.

COUPLED LAMINAR AND TURBULENT FLOW SOLUTIONS FOR JOVIAN ENTRY

A. Kumar*
Old Dominion University, Norfolk, Va.

R. A. Graves Jr.† and K. J. Weilmuenster‡
NASA Langley Research Center, Hampton, Va.

and

S. N. Tiwari#
Old Dominion University, Norfolk, Va.

Abstract

Laminar and turbulent flowfield solutions with coupled carbon-phenolic mass injection are presented for the forebody of a probe entering a nominal Jupiter atmosphere. Solutions are obtained for a 35-deg hyperboloid and for a 45-deg spherically blunted cone using a time-dependent finite-difference method. The radiative heating rates for the coupled laminar flow are significantly reduced as compared to the corresponding no-blowing case; however, for the coupled turbulent flow, it is found that the surface radiative heating rates are sub-

---

Presented as Paper 80-0288 at the AIAA 18th Aerospace Sciences Meeting, Pasadena, Calif., Jan. 14-16, 1980.

*Research Associate Professor, Department of Mechanical Engineering and Mechanics (presently Aerospace Engineer, Hypersonic Propulsion Branch, High-Speed Aerodynamics Division, NASA Langley Research Center).

†Research Leader, Aerothermodynamics Branch, Space Systems Division.

‡Research Engineer, Aerothermodynamics Branch, Space Systems Division.

#Professor, Department of Mechanical Engineering and Mechanics.

stantially increased when compared with the laminar coupled results and often exceed the corresponding no-blowing values. Turbulence is found to have no effect on the surface radiative heating rates for the no-blowing solutions. The present results are compared with the other available solutions, and some additional solutions are presented.

Nomenclature

C_A = ablator mass fraction
C_i = mass fraction of species i, ρ_i/ρ
C_ℓ = mass fraction of element ℓ
H = total enthalpy, $\bar{H}/\bar{V}_\infty^2$
h = enthalpy of mixture, $\sum_{i=1}^{N} C_i h_i$
$\bar{h}_A$ = enthalpy of undecomposed ablation material
h_i = enthalpy of species i, $\bar{h}_i/\bar{V}_\infty^2$
k = thermal conductivity of mixture, $\bar{k}/(\bar{\mu}_\infty \bar{V}_\infty^2/\bar{T}_\infty)$
Le = Lewis number
Le_T = turbulent Lewis number
$\bar{M}$ = molecular weight of mixture
$\bar{M}_i$ = molecular weight of species i
$\bar{M}_\ell$ = molecular weight of element ℓ
$\dot{m}$ = mass injection rate, $\bar{\dot{m}}/\bar{\rho}_\infty \bar{V}_\infty$
N = number of species
n = coordinate normal to the body, $\bar{n}/\bar{R}_n$
N_H = number density of atomic hydrogen (particles/cm^3)
p = pressure, $\bar{p}/\bar{\rho}_\infty \bar{V}_\infty^2$
q^R = net radiative heat flux in n-direction, $\bar{q}^R/\bar{\rho}_\infty \bar{V}_\infty^3$
q^{R+} = component of radiative flux toward the shock
q^{R-} = component of radiative flux toward the wall
$-q_w^C$ = convective heat flux to the wall
$\underline{R}^*$ = universal gas constant (J/mole K)
$\bar{R}_n$ = nose radius (m)
Re = freestream Reynolds number, $\bar{\rho}_\infty \bar{V}_\infty \bar{R}_n/\bar{\mu}_\infty$
r = radius measured from axis of symmetry to point on body surface, $\bar{r}/\bar{R}_n$
s = coordinate measured along the body, $\bar{s}/\bar{R}_n$
Sc = Schmidt number
T = temperature, $\bar{T}/\bar{T}_\infty$
$\bar{T}_\infty$ = freestream temperature (K)
t = nondimensional time, $\bar{t}\bar{V}_\infty/\bar{R}_n$
u = tangential velocity, $\bar{u}/\bar{V}_\infty$

u_s = tangential velocity at shock, $\bar{u}_s/\bar{V}_\infty$
v = normal velocity, $\bar{v}/\bar{V}_\infty$
$\bar{V}_\infty$ = freestream velocity (m/sec)
β = $r + n \cos \theta$
$\beta_{i,j}$ = coefficients for sublimation temperature [Eq. (5)]
δ_s = shock standoff distance, $\bar{\delta}_s/\bar{R}_n$
$\delta_{i\ell}$ = number of atoms of the ℓth element in species i
ε = emissivity
ε^+ = normalized eddy viscosity, μ_T/μ
θ = body angle measured from the body axis
κ = local body curvature, $\bar{\kappa}\bar{R}_n$
λ = $1 + n\kappa$
μ = viscosity, $\bar{\mu}/\bar{\mu}_\infty$
μ_T = eddy viscosity, $\bar{\mu}_T/\bar{\mu}_\infty$
$\bar{\mu}_\infty$ = freestream viscosity (N sec/m^2)
ρ = density, $\bar{\rho}/\bar{\rho}_\infty$
$\bar{\rho}_\infty$ = freestream density (kg/m^3)
σ = Prandtl number
σ_T = turbulent Prandtl number
$\bar{\sigma}$ = Stefan-Boltzmann constant

Superscript

$(\bar{\ })$ = dimensional quantities

Subscript

i = i^{th} species
ℓ = ℓ^{th} element
s = shock values
sub = sublimation
w = wall values
∞ = freestream values
- = values for the solid ablation material at the surface
o = stagnation point

Introduction

The entry probe for the NASA Galileo mission to the planet Jupiter will encounter very severe aerodynamic heating owing to exceedingly high entry velocities. A massive heat shield will be required on the probe's surface to protect the scientific instruments from this aerodynamic heating. To reduce the uncertainties in the design of the heat shield, it is necessary to predict accurately the entry aerothermal environment. Since it is extremely difficult to duplicate the entry environment in an experimental facility, the design of the heat shield for the Galileo probe relies primarily on analytical predictions.

To predict the flowfield accurately, one must model all the physical phenomena occurring in the flow. For the Galileo probe, the gases surrounding the probe forebody will have temperatures in the range of 10,000 to 20,000 K and the surface heating will be primarily due to radiative heating. This will result in massive ablation of the probe's heat shield. Moreover, the flow is expected to be mostly turbulent. The analytical technique should include all the above-mentioned phenomena to predict accurately the surface heating rates and other flow characteristics.

Reference 1 presented laminar and turbulent flowfield solutions for the analytic-shaped bodies (hyperboloids) using a nominal entry trajectory for Jupiter. The viscous shock-layer analysis included coupled ablation injection, detailed spectral calculation of radiative heating, and equilibrium chemistry. In Ref. 2, this analysis was further used to obtain some coupled turbulent flowfield solutions for the forebody of a 45-deg spherically blunted cone (Galileo-probe configuration) only at certain times for a Jupiter entry trajectory. No solutions could be obtained for the coupled laminar flowfield with massive blowing conditions. Reference 3 has presented coupled turbulent solutions for the Galileo probe using a method that assumes a shock shape and then predicts the body shape. There still is a great need for an analytical technique that can accurately predict the coupled laminar and turbulent flowfields for the Galileo probe over the entire range of entry trajectory. The purpose of the present study was to develop such a technique.

In Ref. 4, a time-asymptotic finite-difference method was developed to predict the forebody flowfield past blunt axisymmetric bodies, such as spherically blunted cones, hyperboloids, etc., at zero and small angles of attack. The method uses time-dependent, viscous shock-layer-type equations to describe the flowfield. The analysis was modified in Ref. 5 for the laminar and turbulent flow of a radiating and reacting gas under chemical equilibrium. It is found in Ref. 6 that the method works well for both the laminar and turbulent flows even in the presence of massive surface blowing. The present investigation uses the above method for obtaining the laminar and turbulent flowfield solutions for a 35-deg hyperboloid and a 45-deg spherically blunted cone (Galileo probe shape). The analysis includes coupled ablation injection, detailed spectral calculation of radiative heating, and equilibrium chemistry. The radiative heat transfer and equilibrium chemistry calculations are made using the same techniques as in Ref. 1. The results of the present analysis are compared with those of Refs. 2 and 3 and additional solutions are presented for the conditions for which Ref. 2 presents no solution.

Analysis

Governing Equations

The equations of motion for the laminar and turbulent flows of reacting gas mixtures in chemical equilibrium are given in Ref. 5. These viscous shock-layer-type equations are modified for the present analysis to include the elemental continuity equation that is required here owing to the surface ablation injection. The governing equations in the body-oriented coordinate system (Fig. 1) for the flow at zero angle of attack are expressed as

$$\frac{\partial U}{\partial t} + \frac{\partial M}{\partial s} + \frac{\partial \hat{N}}{\partial n} + Q = 0 \tag{1}$$

where

$$U = \lambda \begin{bmatrix} \rho \\ \rho u \\ \rho v \\ \rho H - p \\ \rho C_\ell \end{bmatrix}, \quad M = \begin{bmatrix} \rho u \\ p + \rho u^2 \\ \rho uv \\ \rho uH \\ \rho uC_\ell \end{bmatrix}$$

$$\hat{N} = \lambda \begin{bmatrix} \rho v \\ \rho uv - \tau \\ p + \rho v^2 \\ \left\{ \rho vH - \frac{\mu}{\sigma Re}\left(1 + \varepsilon^+ \frac{\sigma}{\sigma_T}\right)\frac{\partial h}{\partial n} - u\tau \right. \\ \left. - \frac{\mu}{\sigma Re}\left[Le - 1 + \varepsilon^+ \frac{\sigma}{\sigma_T}\left(Le_T - 1\right)\right] \sum_{i=1}^{N} h_i \frac{\partial C_i}{\partial n} + q^R \right\} \\ \rho vC_\ell - \frac{\mu}{\sigma Re}\left(Le + \varepsilon^+ \frac{\sigma}{\sigma_T} Le_T\right)\frac{\partial C_\ell}{\partial n} \end{bmatrix}$$

$$Q = \frac{\lambda \sin\theta}{\beta} M + \frac{\cos\theta}{\beta} \hat{N}$$

$$+ \frac{1}{\beta} \begin{bmatrix} 0 \\ \beta\kappa(\rho uv - \tau) - p\lambda \sin\theta \\ -\beta\kappa\rho u^2 - p(\lambda\cos\theta + \beta\kappa) \\ 0 \\ 0 \end{bmatrix}$$

Here

$$C_\ell = \sum_{i=1}^{N} \delta_{i\ell} \frac{\bar{M}_\ell}{\bar{M}_i} C_i$$

and

$$\tau = \mu(1 + \varepsilon^+)/Re(\partial u/\partial n - u\kappa/\lambda)$$

The above equations are not valid at the axis of symmetry where $\beta = 0$. A limiting form of the governing equations is obtained at $s = 0$ by differentiating Eq. (1) with respect to s and taking the limit as $s \to 0$. The following equations are

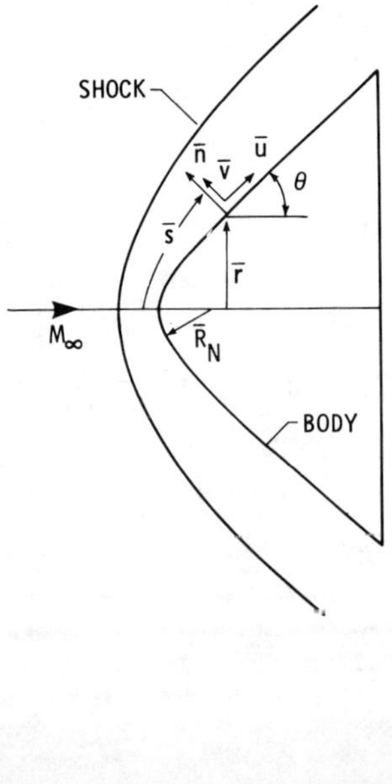

Fig. 1 Coordinate system.

obtained at the axis of symmetry

$$\frac{\partial U_o}{\partial t} + \frac{\partial M_o}{\partial s} + \frac{\partial \hat{N}_o}{\partial n} + Q_o = 0 \tag{2}$$

where

$$U_o = \lambda \begin{bmatrix} \rho \\ \rho u \\ \rho v \\ \rho H - p \\ \rho C_\ell \end{bmatrix}, \quad M_o = \begin{bmatrix} 2\rho u \\ p + 2\rho u^2 \\ 2\rho uv \\ 2\rho uH \\ 2\rho uC_\ell \end{bmatrix}$$

$$\hat{N}_o = \lambda \begin{bmatrix} \rho v \\ \rho uv - \tau \\ p + \rho v^2 \\ \left\{\rho vH - \frac{\mu}{\sigma Re}\left(1 + \varepsilon^+ \frac{\sigma}{\sigma_T}\right)\frac{\partial h}{\partial n} - u\tau \right. \\ \left. - \frac{\mu}{\sigma Re}\left[Le - 1 + \varepsilon^+ \frac{\sigma}{\sigma_T}\left(Le_T - 1\right)\right] \sum_{i=1}^{N} h_i \frac{\partial C_i}{\partial n} + q^R \right\} \\ \rho vC_\ell - \frac{\mu}{\sigma Re}\left(Le + \varepsilon^+ \frac{\sigma}{\sigma_T} Le_T\right)\frac{\partial C_\ell}{\partial n} \end{bmatrix}$$

$$Q_o = \kappa \begin{bmatrix} \rho v \\ 2(\rho uv - \tau) \\ \left[-p + \rho(v^2 - u^2)\right] \\ \left\{\rho vH - \frac{\mu}{\sigma Re}\left(1 + \varepsilon^+ \frac{\sigma}{\sigma_T}\right)\frac{\partial h}{\partial n} - u\tau \right. \\ \left. - \frac{\mu}{\sigma Re}\left[Le - 1 + \varepsilon^+ \frac{\sigma}{\sigma_T}\left(Le_T - 1\right)\right] \sum_{i=1}^{N} h_i \frac{\partial C_i}{\partial n} + q^R \right\} \\ \rho vC_\ell - \frac{\mu}{\sigma Re}\left[Le + \varepsilon^+ \frac{\sigma}{\sigma_T} Le_T\right]\frac{\partial C_\ell}{\partial n} \end{bmatrix}$$

The equation of state is given by

$$p = \left(\frac{R^* \bar{T}_\infty}{\bar{M}\ \bar{V}_\infty^2} \right) \rho T \tag{3}$$

Two independent variable transformations are applied to the governing equations. The first transformation maps the computational domain into a rectangular region in which both the shock and the body are made boundary mesh lines. The second transformation further maps the computational region into another plane to allow higher resolution near the body surface without any significant increase in the number of mesh points in the normal direction. The details of these transformations are given in Ref. 4.

Boundary Conditions

No-slip boundary conditions are used at the surface. The wall temperature and mass injection rate are either specified or calculated. The boundary conditions at the shock are calculated by using the shock relations.

For the calculated mass injection conditions, the ablation process is assumed to be quasisteady and the wall temperature is the sublimation temperature of the ablator surface. The coupled mass injection rate is then given by

$$\dot{m} = \left(\frac{-\bar{q}_w^C - \bar{q}_w^R}{\sum_{i=1}^{N} (C_i \bar{h}_i)_w - \bar{h}_A} \right) / \bar{\rho}_\infty \bar{V}_\infty \tag{4}$$

The sublimation temperature for the carbon-phenolic ablator is given by

$$\bar{T}_{sub} = \sum_{j=1}^{5} \beta_{1,j} C_A^{j-1} + \log \bar{p}_w \sum_{j=1}^{5} \beta_{2,j} C_A^{j-1} + (\log \bar{p}_w)^2 \sum_{j=1}^{5} \beta_{3,j} C_A^{j-1} \tag{5}$$

where $\bar{p}_w$ is the wall pressure in atmospheres and C_A is the ablator mass fraction at the wall. The values of $\beta_{i,j}$ are given in Table 1. These are applicable for a freestream gas

Table 1 Sublimation temperature coefficients

Coefficients	j				
	1	2	3	4	5
$\beta_{1,j}$	5552	-20184	53058	-57933	23243
$\beta_{2,j}$	1798	-12049	30145	-32045	12457
$\beta_{3,j}$	322	-2208	5270	-5450	2092

composition of 89% H_2, 11% He by volume and for ablator mass fractions of 0.4 to 1.0.

For ablation injection, the elemental concentrations at the wall are governed by convection and diffusion as given by the equation

$$\left(\frac{\partial C_\ell}{\partial n}\right)_w - \frac{1}{Re}\left(\frac{\dot{m}\ Sc}{\mu}\right)_w [(C_\ell)_w - (C_\ell)_-] = 0 \tag{6}$$

The net radiative flux, q^R, can be represented as

$$q^R = q^{R+} - q^{R-} \tag{7}$$

At the surface

$$\bar{q}_w^{R+} = \varepsilon\sigma\bar{T}_w^4 \tag{8}$$

The surface is treated as a gray surface with an assumed emissivity, ε, of 0.8 and a reflectivity of 0.0. The heat transferred to the wall because of conduction and diffusion is

$$-q_w^C = \frac{1}{Re}\left(k\frac{\partial T}{\partial n} + \frac{\mu}{Sc}\sum_{i=1}^{N} h_i \frac{\partial C_i}{\partial n}\right)_w \tag{9}$$

Radiative Transport

The radiative flux, q^R, is calculated with the radiative transport code RAD[7,8] which has been incorporated into the present computer code.

The RAD accounts for the effects of nongray self-absorption. Molecular band, continuum, and atomic line transitions are included. A detailed frequency dependence of the absorption coefficients is used for integrating over the radiation frequency spectrum, and the tangent slab approximation is used for integrating over physical space. The chemical species considered in the present study for determining the radiative transport are H, H_2, H^+, e^-, C, C_2, C_3, C^+, C^-, CO, O, O_2, O^+, and O^-. The radiation properties used in this study for C_2 and C_3 are those reported in Refs. 9 and 10, respectively.

Thermodynamic and Transport Properties

The equilibrium composition is determined by a free-energy minimization calculation as developed in Ref. 11 and written for the present code in Ref. 12. Thermodynamic properties for specific heat, enthalpy, and free energy and transport properties for viscosity and thermal conductivity are required for each species considered. Values for the thermodynamic and transport properties are obtained by using polynomial curve fits. Mixture viscosity is obtained by using the semiempirical formula of Wilke.[13]

Six chemical species are used to describe the hydrogen-helium gas mixture: e^-, H, H_2, H^+, He, and He^+. With carbon-phenolic injection, 13 additional species are used which are C, C_2, C_3, C^+, C_2H, C_3H, C_4H, C_2H_2, O, O_2, O^+, CO, and CO_2. The Lewis number and Prandtl number of the mixture are set equal to 1.1 and 0.64, respectively.

Eddy-Viscosity Approximations

A two-layer eddy-viscosity model consisting of an inner law based on Prandtl's mixing-length concept and the Clauser-Klebanoff expression (based on Refs. 14 and 15) for the outer law is used in the present investigation. This model, introduced by Cebeci,[16] assumes that the inner law is applicable for the flow from the wall out to the location where the eddy viscosity given by the inner law is equal to that of the outer law. The outer law is then assumed applicable for the remainder of the viscous layer. It is noted that the eddy viscosity degenerates to approximately zero in the inviscid portion of the shock layer. The degeneracy is expressed in terms of the normal intermittency factor given by Klebanoff.[15] Reference 2 gives a more detailed description of the turbulence model and various expressions for it. The only difference between the models used in the present calculations and that used in Ref. 2 is in the boundary-layer edge definition. The present analysis uses the definition given in Ref. 17, which is based on an index of diffusion, conduction, and dissipation.

The turbulent Prandtl number and Lewis number are assumed to be 0.9 and 1.0, respectively.

Method of Solution

A time-asymptotic, two-step, finite-difference method due to MacCormack[18] is used to solve the governing equations. The details of the method are given in Ref. 4.

The calculations of equilibrium chemistry, radiative heat flux, and eddy viscosity required a significant amount of computing time, and it is not feasible to perform these calculations in each time-step. In the present analysis, the eddy viscosity is calculated after every 25 time-steps, the equilibrium chemistry after every 200 time-steps, and the radiative heat flux after every 1000 or 2000 time-steps.

Results and Discussion

Forebody flowfield solutions and surface-heating rates are presented for a probe entering a nominal Jupiter atmosphere. Most of the results are obtained for the flow past a 45-deg sphere cone (the current Galileo probe configuration), but a sample calculation is made for the flow past a 35-deg (asymptotic half angle) hyperboloid for comparison with the existing results. The freestream conditions correspond to an entry trajectory into the Jupiter atmosphere where the atmospheric gas model is the Orton nominal atmosphere consisting of 89% H_2 and 11% He by volume (see Ref. 2 for further details). These freestream conditions are tabulated in Table 2. The nose radius of the probe is 0.3112 m.

In this study, solutions are obtained for both laminar and turbulent flows with and without coupled ablation injection. In all the turbulent solutions, transition from laminar to turbulent flow is assumed to occur instantaneously at the first mesh point away from the stagnation point which is at $s = 0.157$. Both laminar and turbulent solutions are compared with the existing results for the hyperboloid, but only turbulent solutions are available for the sphere cone.

Figures 2 and 3 show the surface heating rates with and without coupled ablation injection for the sphere cone at 111.3 sec. It is seen from Fig. 2 that the turbulence has no effect on the radiative heating rates for no-injection, but the turbulent convective heating rates are much higher than the laminar convective heating rates, as is seen from Fig. 3. The radiative heating rates for the coupled laminar flow are

Table 2 Freestream conditions

Time from entry point, sec	Altitude, km	$\bar{V}_\infty$, km/sec	$\bar{\rho}_\infty$ kg/m^3
99.9	195.56	48.07	$2.65*10^{-5}$
103.9	169.57	46.96	$7.19*10^{-5}$
107.2	149.10	44.83	$1.64*10^{-4}$
109.0	138.60	42.88	$2.54*10^{-4}$
110.2	131.90	41.16	$3.38*10^{-4}$
111.3	126.05	39.29	$4.36*10^{-4}$
112.2	121.49	37.52	$5.34*10^{-4}$
113.5	115.31	34.67	$7.02*10^{-4}$
114.2	112.20	33.01	$8.05*10^{-4}$
115.3	107.63	30.31	$9.89*10^{-4}$
116.4	103.45	27.54	$1.20*10^{-3}$
117.4	100.00	25.07	$1.41*10^{-3}$

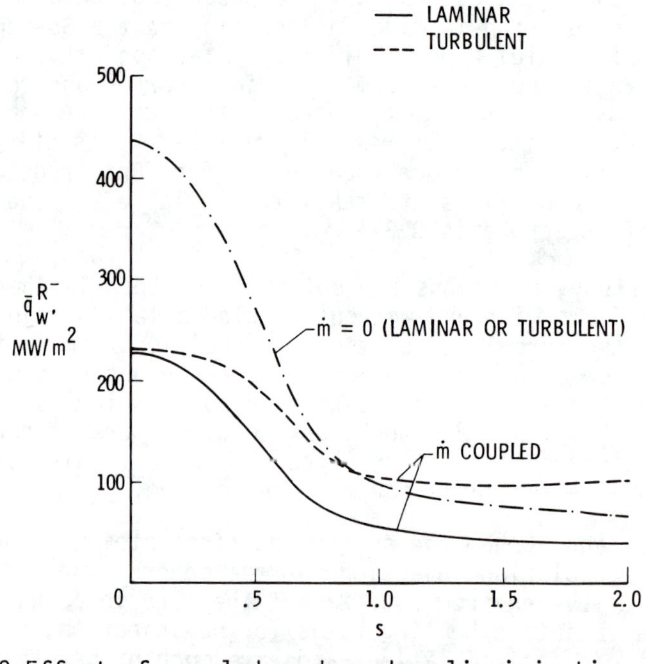

Fig. 2 Effect of coupled carbon-phenolic injection on radiative heating rates for a 45-deg sphere cone at 111.3 sec.

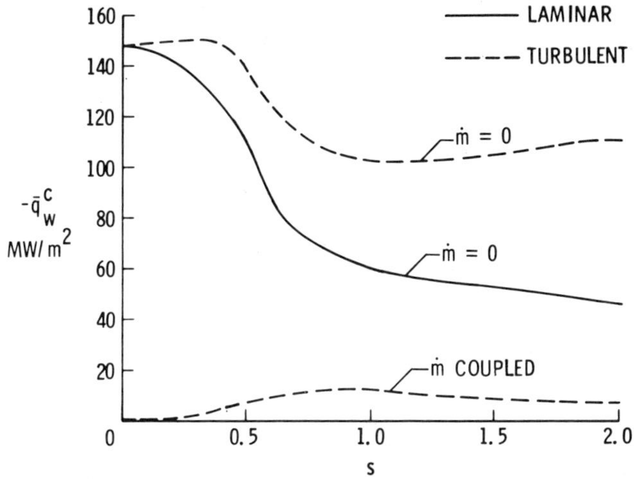

Fig. 3 Effect of coupled carbon-phenolic injection on convective heating rates for a 45-deg sphere cone at 111.3 sec.

significantly reduced all over the body and the convective heating rates are negligible, but the radiative heating rates for the coupled turbulent flow are seen to increase again as compared to the coupled laminar flow and, in fact, are even higher than the no-injection radiative heating rates on the conical portion of the probe. This adverse effect of turbulence on the radiative heating rates was first reported in Ref. 2. The coupled turbulent convective heating rates also increase but still are only a small fraction of the corresponding radiative heating rates. Thus, it is concluded from Fig. 2 that the coupled turbulent radiative heating rates are critical for the design of the probe's heat shield. The enhancement of the radiative heating rates due to turbulence will be discussed later.

Figure 4 shows a comparison of coupled radiative heating rates for the laminar and turbulent flows over the 35-deg hyperboloid with Ref. 2. It is seen that the present laminar radiative heating rates are about 5 to 10% higher and the present turbulent radiative heating rates are about 20% higher than those predicted in Ref. 2. This difference partially is due to the fact that in Ref. 2, the mass injection rate at a particular body station is calculated by using the heating rate at the previous body station. Since the heating rates over the body are seen to decrease with increasing body distance, the mass injection rate at a particular body station (used in Ref. 2) is higher than it should be if the heating rate at that same body station is used. This higher mass

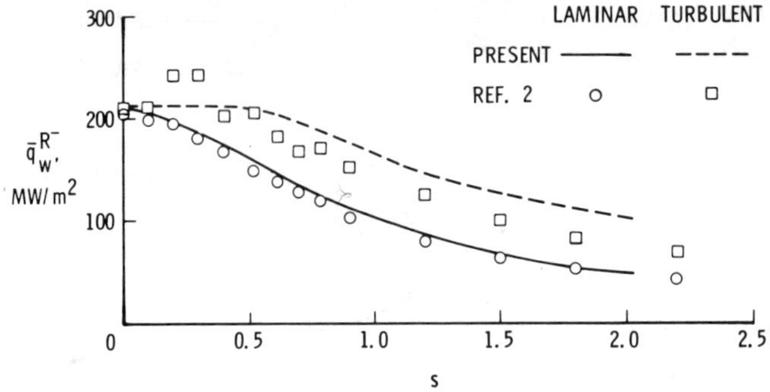

Fig. 4 Comparison of coupled radiative heating rates for a 35-deg hyperboloid at 111.3 sec.

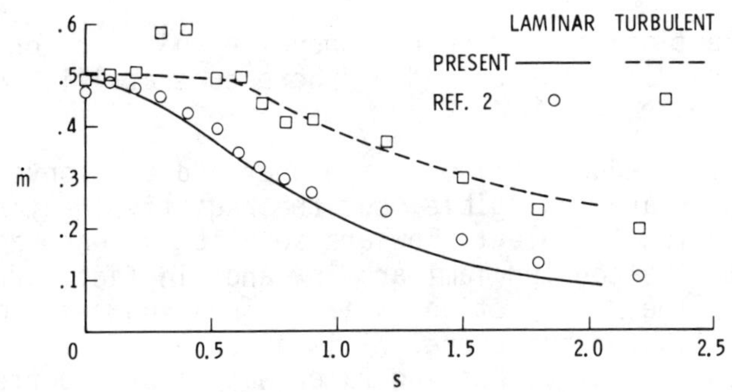

Fig. 5 Comparison of coupled carbon-phenolic mass injection rates for a 35-deg hyperboloid at 111.3 sec.

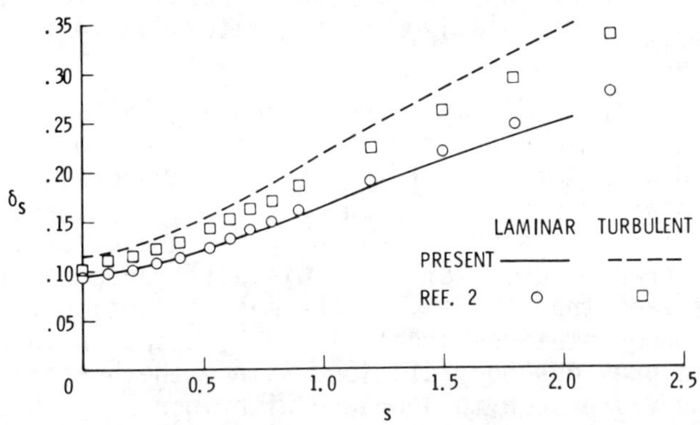

Fig. 6 Comparison of shock shapes for a 35-deg hyperboloid at 111.3 sec.

injection rate results in a slight decrease in the radiative heating rate. The present analysis uses the heating rate at the same body station at which the mass injection rate is being calculated. Another difference between the present results and those of Ref. 2 is that the present analysis does not predict any overshoot in the turbulent radiative heating rates near the stagnation region.

Figure 5 shows the comparison of the corresponding surface mass injection rates. It is seen that although the present laminar radiative heating rates are slightly higher than those in Ref. 2, the present surface mass injection rates are still lower. This is directly due to the reason given earlier. The present turbulent mass injection rates are higher on the downstream part of the body, but the difference is not as much as it was between the radiative heating rates. This decrease in the difference in the mass injection rates is also due to the same reason as discussed earlier.

Shock stand-off distances are compared in Fig. 6. The present analysis predicts about 10% higher shock stand-off distances for the turbulent flow, but the shock stand-off distances for the laminar flow compare well.

All the results presented below are now for the flow past a sphere cone.

Figures 7 and 8 show the coupled laminar and turbulent radiative heating rates and surface mass injection rates at 111.3 sec. The results of Ref. 2 are also plotted for comparison. It is seen that the present results are about 10 to 15% higher on the conical portion of the probe. Unlike Ref. 2, the present heating rates and mass injection rates gradually decrease on the nose portion of the probe. No comparison could be made for the coupled laminar flow as the method of Ref. 2 does not work for the coupled laminar flow past a sphere cone with massive blowing conditions.

Figure 9 shows the surface pressure distribution for the coupled laminar and turbulent flows at 111.3 sec along with the surface pressure distribution predicted in Ref. 2 for the coupled turbulent flow. It is seen from the present results that the surface blowing smooths out the effect of curvature discontinuity at the juncture point of the probe. The surface pressures for the turbulent flow as predicted by the two analyses compare well in the nose region, and on the conical

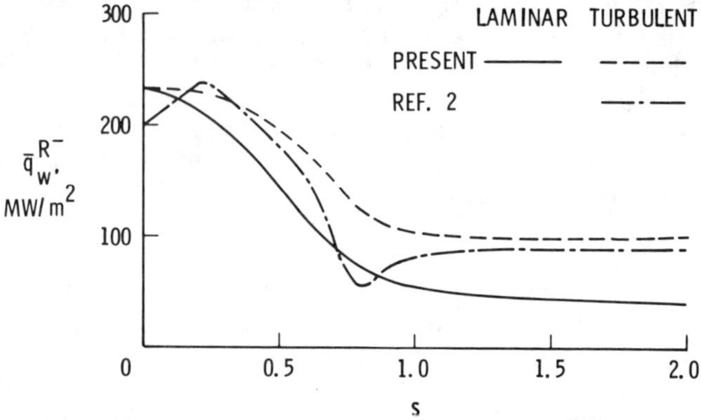

Fig. 7 Comparison of coupled radiative heating rates for a 45-deg sphere cone at 111.3 sec.

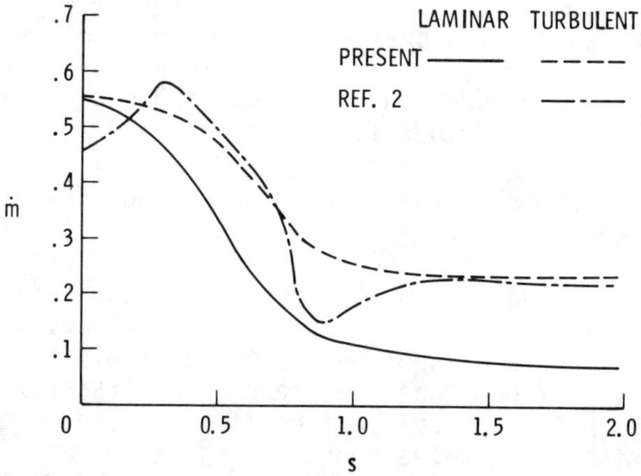

Fig. 8 Comparison of coupled carbon-phenolic mass injection rates for a 45-deg sphere cone at 111.3 sec.

portion of the probe, but the results of Ref. 2 show a sharp decrease and then increase in the pressure around the juncture point. This behavior of the surface pressure in the presence of massive surface blowing is not physically expected.

Figures 10 and 11 show the coupled laminar and turbulent radiative heating rates at 109 and 107.2 sec. The method of Ref. 2 fails to work for these trajectory points, but coupled turbulent radiative heating rates are given in Ref. 3.

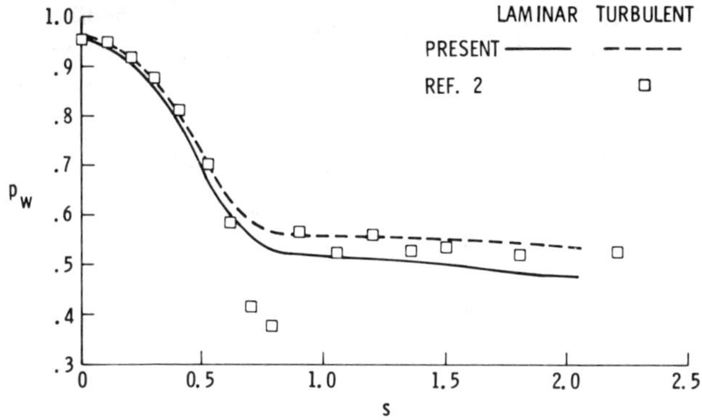

Fig. 9 Comparison of surface pressure distribution for a 45-deg sphere cone at 111.3 sec.

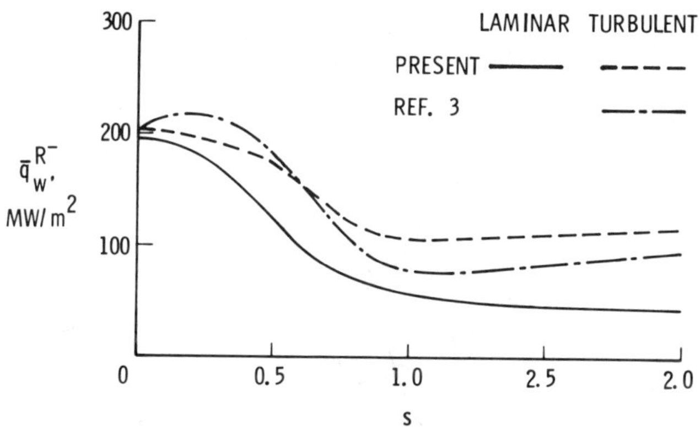

Fig. 10 Comparison of coupled radiative heating rates for a 45-deg sphere cone at 109 sec.

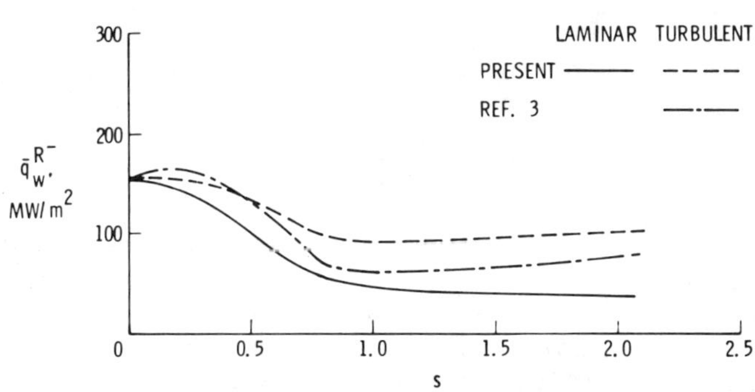

Fig. 11 Comparison of coupled radiative heating rates for a 45-deg sphere cone at 107.2 sec.

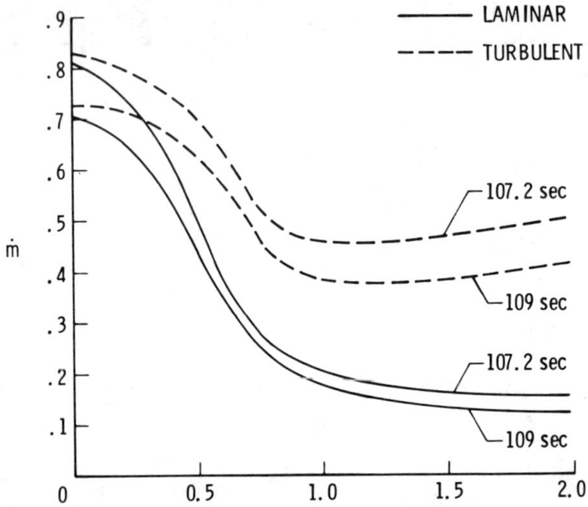

Fig. 12 Coupled carbon-phenolic mass injection rates for a 45-deg sphere cone.

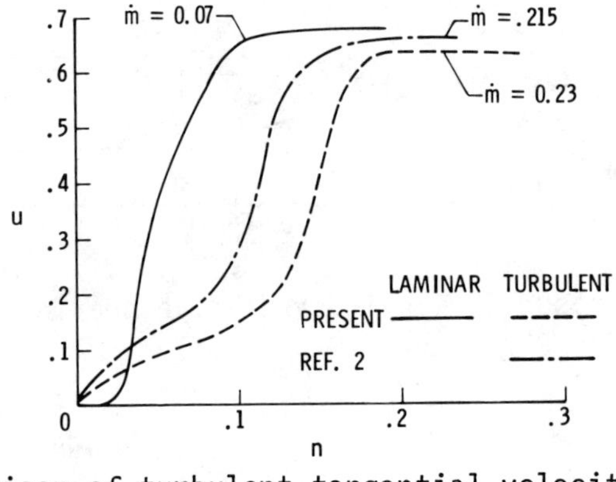

Fig. 13 Comparison of turbulent tangential velocity profiles at $s = 1.8$ for a 45-deg sphere cone at 111.3 sec.

The present radiative heating rates are about 25% higher than those predicted in Ref. 3. Figure 12 shows the corresponding mass injection rates. The nondimensional mass injection rate, $\dot{m}$, is maximum for 107.2 sec of the three trajectory points considered in the present analysis. For other trajectory points, $\dot{m}$ is also expected to be lower than for 107.2 sec. Since the present method worked well for 107.2 sec,

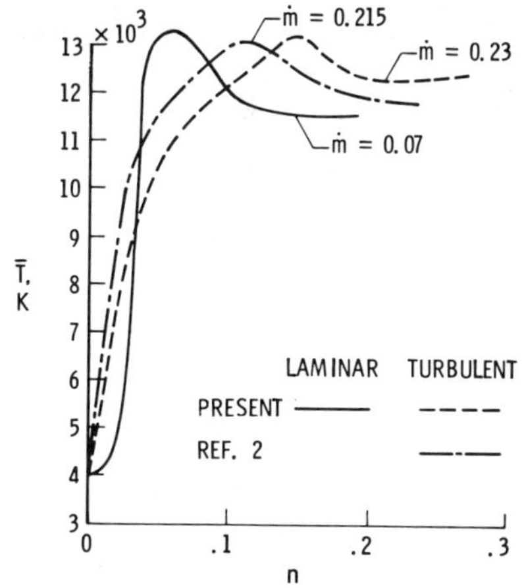

Fig. 14 Comparison of turbulent temperature profiles at s = 1.8 for a 45-deg sphere cone at 111.3 sec.

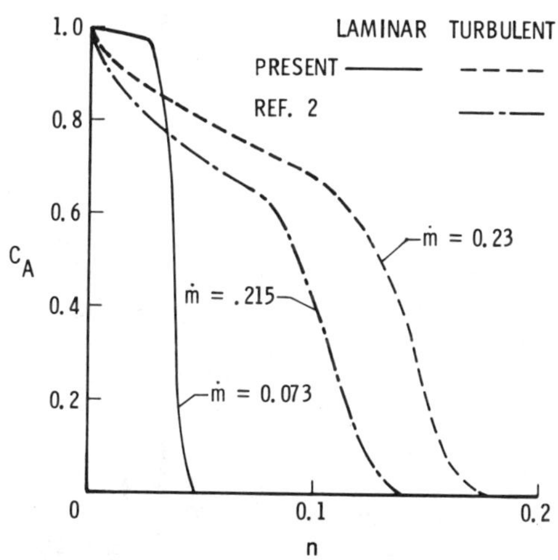

Fig. 15 Effect of turbulence on ablation mass fraction profile at s = 1.8 for a 45-deg sphere cone at 111.3 sec.

it should be able to provide solutions at any point on the trajectory given in Table 2.

Figures 13 through 16 show the effect of turbulence on the shock layer. It is seen from these figures that the turbulence significantly changes the structure of the shock layer. The profiles of Ref. 2 for the turbulent flow are

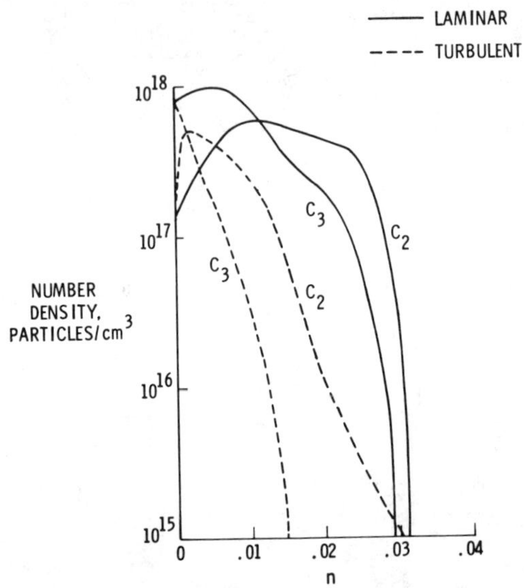

Fig. 16 Effect of turbulence on concentrations of major radiation absorbers at s = 1.8 for a 45-deg sphere cone at 111.3 sec.

also plotted for comparison. In general, the present flow profiles show the similar distributions as in Ref. 2. Tangential velocity and temperature profiles are shown in Figs. 13 and 14. These profiles show the same behavior as observed in Ref. 6 in the presence of massive surface blowing. The gradients near the surface significantly increase for the turbulent flow and the turbulence brings the high-temperature gases closer to the surface. Figure 15 shows that for the turbulent flow, the ablation layer (the portion of the shock layer in which the ablation products are dominantly present) is much thicker but the concentration of the ablation products is decreased because of enhanced diffusion. Figure 16 shows the number density distributions of C_2 and C_3 for laminar and turbulent flows. It is seen that the number densities of C_2 and C_3 for the turbulent flow are significantly reduced as compared to the corresponding laminar values. The high temperatures near the surface for the turbulent flow dissociate these molecules, thus reducing their concentrations. It is shown in Ref. 10 that C_2 and C_3 are the main radiation-absorbing molecules for Jovian entry conditions. Thus, the turbulence has an adverse effect on the effectiveness of the ablation species to absorb radiation.

Conclusion

A time-dependent finite-difference method is used to develop a code for solving the coupled laminar and turbulent flows over the forebody of a probe entering a nominal Jupiter atmosphere. Solutions are obtained for both a 35-deg hyperboloid and a 45-deg sphere cone. Detailed comparisons are made with the existing results. It is found that the present code works well for both the coupled laminar and turbulent flows at all the trajectory points. The code used in Ref. 2 works well for both the coupled laminar and turbulent flows over a hyperboloid, but it works only for the coupled turbulent flows over sphere cones. Even for the coupled turbulent flows, it does not work at several trajectory points where the non-dimensional injection rates are very high, such as at 109 sec or 107.2 sec. The present code uses the heating rate at the same point for which the injection rate is being calculated, whereas in Ref. 2 the heating rate at the previous mesh point is used.

The turbulence is found to alter the structure of the shock layer in such a way that the effectiveness of the ablation layer is significantly reduced. The present analysis supports the finding of Ref. 2 in which the coupled radiative heating rates exceed even the no-injection values on the conical portion of the body. The present coupled turbulent radiative heating rates have the maximum value at the stagnation point whereas the results of Refs. 2 and 3 show the maximum slightly downstream of the stagnation point. The surface pressures predicted by the present analysis appear to be more physically plausible in the presence of massive surface blowing than those in Ref. 2 where a very sharp dip occurs around the sphere-cone juncture point. Detailed flowfield profiles are similar to those in Ref. 2, but there are differences in the magnitude.

Acknowledgment

The authors would like to thank Dr. James N. Moss for his help during the course of this effort, without which it would have taken considerably longer to set up both the chemistry and radiation codes.

References

[1]Moss, J. N., Anderson, E. C., and Simmonds, A. L., "The Impact of Turbulence on a Radiating Shock Layer with Coupled Ablation Injection," AIAA Paper No. 78-1186, July 1978.

[2]Moss, J. N., "A Study of the Aerothermal Entry Environment for the Galileo Probe," AIAA Paper No. 79-1081, June 1979.

[3]Nicolet, W. E. and Balakrishnan, A., "Methods for Predicting Off-Stagnation Point Flowfields for Planetary Entry Probes," AIAA Paper No. 79-1083, June 1979.

[4]Kumar, A. and Graves, R. A., Jr., "Numerical Solution of the Viscous Hypersonic Flow Past Blunted Cones at Angle of Attack," AIAA Paper No. 77-172, January 1977; also AIAA Journal, Vol. 15, Aug. 1977, pp. 1061-1062.

[5]Kumar, A., Tiwari, S. N., and Graves, R. A., Jr., "Effects of Small Angle of Attack on the Radiating Viscous Shock Layer Solutions for Jovian Entry," AIAA Paper No. 78-909, May 1978; also Outer planet Entry Heating and Thermal Protection, Progress in Astronautics and Aeronautics, Vol. 64, edited by R. Viskanta, AIAA, New York, 1979, pp. 147-164.

[6]Kumar, A., Graves, R. A., Jr., and Tiwari, S. N., "Laminar and Turbulent Flows Over a Spherically Blunted Cone with Massive Surface Blowing," AIAA Journal, Vol. 17, Dec. 1979, pp. 1326-1331.

[7]Nicolet, W. E., "Advanced Methods for Calculating Radiation Transport in Ablation-Product Contaminated Boundary Layers," CR-1656, 1970, NASA.

[8]Nicolet, W. E., "User's Manual for the Generalized Radiation Transfer Code (RAD/EQUIL)," CR-116353, 1969, NASA.

[9]Sutton, K. and Moss, J. N., "Radiation Absorption by the C_2 Band Systems for Jupiter Entry Conditions," AIAA Paper No. 79-0033, Jan. 1979.

[10]Moss, J. N., Jones, J. J., and Simmonds, A. L., "Radiative Flux Penetration through a Blown Shock Layer for Jupiter Entry," AIAA Paper No. 78-908, May 1978; also Outer Planet Entry Heating and Thermal Protection, Progress in Astronautics and Aeronautics, Vol. 64, edited by R. Viskanta, AIAA, New York, 1979, pp. 22-41.

[11]Stroud, C. W. and Brikley, K. L., "Chemical Equilibrium of Ablation Materials Including Condensed Species," TN D-5391, 1969, NASA.

[12]Kumar, A., Graves, R. A., Jr., and Weilmuenster, K. J., "User's Guide for the Vectorized Code 'EQUIL' for Calculating Equilibrium Chemistry on CDC-STAR-100 Computer," TM 80193, 1980, NASA.

[13]Wilke, C. R., "A Viscosity Equation for Gas Mixtures," Journal of Chemical Physics, Vol. 18, No. 4, April 1950, pp. 517-519.

[14]Clauser, F. H., "The Turbulent Boundary Layer," Vol. IV of Advances in Applied Mathematics, H. L. Dryden and Th. Von Kármán, eds., Academic Press, New York, 1956, pp. 1-51.

[15]Klebanoff, P. S., "Characteristics of Turbulence in a Boundary Layer With Zero Pressure Gradient," Rep. 1247, 1955, NACA.

[16]Cebeci, T., "Behavior of Turbulent Flow Near a Porous Wall with Pressure Gradient," AIAA Journal, Vol. 8, No. 12, 1970, pp. 2152-2156.

[17]Anderson, E. C. and Wilcox, D. C., "Vorticity Interaction Effects on Blunt Bodies," CR-2778, 1977, NASA.

[18]MacCormack, R. W., "The Effect of Viscosity in Hypervelocity Impact Cratering," AIAA Paper No. 69-354, April 1969.

PRELIMINARY AEROTHERMAL ANALYSIS FOR SATURN ENTRY

E. V. Zoby* and J. N. Moss *

NASA Langley Research Center, Hampton, Va.

Abstract

A preliminary thermal analysis based on recently defined Saturn entry conditions has been conducted. The study, using viscous-shock-layer and engineering codes employed for Project Galileo, investigated nonequilibrium chemistry effects on the Saturn thermal environment, defined the primary heat-transfer mode for heat-shield design, delineated some problem areas for future thermal studies, and validated BIRCHES (Blunt Body Inviscid Radiative and Convective Heating Engineering Solutions) for parametric or design studies. The effect of nonequilibrium chemistry appears to significantly influence only the radiative fluxes with effects localized to the stagnation region. However, the heat-transfer mode pertinent to the overall heat-shield design is convection. The convective results of BIRCHES and a detailed code are in good agreement. The resulting mass-loss rates for the currently prescribed nominal Saturn entry conditions are small when compared with the values for nominal Jupiter entry conditions. With coupled carbon-phenolic ablation injection, the convective heating rates are reduced substantially while the radiative heating rates are increased when compared with the corresponding no-injection results.

Nomenclature

C_{h,U_t} = transformed enthalpy proportionality constant

C_{T,U_t} - transformed temperature proportionality constant

$\dot{m}$ = mass-loss rate

$\dot{q}_c$ = convective heat transfer

Presented as Paper 80-0359 at the AIAA 18th Aerospace Sciences Meeting, Pasadena, California, Jan. 14-16, 1980.

*Aero-Space Technologist.

$\dot{q}_{c,L}$ = convective laminar heat transfer
$\dot{q}_r$ = radiative heat transfer
R_B = base radius of probe
R_N = nose radius of probe
S = body wetted length
T_W = wall temperature
$T_{W,sub}$ = sublimation wall temperature
U_t = transformed freestream velocity
u = freestream velocity
= freestream density
U_∞ = freestream velocity
ρ_∞ = freestream density
θ_c = sphere-cone half angle
θ_s = shock angle

Introduction

The planetary exploration program of the National Aeronautics and Space Administration (NASA) includes missions to the planet Saturn. One recent mission study[1] consisted of a Saturn orbiter and probe and a probe to explore Titan, a satellite of Saturn. Such a mission is referred to as the Saturn Orbiter Dual Probe (SO2P) mission. A prospective launch date is the late 1980's with an approximate transit time of 7 years. The SO2P project proposes to use much of the software technology and probe design developed for Project Galileo.

For a successful mission, the aerothermodynamic environment experienced by the entry probe must be defined. Since experimental facilities cannot adequately duplicate the aerodynamic and thermal conditions expected during Saturn entry, much of the required design information must be obtained from computational analyses. The thermal environment experienced by a probe entering the Saturn atmosphere has been investigated in previous preliminary studies.[2-5] These analytical survey studies considered a wide range of entry conditions, atmospheric gas composition, and probe configurations, as well as the chemical state of the gas mixture. The studies were based primarily on approximate viscous or inviscid analyses.

The purpose of this investigation is first to define the Saturn thermal environment using numerical procedures developed for the Galileo probe design. Particular factors that require further investigation are defined. Thus, the present investigation, while a preliminary study, is intended to serve as a baseline for future Saturn probe design studies and to provide the first detailed coupled viscous flowfield calculations[6,7] for the Saturn environment. Discussions of the nonequilibrium chemistry effect on the thermal environment, of the significance of the different heat-transfer modes, and of peak heat-shield mass-loss rates compared to peak values for Jovian conditions are included. Secondly, the validity of an approximate engineering method developed for computing the radiative[8] and convective[9] heat-transfer rates about blunt bodies entering the atmospheres of the outer planets is demonstrated by comparison with results of a detailed viscous-shock-layer (VSL) analysis. The code is hereafter referred to as BIRCHES (Blunt Body Inviscid Radiative and Convective Heating Engineering Solutions). These major topics are addressed, herein, by basing the numerical calculations on entry conditions resulting from the SO2P study.[1]

Analysis

Three different analyses are used to compute the heating rates and the flowfield about a probe entering a proposed Saturn atmosphere. Two of the analyses are VSL analyses,[6,7] one of which employs a detailed equilibrium chemistry code[6] and the other incorporates both an approximate equilibrium and a finite-rate chemistry option[7] to compute the flowfield. Preliminary studies[3,4] indicated the presence of a significant chemical nonequilibrium layer behind the shock which may affect significantly the predicted heating rates for some of the outer planet missions. The present study includes the detailed chemical nonequilibrium flowfield analysis to investigate the impact of the flow condition on surface heating rates for Saturn entry conditions. The third method is BIRCHES which provides an approximate inviscid radiative flowfield analysis[8] with convective heating capability.[9]

The VSL analysis of Moss[6] includes a detailed description of the radiative transport, equilibrium chemistry, and transport properties. The analysis provides for ablation injection at the surface and the flow may be either laminar or turbulent. The set of VSL equations is obtained from the steady-state Navier-Stokes equations by retaining terms up to second order in the inverse square root of the Reynolds number. The set of governing equations is then solved as a parabolic set of equations using an implicit finite-difference numerical procedure. This analysis[6] was used to provide "benchmark" calculations for the design of the Galileo probe.

The VSL analysis of Tiwari and Szema[7] employs the same solution procedure as used by Moss[6] to solve the steady-state VSL equations. The analysis is currently restricted to a nonblowing laminar flowfield about a hyperboloid. Equilibrium and nonequilibrium chemistry options are included. The equilibrium calculations are based on an approximate equation of state[10] while the finite-rate calculations use rate constants based on results of Leibowitz.[4,11] The radiative transport calculations use a 58-step absorption model[8] for the hydrogen-helium gas mixture.

Since the detailed analyses require large computer run times and storage, BIRCHES, an approximate inviscid radiative flowfield analysis[8] with a convective heating-prediction capability,[9] has been developed. Important aspects of BIRCHES, including the computational technique, the thermodynamic and transport-property correlations, the step model for absorption coefficients, and the analytic shock-shape equations, have previously been discussed.[8,10,12] Briefly, BIRCHES uses the inverse flowfield procedure of Maslen.[13] The thermodynamic[10] and transport[12] properties correlations are developed over a range of temperatures, pressures, and hydrogen-helium gas compositions. The step model consists of 58 steps[8] and the absorption coefficients are computed by summing the product of the species absorption cross-section and number density. The number density is based on an equilibrium calculation, and hence, the radiative calculation is not restricted to one gas mixture. Two analytic expressions are utilized that are capable of generating body shapes closely approximating either a spherically blunted cone or a hyperboloid. The approximate convective heating equations[9] are applicable to nonreacting or reacting gas flows at conditions of constant or variable edge entropy. BIRCHES has been demonstrated for Jovian conditions to provide reliable comparisons with radiative and convective heating rates predicted by the detailed results of Moss[6] and was used in parametric or tradeoff studies for the Galileo probe. Such a procedure should prove useful for future Saturn design studies. Note that for current Saturn entry conditions, the probe enters at velocities much lower than those for the Galileo mission. Thus, the freestream velocity range on the thermodynamic correlations[10] must be expanded to include adequately the Saturn conditions. This effort has been accomplished and the revised thermodynamic correlations are given in the Appendix.

Discussion

The results of the numerical procedures are used to illustrate factors that may impact the design of a probe entering the Saturn atmosphere. The importance of chemical nonequilibrium effects for both the radiative and convective nonblowing surface heating rates, as well as the relative significance of nonblowing radiative and convective heating modes, is demonstrated for prescribed entry conditions. Also, the heating results for BIRCHES and a detailed code are compared. All calculations are for 0 deg angle-of-attack conditions.

Entry Conditions

Freestream conditions resulting from four different entry trajectories are considered. The four trajectories are referred to herein as trajectory I, II, III, or IV. These four trajectories include variations in entry conditions and gas composition. Table 1 presents the pertinent values associated with each trajectory. All four trajectories are for a 45 deg spherically blunted cone (R_N = 0.3112 m, and R_B/R_N = 2.0) entering the Saturn atmosphere.

Trajectory I presents the entry conditions resulting from the SO2P mission study[1] with the recommended nominal atmosphere[14] for Saturn. The entry is off the equatorial plane where the trajectory calculation is initiated (zero time) at an altitude of 450 km, a latitude of -3 deg, and a heading of S. 114.97 deg E.

Trajectory II is selected to represent an upper bounds on the entry velocity based on previous and current considerations. The entry conditions are for an equatorial entry where the trajectory calculations are initiated at an altitude of 1000 km. Trajectories III and IV are similar to trajectories I and II, respectively, except the gas composition is for a 0.75 H_2 - 0.25 He gas mixture by volume. This gas composition closely approximates the recommended cool atmosphere[14] and corresponds to the limiting gas composition range for the thermodynamic and transport property correlations. Thus, trajectories II, III, and IV are used in this preliminary investigation to provide information on the effects of a change in entry velocity or gas composition on the thermal environment.

The trajectories are calculated with the POST[15] computer program. For these calculations, the ballistic coefficient is held constant at its initial value. Hence, the effects of mass loss and shape change are not included in the present trajectory calculations.

Table 1 Probe characteristics and entry conditions

Parameters	Trajectory I	II	III	IV
Body half angle	45	45	45	45
Nose radius, m	0.3112	0.3112	0.3112	.3112
Base-to-nose radius ratio	2	2	2	2
Drag coefficient	1.05	1.05	1.05	1.05
Mass, kg	242.0	242.0	242.0	242.0
Ballistic coefficient, kg/m^2	189.5	189.5	189.5	189.5
Atmosphere (by volume)	89%H_2-11%He	89%H_2-11%He	75%H_2-25%He	75%H_2-25%He
Entry altitude, km	450	1000[a]	450	1000[a]
Inertial (and relative) entry velocity, km/s	35.93 (28.1)	41.11 (32)	35.93 (28.1)	41.11 (32)
Inertial (and relative) entry angle, deg	-23.02 (-30)	-22.91 (-30)[b]	-23.02 (-30)	-22.91 (-30)[b]
Entry latitude, deg	-3	0	-3	0
Azimuth angle-of-inertial velocity vector, deg	114.97	90	114.97	90

[a] Time at 450 km is 35 s.

[b] Inertial angle at 450 km is -22.11 deg.

Effects of Nonequilibrium Chemistry

Results of previous investigations[3,4] showed that a consideration of chemical nonequilibrium effects impacts the thermal design of probes entering the atmospheres of the outer planets. Primarily, these earlier investigations concluded that the presence of a chemical nonequilibrium layer behind the shock layer reduces the radiative heating rates to the probe surface when compared to a corresponding equilibrium prediction. A recent investigation of Tiwari and Szema[7] showed that the radiative heating rates for chemical nonequilibrium were approximately 10% greater than the corresponding equilibrium solutions at Jupiter entry conditions. This apparent contradiction in the overall effects of chemical nonequilibrium on the radiative heating rates would seem to warrant further clarification but is not addressed in the present paper.

For this preliminary investigation, the analysis of Tiwari and Szema[7] is used to assess the impact of nonequilibrium chemistry on a thermal analysis for a Saturn entry probe. As mentioned in the "Analysis" section, the method is based on a detailed VSL solution, uses the finite-rate chemistry model of Leibowitz[4,11] and is restricted to a laminar flowfield about a hyperboloid. The impact of the nonequilibrium chemistry effects on the Saturn thermal environment that are computed for the hyperboloid should be representative in general of those corresponding results that would be computed for the baseline 45 deg spherically blunted cone. A 35 deg hyperboloid (R_N = 0.3112 m, R_B/R_N = 2.0) is used for the current calculations since it has the same drag coefficient as a 45 deg spherically blunted cone with the same base and nose radii.

A typical comparison of the radiative heating-rate distribution, based on nonequilibrium and equilibrium chemistry calculations, is shown in Fig. 1a) and b) for trajectories I and II, respectively. The freestream conditions correspond to the time of peak equilibrium radiative heating for both trajectories. The radiative stagnation-point heating-rate results of Tiwari and Szema[7] are approximately a factor of 2 greater than the results of Moss.[6] The difference between the equilibrium and nonequilibrium radiative heating rates decreases rapidly away from the stagnation point such that at a body station of $S/R_N \approx 1$, the differences are negligible. Thus, the impact of the nonequilibrium radiative heating on the probe design would seem to be confined to the stagnation region.

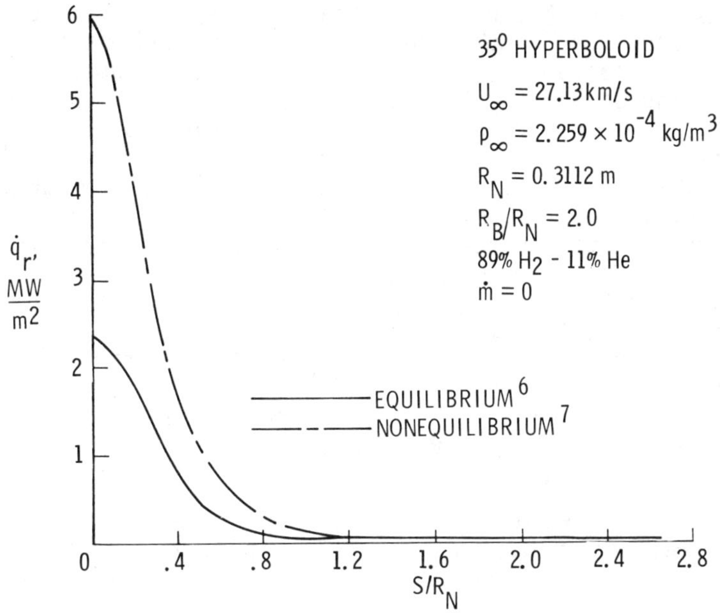

a) Trajectory I

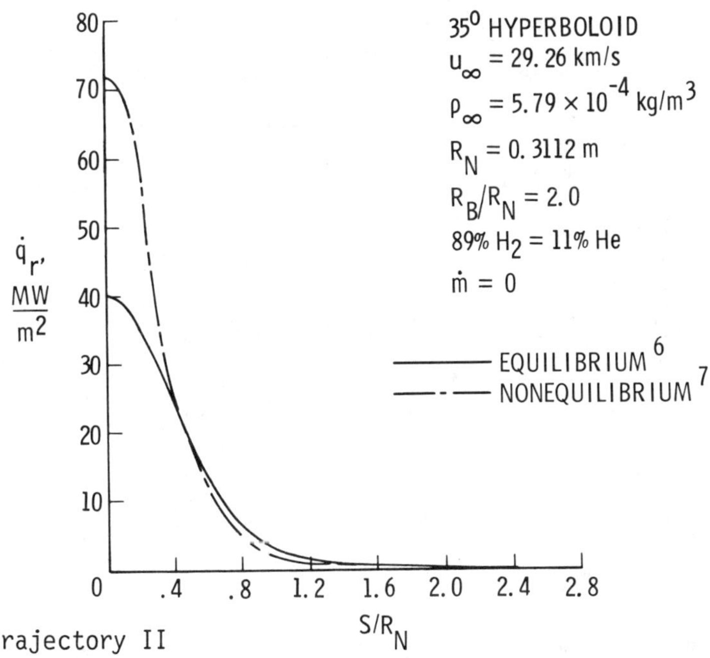

b) Trajectory II

Fig. 1 Nonequilibrium and equilibrium laminar radiative heat-transfer distributions ($\dot{m} = 0$).

A summary of the radiative stagnation-point heating-rate calculations based on the analyses of Tiwari and Szema[7] and Moss[6] for trajectories I and II is presented in Table 2. The computed radiative rates for trajectory II (higher entry velocity) are greater than the computed fluxes for trajectory I. Note that the nonequilibrium values peak at a time earlier (higher altitude) than peak equilibrium values. Furthermore, the ratio of the peak chemical nonequilibrium radiative flux to the peak equilibrium value for trajectory I is 2.2 while the corresponding ratio for trajectory II conditions is 2.6. Based on present calculations, preliminary results indicate that the relative significance of nonequilibrium chemistry on the stagnation radiative heating rates is less for trajectory III and IV calculations (not presented herein) for which the freestream concentration of H_2 is lower than trajectory I conditions.

A typical comparison of the laminar nonblowing convective heating-rate distribution based on nonequilibrium and equilibrium chemistry calculations and an equilibrium wall

Table 2 Nonequilibrium and equilibrium radiative stagnation-point heating rates[a]

Time, sec	Equilibrium[6]	Nonequilibrium[7]
	Trajectory I	
13.00	0.43	4.20
15.00[c]	1.27	6.11
16.75[b]	2.31	5.94
17.75	2.25	4.12
19.00	1.14	1.31
	Trajectory II	
47.00	4.24	30.79
49.00	15.86	62.97
50.50[c]	32.53	87.54
51.75[b]	40.00	71.54
52.50	30.30	41.51

[a] $\dot{q}_r$ = MW/m^2; 35 deg hyperboloid, $\dot{m} = 0$, laminar flow.

[b] Time of peak equilibrium radiative heating.

[c] Time of peak nonequilibrium radiative heating.

temperature of 4000 K is shown in Fig. 2a) and b) for trajectories I and II, respectively. A comparison of the laminar heating-rate results of Tiwari and Szema[7] and Moss[6] indicates that nonequilibrium chemistry effects on the convective heating rates appear to be small (<15%). The relative significance of nonequilibrium chemistry is less for convective heating than for radiative heating due to the assumption of an equilibrium wall condition and that influences on convective heating are confined to the near-wall region. Radiative heating is a function of the thermodynamic properties, especially the temperature, through the flowfield; the temperature distributions based on equilibrium and nonequilibrium chemistry can differ significantly. In addition, a comparison of the respective laminar convective and radiative distributions for trajectories I and II shows that the laminar convective rates are, in general, much greater than the corresponding radiative values, particularly downstream of the stagnation region where most of the surface area is located.

Heat-Transfer Modes

Typical comparisons of the radiative and convective power ($\int \dot{q} dA$) for Saturn entry conditions are presented in Table 3. The results are computed for trajectory I and II conditions for flow about a 35 deg hyperboloid and a surface temperature of 4000 K. The results indicate that the dominant heat-transfer mode for the currently prescribed Saturn conditions is convection. Also note that the convective results presented in Table 3 are based only on the presence of a laminar flowfield. However, the point is made that the computed stagnation-point radiative heating rates based on either nonequilibrium or equilibrium chemistry for trajectory II, III, or IV conditions, higher entry velocities, or the cool atmosphere (lower concentrations of H_2) are at least as great as the corresponding laminar convective rates. This result would appear to also impact future probe design studies.

Approximate Code

The radiative and convective heating rates computed with BIRCHES are validated by comparison with the detailed nonblowing results of Moss.[6] All calculations are performed for the 45 deg spherically blunted cone (R_N = 0.3112 m, R_B/R_N = 2.0), the current baseline configuration for the Saturn probe. Heating-rate results are computed based on trajectories I, II, III, and IV. These comparisons are

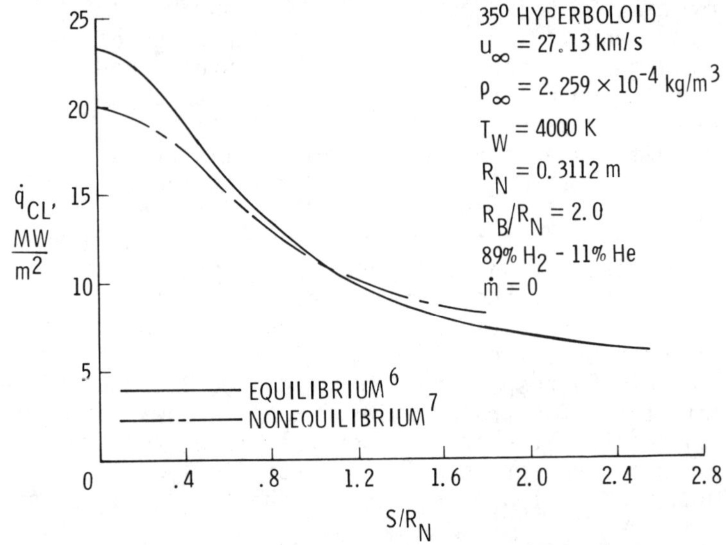

a) Trajectory I

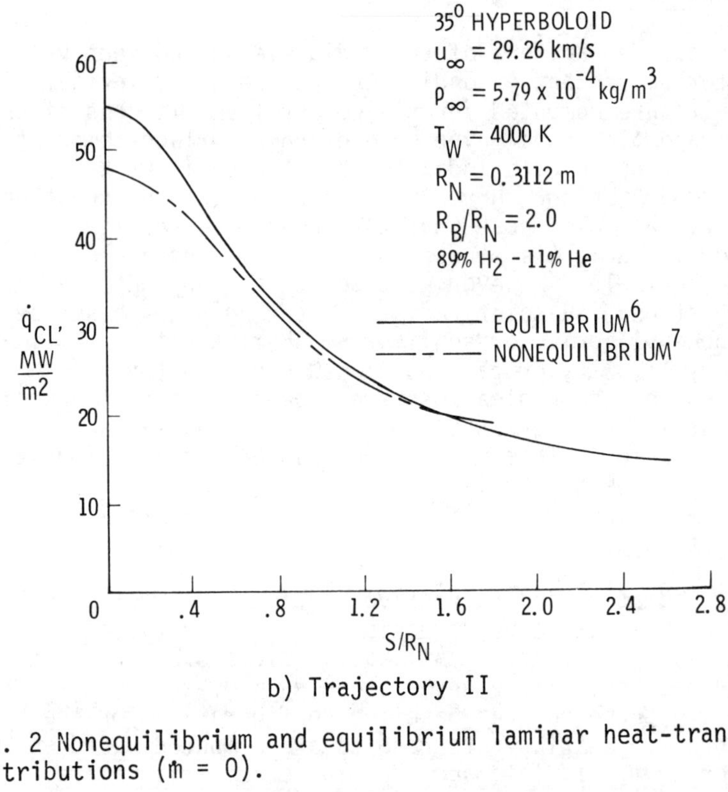

b) Trajectory II

Fig. 2 Nonequilibrium and equilibrium laminar heat-transfer distributions ($\dot{m}$ = 0).

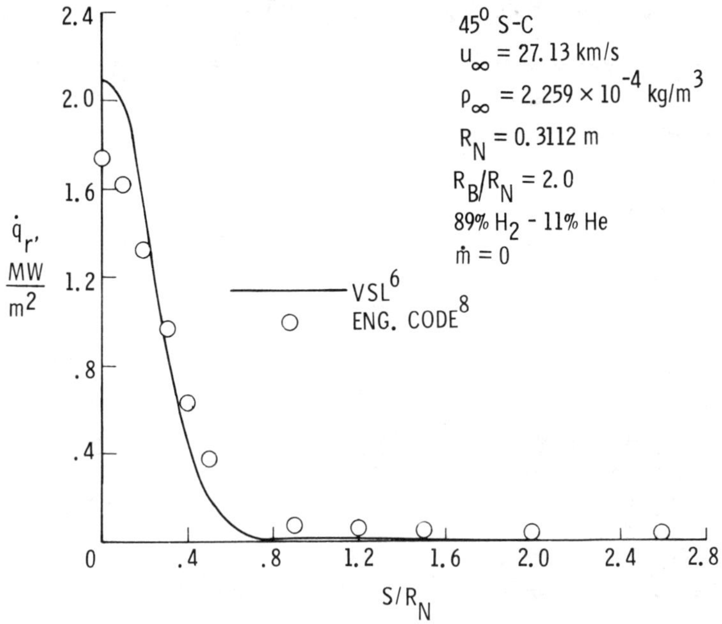

a) Trajectory I

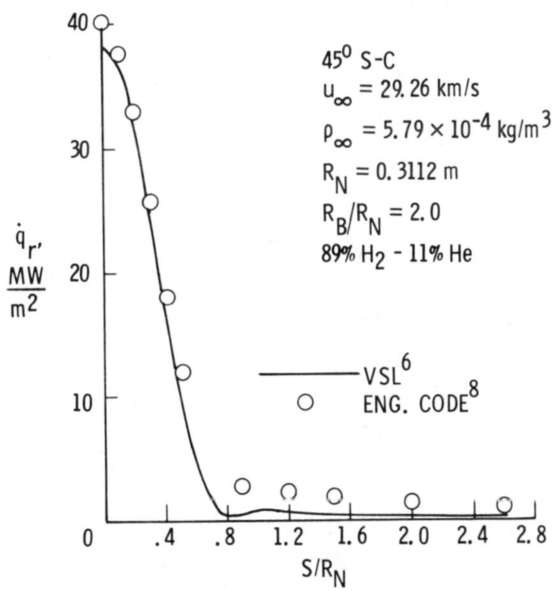

b) Trajectory II

Fig. 3 Comparison of laminar radiative heat-transfer distributions (ṁ = 0).

Table 3 Equilibrium radiative and convective power[a]

Time, s	$\int \dot{q}_r \, dA$	$\int \dot{q}_c \, dA$[b]
	Trajectory I	
13.00	0.04	6.4
15.00	0.10	10.3
16.75	0.16	15.1
17.75	0.16	17.2
19.00	0.12	20.9
	Trajectory II	
47.00	0.95	13.9
49.00	3.10	22.3
50.50	5.30	30.5
51.75	4.80	36.8
52.50	2.80	39.7

[a]35 deg hyperboloid; power = MW; $\dot{m}$ = 0, laminar flow.

[b]Wall temperature = 4000 K.

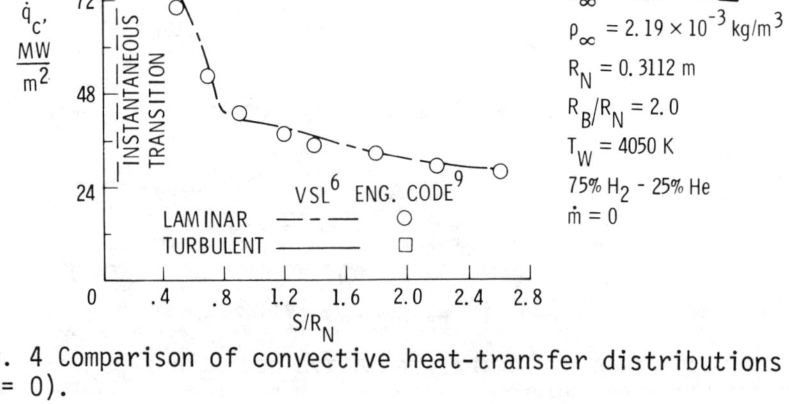

Fig. 4 Comparison of convective heat-transfer distributions ($\dot{m}$ = 0).

intended to demonstrate the reliability of the approximate heating predictions for various entry conditions and gas compositions.

Radiative Comparisons. Typical comparisons for radiative heating-rate distributions, based on detailed and approximate calculations, are presented in Fig. 3a) and b). The freestream conditions correspond to the times of peak equilibrium radiative heating for trajectories I and II, respectively. Discrepancies of less than 25% are noted for comparisons over the hemispherical portion of the probes. However, on the conical portion of the forebody where the radiative levels based on the detailed calculations are of the order of 0.1 MW, the approximate results overpredict significantly (factors as large as 7) the detailed results. The poor agreement at these low radiative heating levels may result from several of the approximations incorporated in BIRCHES. For the more severe Jovian entry conditions, the approximate radiative calculations were in good agreement with the detailed radiative calculations with discrepancies of less than 20%.[8]

Convective Comparisons. Preliminary detailed calculations indicated that the presence of a radiating flowfield had little effect (<5%) on the convective heat-transfer results. These findings were true for both laminar and turbulent flows. Hence, all convective heating-rate comparisons are based on a nonradiating flowfield condition.

A typical comparison of the approximate and detailed convective heating-rate distributions about the probe is presented in Fig. 4. The freestream conditions correspond to the time of peak stagnation-point heating for trajectory IV. Both laminar and turbulent results are shown. Instantaneous transition was assumed at a body station where $S/R_N = 0.1$. In general, discrepancies of less than 10% are observed. However, the approximate stagnation-point heating-rate value is approximately 15% lower than the result of Moss.[6]

A summary of the convective heating results for the four trajectories is presented in Table 4 in terms of the convective power ($\int \dot{q}_c \, dA$). Calculations were made for all trajectories with the wall temperature equal to the stagnation-point sublimation temperature for a carbon-phenolic ablator (assuming massive blowing). This wall-temperature condition is noted in Table 4 as $T_{W,sub}$ and represents ranges of wall temperatures from 3500 to 4100 K. For the heating calculations, the approximate and detailed results are in good agreement with discrepancies of less than 15% noted (<10% for

Table 4 Equilibrium convective power[a]

Time, s	Laminar, $T_{W,sub}$, VSL[6]	Laminar, $T_{W,sub}$, Eng. Code[9]	Laminar, T_W = 2000 K, VSL[6]	Laminar, T_W = 2000 K, Eng. Code[9]	Turbulent[b], $T_{W,sub}$, VSL[6]	Turbulent[b], $T_{W,sub}$, Eng. Code[9]	Turbulent[b], T_W = 2000 K, VSL[6]	Turbulent[b], T_W = 2000 K, Eng. Code[9]
				Trajectory I[c,d]				
13.00	8.3	7.2	15.4	14.1	11.6	10.6	24.7	23.8
15.00	12.3	11.2	22.5	20.5	20.5	19.9	40.7	41.9
17.75	19.6	18.4	35.7	32.6	43.4	40.2	84.6	79.7
19.00	21.8	20.9	40.7	37.3	55.1	50.1	107.5	101.3
21.00	17.9	17.7	39.3	36.8	53.3	48.5	119.1	116.5
24.00		-1.3	16.9	16.2		-3.7	56.2	56.1
				Trajectory II[c,e]				
44.00		7.6	12.6	12.3		8.3	15.8	15.5
47.00	16.4	15.5	25.0	23.0	21.3	23.5	37.7	39.7
49.00	25.3	23.9	39.0	34.4	39.1	44.9	72.2	71.7
51.75	38.9	36.6	62.4	54.5	90.9	90.2	156.4	148.1
52.50	40.0	37.9	65.8	58.0	106.5	100.2	181.4	170.4
55.00	20.7	20.5	46.4	42.9	71.0	63.1	157.3	154.7
57.00		-10.3	18.7	17.8		-3.1	65.0	65.0
				Trajectory III[d,f]				
18.50		5.8				5.9		
20.00		10.2				13.6		
22.00	21.8	20.4			33.0	38.2		
23.50	32.7	31.0			65.6	70.9		
25.00	37.2	35.7			107.9	99.9		
26.00	27.5	26.8			94.0	83.2		
27.00	11.3	11.2			40.5	34.5		
				Trajectory IV[e,f]				
53.00		20.0				26.0		
54.50		35.4				59.0		
56.00	55.3	54.7			103.5	118.9		
56.75	63.0	60.5			145.7	152.7		
57.75	59.0	56.1			187.5	171.9		
59.00	26.6	26.2			103.6	89.2		

[a]Convective power, MW = $\int \dot{q}_c \, dA$, $\dot{m} = 0$.

[b]Instantaneous transition at $S/R_N = 0.1$.

[c]Gas composition 89% H_2 - 11% He by volume.

[d]Relative entry velocity, 28.1 km/s.

[e]Relative entry velocity, 32.0 km/s.

[f]Gas composition 75% H_2 - 25% He by volume.

laminar calculations). Further inspection of the results presented in Table 4 shows that negative flux values are computed for trajectories I and II at times after peak heating for the $T_{W,sub}$ condition. For these conditions, the assumed wall temperature is actually greater than the shock temperature. These results indicate (as is discussed in a subsequent section) that the Saturn entry conditions do not sustain the thermal environment required for large heat-shield mass-loss rates. Thus, the effect of a possible large variation in wall temperature on the validity of the approximate results was investigated by computing the convective flux for trajectories I and II at a wall temperature of 2000 K. The approximate and detailed results are presented in Table 4 with discrepancies of less than 15%.

Transition

Since the thermal design for current Saturn entry conditions appears to be convective dominated, selection of a transition criterion is critical to heat-shield design studies. For the radiative-dominated Jovian entry conditions, massive-surface-blowing rates prompted the assumption of instantaneous transition at a S/R_N value of 0.1. This transition criterion has been used herein for the preliminary study.

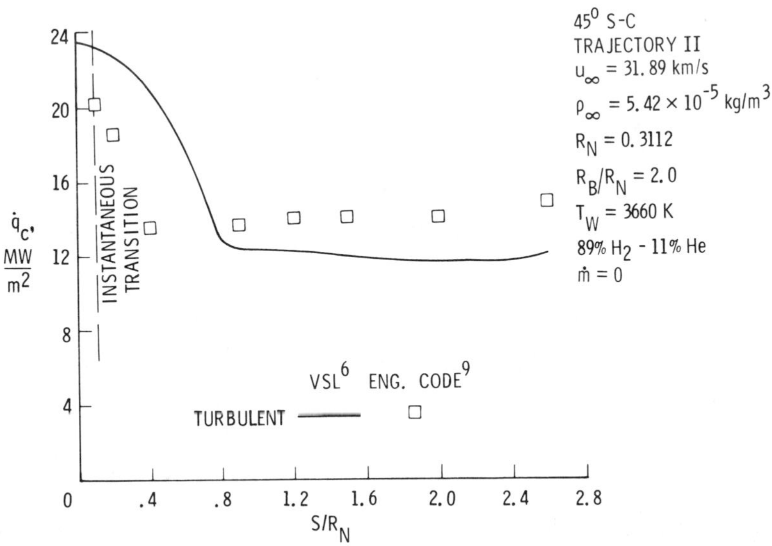

Fig. 5 Comparison of turbulent convective heat-transfer distributions at low Reynolds number conditions ($\dot{m} = 0$).

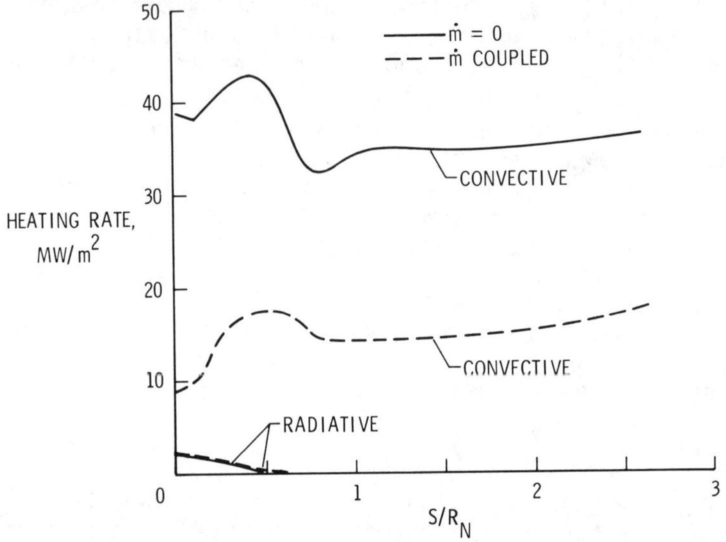

a) Trajectory I (U_∞ = 27.13 km/s)

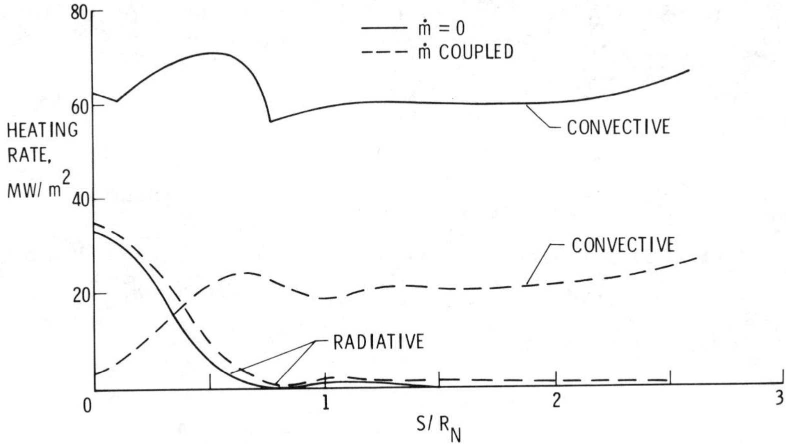

b) Trajectory II (U_∞ = 29.26 km/s)

Fig. 6 Effect of coupled carbon-phenolic injection on heating (equilibrium, instantaneous transition - (S/R_N) = 0.1).

The effect of using such a transition criterion for times early in a trajectory is demonstrated in Fig. 5. The approximate and detailed heating rates computed for trajectory II and a wall temperature of 3660 K are in poor agreement. The poor agreement can be attributed in part to using an unrealistically low Reynolds number in the approximate turbulent heating-rate equations. The local Reynolds number at the end of the forebody and the corresponding momentum-thickness Reynolds number are less than 65,000 and 120, respectively. Conservative values for transition onset have been quoted in recent Venusian studies as 150,000 for the local Reynolds number[16] and 200 for the momentum-thickness Reynolds number.[17] Thus, turbulent flow based on these Reynolds number criteria would not exist for the conditions shown in Fig. 5.

Coupled Solutions

Adequate heat-shield design requires reliable estimates of the effectiveness of the gaseous products of ablation in reducing the surface heating rates and thus the heat-shield mass loss. The heating-rate distributions, both with and without carbon-phenolic mass injection, have been computed near the time of peak heating for both trajectories I and II and the results are presented in Fig. 6. Instantaneous transition to turbulent flow is arbitrarily initiated at an S/R_N value of 0.1. The no-injection heating results were obtained using the same wall temperature as the corresponding solutions with coupled carbon-phenolic injection. With coupled injection, a slight increase in the radiative heating is observed (large percentage increase on the conical portion of the probe) when compared with the no-injection values. This phenomenon, at least for the conical portion of the probe, has been obtained in Galileo design calculations.[6,18] Coupled injection produces a substantial reduction in convective heating as evidenced by the results presented in Fig. 6. The extent of the reduction is a function of entry conditions. The carbon-phenolic mass-loss rates, assuming quasisteady sublimation, are 0.60 kg/s and 1.15 kg/s for the heating rates shown in Figs. 6a and 6b, respectively. These mass-loss rates are substantially less than the 7.1 kg/s predicted[6] for a peak heating condition during a nominal Galileo entry. The present results show that for nominal Saturn entry conditions, such as those for trajectory I, the heat-shield mass loss will be dependent almost entirely on the convective heating and the ablative mass loss will be modest when compared with that predicted for Jupiter entry. For higher speed entries, such as represented by trajectory II, the solutions with coupled mass injection show the flank region of the probe to be dominated by convective heating, whereas the stagnation region is dominated by radiative heating.

Conclusion

A preliminary thermal analysis based on recently defined Saturn entry conditions has been conducted. The present thermal analysis exercised VSL and engineering (BIRCHES) methods developed for Project Galileo.

The nominal trajectory that is used as a model for the purposes of this investigation is based on a 45 deg spherically blunted cone entering the Saturn atmosphere consisting of a 0.89 H_2 - 0.11 He gas mixture by volume at a relative entry velocity of 28.1 km/s with a relative entry angle of -30 deg. Off-nominal conditions considered in this investigation provide information on the effects of a change in entry velocity or gas composition on the thermal environment.

The analysis included investigations of nonequilibrium chemistry effects on the probe-surface heating and of the relative significance of the heat-transfer modes; a discussion of problem areas that should be addressed in future thermal design studies; and a validation of BIRCHES for Saturn entry conditions.

According to the analysis used in this investigation, the apparent effect of nonequilibrium chemistry on the Saturn thermal environment is to increase the stagnation-region radiative rates compared to corresponding fluxes computed for equilibrium chemistry conditions. For the peak heating times of the trajectories, ratios of the chemical nonequilibrium value approximately as large as 3 are computed. However, the relative significance of nonequilibrium on the radiative fluxes was found to decrease for higher entry velocity conditions and for lower concentrations of hydrogen in the freestream.

The nonequilibrium chemistry effect on the convective rates is relatively small. In addition, the convective rates are much greater than the corresponding radiative fluxes, especially over the conical portion of the forebody. Thus, the dominant heating mode for the prescribed Saturn entry conditions is convection.

The convective heating-rate calculations based on results of BIRCHES and a detailed code are in good agreement, but the approximate radiative calculations overpredicted significantly flank region heating where the predicted detailed radiative fluxes are extremely low; however, BIRCHES is well suited for parametric or preliminary Saturn probe design studies. Comparison of detailed and approximate turbulent heating-rate calculations demonstrated that realistic selection of a transition criterion will be critical for Saturn thermal studies.

The resulting mass-loss rates for the currently prescribed nominal Saturn entry conditions are small when compared with the values for nominal Jupiter entry conditions. With coupled carbon-phenolic ablation injection, the convective heating rates are reduced substantially while the radiative heating rates are increased when compared with the corresponding no-injection results.

Appendix: Revised Thermodynamic Correlations

A method for correlating the enthalpy and temperature as functions of the pressure and density over a range of temperatures, pressures, and hydrogen-helium gas mixtures has been reported.[10] In that investigation, two methods are proposed for calculating both the enthalpy and temperature proportionality constants. The method used for the outer-planet flowfield calculations is referred to as method II.[10] In the method II technique, the transformed proportionality constant for both the enthalpy and temperature is represented as a fifth-order least squares polynomial in terms of the freestream velocity. The results of the polynomial expressions are implied to be applicable over a velocity range, given as $U_\infty \sin\theta_s$, from 20 to 80 km/s.

As a result of the lower limit on $U_\infty \sin\theta_s$, a thermal analysis using the engineering code is not possible for significant portions of currently prescribed Saturn entry conditions. For the present investigation, results of additional equilibrium shock calculations have been included in the method II correlation technique[10] to extend the range for $U_\infty \sin\theta_s$ to 10 km/s. Two fifth-order least squares polynomials are used herein to represent either the transformed enthalpy or temperature proportionality constants over the velocity range, $U_\infty \sin\theta_s$, from 10 to 80 km/s. The new polynomial expressions are presented in the nomenclature of the initial investigation[10] (note that the additional pertinent relations in that reference are unchanged).

For $U_t > 26.76$ km/s,

<u>enthalpy coefficient</u>

$$C_{h,U_t} = 5.6761943 - 0.52805436 U_t + 0.02042713\, U_t^2 - 0.00037944577 U_t^3 + 0.0000034328026 U_t^4 - 0.00000001222264 U_t^5 \quad \text{(A1)}$$

temperature coefficient

$$C_{T,U_t} = -546.33326 + 61.723131U_t - 2.25117U_t^2 + 0.040035884U_t^3 - 0.00035267023U_t^4 + 0.0000012408732U_t^5 \quad \text{(A2)}$$

For $U_t < 26.76$ km/s,

enthalpy coefficient

$$C_{h,U_t} = -3.6968734 + 1.2003352U_t - 0.14153049U_t^2 + 0.0083260905U_t^3 - 0.00023753921U_t^4 + 0.0000026013019U_t^5 \quad \text{(A3)}$$

temperature coefficient

$$C_{T,U_t} = 623.14894 - 164.99894U_t + 20.347507U_t^2 - 1.2118144U_t^3 + 0.034511891U_t^4 - 0.00037612162U_t^5 \quad \text{(A4)}$$

where $U_t = U_\infty \sin\theta_s$ 1. + 0.7467 (1. - XH2) and XH2 is the hydrogen mole fraction at freestream conditions.

Acknowledgments

The authors wish to acknowledge the contributions of Richard Powell, who provided the trajectory data; S. V. Subramanian, who provided nonequilibrium calculations; and Ann Simmonds, who assisted in data reduction.

References

[1]Wright, J. L. and Hibbs, A. R., "Saturn Orbiter Dual Probe Mission Concept," JPL 710-20, Sept. 1978.

[2]Nicolet, W. E., Howe, J. T., and Mezines, S. A., "Outer Planet Probe Entry Thermal Protection, Part I: Aerothermodynamic Environment," AIAA Paper 74-700, July 1974.

[3]Howe, J. T., "Hydrogen Ionization in the Shock Layer for Entry Into the Outer Planets," AIAA Journal, Vol. 12, June 1974, pp. 875-876.

[4]Leibowitz, L. P. and Kuo, T., "Outer Planet Entry Nonequilibrium Heating," AIAA Paper 75-1149, Sept. 1975.

[5]Sutton, K., "Radiative Heating About Outer Planet Entry Probes," AIAA Paper 75-183, Jan. 1975, New York.

[6]Moss, J. N., "A Study of the Aerothermal Entry Environment for the Galileo Probe," AIAA Paper 79-1081, June 1979.

[7]Tiwari, S. N. and Szema, K. Y., "Effects of Precursor Heating on Radiating and Chemically Reacting Viscous Flow Around a Jovian Entry Body," NASA CR-3186, Sept. 1979.

[8]Zoby, E. V., Sutton, K., Olstad, W. B., and Moss, J. N., "An Approximate Inviscid Radiating Flow Field Analysis for Outer Planet Entry Probes," AIAA Paper 78-189, Jan. 1978.

[9]Zoby, E. V., Moss, J. N., and Sutton, K., "Approximate Convective Heating Equations for Hypersonic Flows," AIAA Paper 79-1078, June 1979.

[10]Zoby, E. V., Gnoffo, P. A., and Graves, R. A., Jr., "Correlations for Determining Thermodynamic Properties of Hydrogen-Helium Gas Mixtures at Temperatures From 7000 to 35,000 K," NASA TN D-8262, Aug. 1976.

[11]Leibowitz, L. P., "Measurements of the Structure of an Ionizing Shock Wave in a Hydrogen-Helium Mixture," Physics of Fluids, Vol. 16, Jan. 1973, pp. 59-68.

[12]Zoby, E. V., Graves, R. A., Jr., Moss, J. N., Simmonds, A., and Kumar, A., "Correlations for Determining Transport Properties of Hydrogen-Helium Gas Mixtures at Temperatures From 1000 to 25,000 K," AIAA Paper 79-0034, Jan. 1979.

[13]Maslen, S. H., "Inviscid Hypersonic Flow Past Smooth Symmetric Bodies," AIAA Journal, Vol. 2, June 1964, pp. 1055-1061.

[14]Anon, "The Planet Saturn (1970)," NASA SP-8091, June 1972.

[15]Brauer, G. L., Cornick, D. E., and Stevenson, R., "Capabilities and Applications of the Program to Optimize Simulated Trajectories (POST)," NASA CR-2770, 1977.

[16]Brewer, R. A., "Selection of a Heat Protection System for Venusian Entry," AIAA Paper 75-731, May 1975.

[17]Sutton, K., "Coupled Nongray Radiating Flow About Planetary Entry Bodies," AIAA Journal, Vol. 12, Aug. 1974, pp. 1099-1105.

[18]Nicolet, W. E. and Balakrishnan, A., "Methods for Predicting Off-Stagnation Point Flowfields for Planetary Entry Probes," AIAA Paper 79-1083, June 1979.

THERMOPHYSICAL AND SYSTEMS INTEGRATION CONSIDERATIONS IN AEROBRAKING DESIGN

J. R. French* and A. D. McRonald†

Jet Propulsion Laboratory, California Institute of Technology, Pasadena, Calif.

Abstract

The aerobraking concept for decelerating spacecraft into low-energy orbits is summarized. Data and comparisons are given for aerobraking approaches to Venus, Earth, Mars, and Titan. Calculations are based on adaption of a craft similar to the Venus Orbiting Imaging Radar vehicle to an aerobraking configuration. Special attention is given to integration of the aerobraking parts into the system and protection of the craft from heat loads due to the aerobraking maneuvers. Aerobraking shows substantial reductions in required launch mass for all the targets.

Nomenclature

C_D = drag coefficient
h_ρ = local density scale height
P_o = initial period of orbit
p = pressure
p_{p1} = peak stagnation pressure, first pass
Q_1 = time-integrated heating, first pass
q_p = heat flux at periapsis
t = pass duration
V = velocity
Z_a = apoapsis altitude
Z_p = periapsis altitude
α = entry angle of attack

Presented as Paper 80-1492 at the AIAA 15th Thermophysics Conference, Snowmass, Colo., July 14-16, 1980.

*Manager, Advanced Planetary Study, Technology and Space Program Development.

†Member of Technical Staff, Applied Mechanics Technology, Applied Mechanics Division.

ΔV = change in velocity
ρ = density
ρ_p = periapsis density

Introduction

The use of atmospheric drag to circularize an initially elliptical orbit of a spacecraft at a planet or other body with an atmosphere is a concept that has recently been studied for the proposed Venus Orbiting Imaging Radar (VOIR) mapping mission as a means of reducing launch mass requirements. Studies of its use in Earth orbit for the return of the Space Tug from synchronous orbit were reviewed by Hethcoat,[1] and an experimental verification in Earth orbit was conducted in the Atmospheric Explorer series.[2] First studies of the application of aerobraking to the VOIR mission were reported by McRonald.[3,4]

A further, more detailed study of the VOIR application was included in an aerobraking feasibility study conducted in 1979,[5] in which the use of a large aft deployable aerobraking surface was included to make the spacecraft aerodynamically stable during the atmospheric passes, and questions of thermophysics and systems integration were addressed. This study also analyzed aerobraking parameters at Earth and Mars.

In all cases, the goal is a considerable reduction in spacecraft mass by eliminating propellant mass required for the circularization maneuver.

This paper compares aerobraking at Venus, Earth, and Mars; extends the analysis to Titan; and describes some aspects of thermophysics and system integration in the VOIR mission.

The Aerobraking Concept

Aerobraking is the term applied to very gradual orbit decay of a spacecraft, basically in its final circular orbit configuration, and it differs markedly from the use of an atmosphere for orbit insertion, aerocapture, where the maneuver is performed in one or very few passes, requiring the spacecraft to be encapsulated in an entry vehicle.

Conceptually, the decay of an elliptical orbit by aerobraking is simple. A spacecraft is initially injected by chemical propulsion into a highly elliptical orbit with periapsis just above the planetary atmosphere. Aerobraking is initiated by lowering the periapsis (by a small propulsive maneuver at apoapsis) to a level where the spacecraft makes a brief pass through the atmosphere near periapsis. The resulting small velocity loss (ΔV) causes a slight reduction in apoapsis distance and period, while the periapsis distance scarcely changes. If allowed to proceed, this process will tend to continue until the orbit becomes nearly circular, when it is terminated by performing a small propulsive maneuver (at apoapsis) to raise the periapsis out of the atmosphere. The process is inherently stable and regular, and it is very flexible in that one can intervene to raise or lower periapsis by a few kilometers during the orbit decay, thus controlling the decay rate.

Analysis shows that the limiting factor on the allowable rate of orbit decay is heating, and a front shield must be included to protect the spacecraft. In addition, it is advantageous to reduce the total orbit decay time by increasing the drag surface area, i.e., adding a large drag brake. Because of launch vehicle limitations on size, the large brake must be deployable.

The rate at which the orbit decays is determined by the vehicle drag at each successive pass, and this may vary or be uncertain for several reasons, including:

1) Atmospheric density may not be known precisely at the chosen periapsis altitude.

2) Density may vary with time of day and location of periapsis.

3) Solar gravity may perturb the periapsis distance, especially initially, when the orbit is highly elliptical.

4) Periapsis distance may be perturbed by gravity anomalies of the planet.

5) Aerodynamic characteristics (e.g., drag and lift) of the spacecraft may not be known precisely because the aerodynamic conditions for aerobraking fall in the transitional regime (about which information is incomplete) between free molecular and continuum conditions.

6) Vehicle angle of attack changes during the atmospheric pass, and this "excursion" may vary slightly. (Even an inherently stable vehicle will oscillate about $\alpha = 0$ deg.)

Most of these factors will follow a regular pattern as aerobraking proceeds, but the uncertainties they cause make it prudent for aerobraking to be initiated conservatively at a low rate, stepped up to the design level as experience and confidence in prediction is gained, and monitored by onboard measurements (e.g., accelerometers, thermometers) and Earth-based orbit determination.

There are four further features that become important towards the end of aerobraking, as the orbit nears circular:

1) Period between passes becomes short (<2 h for Venus and Earth).

2) Angle through which the velocity vector rotates in the atmospheric pass, initially only a few degrees, becomes much larger as the portion of the orbit in the atmosphere increases.

3) Rate of fall of periapsis, initially a few meters per pass, becomes as large as several hundred meters per pass.

4) Time-integrated heating per pass increases a little due to the increased pass duration (although the peak heating rate decreases).

The large angle is important because the vehicle must rotate through this angle during the pass to maintain the design condition of near zero angle of attack, and must be reset to the appropriate attitude at the start of the next pass.

The remedies to these endphase problems are straightforward. Raising the periapsis gradually towards the end of aerobraking will reduce factors 3) and 4) (fall of periapsis and heating) and will allow a larger margin of error in defining angular attitude as forces and heating are reduced. The remedy for factor 1) is to temporarily raise periapsis out of the atmosphere for a few orbits on a few occasions in the last day or two of aerobraking to allow time for review or orbit determination (aerobraking can be stopped and started without much loss of time because the period is short).

We note that the angular rotation of the spacecraft as it passes through periapsis (to maintain zero angle of attack) is much simpler when the aft aerobrake is located behind the spacecraft mass center; the vehicle is stable, i.e., aerodynamic forces at nonzero angle of attack tend to reduce the angle of attack to zero. This means that the vehicle will negotiate the atmospheric passes successfully by virtue of its natural stability.

Mission Profile for Aerobraking

The aerobraking concept is implemented as follows:

1) Launch from Earth and interplanetary cruise are the same as for any planetary mission.

2) Orbit insertion is by propulsive means into an elliptical orbit with periapsis just above the atmosphere; initial orbit period is a tradeoff between propulsion mass and circularization time.

3) Once the spacecraft is established in orbit, the drag brake is deployed, the spacecraft is checked, and the orbit is determined accurately.

4) When all is ready, a propulsive maneuver at apoapsis drops periapsis to the chosen altitude for aerobraking (a gradual approach will probably be used, for conservatism). Before entry to each pass, the spacecraft attitude is reset from the exit value at the previous pass or the attitude required for an apoapsis propulsive maneuver (for most cases a change of less than 10 deg).

5) Thereafter, the aerobraking process is monitored by onboard measurements and orbit determination, with occasional adjustments being made, as required, to periapsis altitude (small propulsive burns at apoapsis). These maneuvers may include temporarily raising periapsis to arrest aerobraking if time is required for more accurate orbit determination before proceeding, and may include a gradual tapering off of the aerobraking rate as the orbit becomes nearly circular. (Ideally, aerobraking is ended when the apoapsis approaches the final circular altitude, at which point the minimum propellant is used to raise periapsis and circularize the orbit.)

6) At the end of aerobraking, two propulsive maneuvers are executed to adjust periapsis and apoapsis to the desired circular values. The drag brake and front shield may then be jettisoned.

Comparison of Aerobraking at Venus, Earth, Mars, and Titan

As mentioned previously, a recent study[5] evaluated the aerobraking concept for possible application to the VOIR mission. A nominal case was adopted, of orbit decay in 45 days from an initial orbit of period 12 h, for a spacecraft of mass

Table 1 Parameters linear with periapsis density ρ_p: initial period P_o = 24 h, $\rho_p = 3.0 \times 10^{-8}$ kg/m^3

Symbol	Parameter definition	Probe target			
		Venus	Earth	Mars	Titan
p_{p1}	Peak stagnation pressure, 1st pass, N/m^2	2.90	3.39	0.67	0.15
q_{p1}	Peak heating rate, 1st pass, W/cm^2	1.43	1.80	0.16	0.016
Q_1	Time-integrated heating, 1st pass, J/cm^2	93.4	139.1	20.6	12.0
Q_e	Time-integrated heating, last pass, J/cm^2	183	272	34.2	13.2
$dZ_{a,e}$	Fall in apoapsis altitude, last pass, km	70	100	33	70
dZ_p	Fall in periapsis altitude, 1st pass, km	0.0044	0.0087	0.0042	0.134
$dZ_{p,e}$	Fall in periapsis altitude, last pass, km	0.42	0.90	0.30	0.80

1630 kg, with an aft aerobrake diameter of 9.0 m, at constant periapsis density, ρ_p, of 3.0×10^{-8} kg/m^3, periapsis altitude, Z_p, of ∿132 km (Venus Model I atmosphere[6]).

Subsequently, the second volume of the feasibility study[5] analyzed aerobraking at Venus, Earth (U.S. 1976 Standard Atmosphere[7]), and Mars (1975 Mean Atmosphere[8]). The data for Mars have been recalculated using the measured Viking I density profile,[9] and data for Titan have been added (the 1973 Nominal Model[10]).

For these bodies and atmospheres, aerobraking orbit decay has been evaluated: from initial periods of 12, 24, and 48 h; at constant periapsis density ρ_p of 7, 5, 3, and 1.5×10^{-8} kg/m^3, using the VOIR reference spacecraft (1630 kg mass, 9.0 m diam aerobrake, drag coefficient $C_D = 2$). The range of decay times was 22 to 189 days for Venus, 16 to 133 days for Earth, 17 to 220 days for Mars, and 8 to 110 days for Titan. In addition, a ρ_p value of 15×10^{-8} was analyzed for Mars and Titan.

The analysis requires the density ρ and the local density scale height h_ρ as a function of altitude, shown in Figs. 1 and 2, respectively. Tables 1 and 2 give the aerobraking parameters. Table 1 gives sample parameters which depend linearly on ρ_p (the peak stagnation pressure p_{p1}, the peak heating rate q_{p1}, the time integrated heating Q_1, and the fall of periapsis dZ_{p1} for the first atmospheric pass, for the case with initial period Po = 24 h, $\rho_p = 3.0 \times 10^{-8}$ kg/m^3. Also given are Q, dZ_p, and dZ_a, the fall in apoapsis altitude for the last pass (when the apoapsis approaches the target circular altitude, this being 300 km for Venus, Earth, and Mars, and 1000 km for Titan).

Table 2 gives parameters which are generally independent of ρ_p: the half-pass duration $2_{½}$, the entry angle of attack α_5 (the angle through which the velocity vector rotates), measured at 5% of peak level, for the first and last passes; $\Delta V_{circ,o}$ and $\Delta V_{circ,e}$, the equivalent propulsion ΔV to go circular, initially, and at the end of aerobraking, and P_e, the period at the end. The difference between $\Delta V_{circ,o}$ and $\Delta V_{circ,e}$ is the potential gain from aerobraking. For example, for P_o = 12 h, the ΔV difference is 2370, 2177, 966, and 354 m/s for Earth, Venus, Mars, and Titan, respectively.

Figures 3 and 4 show the aerobraking parameters graphically as a function of periapsis density ρ_p. Figure 3 gives the total decay time T (at constant ρ_p), and Fig. 4 shows the

Table 2 Parameters insensitive to P_o and ρ_p, at P_o = 24 h, $\rho_p = 3.0 \times 10^{-8}$ kg/m^3

Symbol	Parameter definition		Probe target			
			Venus	Earth	Mars	Titan
$t_{1/2,1}$	Half pass duration (5% of peak), first pass, s		64	76	128	721
$t_{1/2,e}$	Half pass duration (5% of peak), last pass, s		320	390	490	1376
$\alpha_{5,1}$	Change in angle of V (5% of peak), first pass, deg		3.2	3.9	5.4	16.1
$\alpha_{5,e}$	Change in angle of V (5% of peak), last pass, deg		21.0	26.5	27.2	37.2
$\Delta V_{circ,o}$	Total propulsive ΔV to go circular, initially, m/s	P_o = 12 h	2297	2489	1061	516
		P_o = 24 h	2542	2748	1193	605
		P_o = 48 h	2694	2908	1274	657
$\Delta V_{circ,e}$	Total propulsive ΔV to go circular, last pass, m/s		120	119	95	162
P_e	Final period, h		1.56	1.51	1.89	4.60

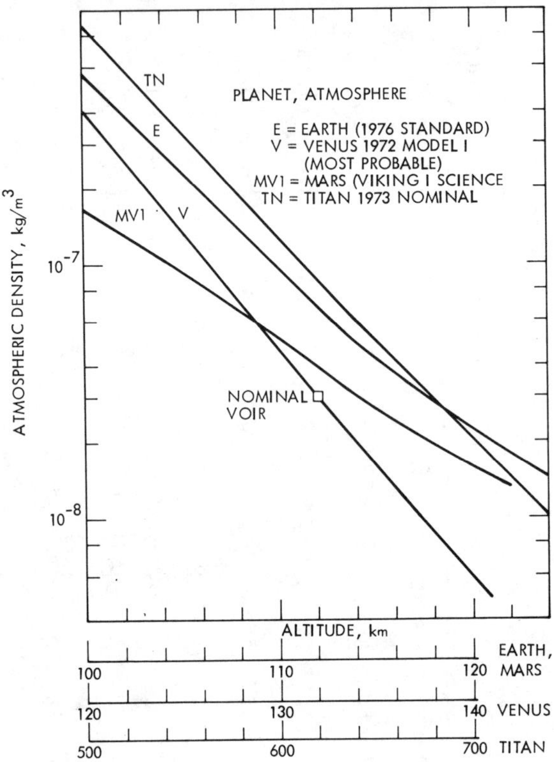

Fig. 1 Atmospheric density as a function of altitude.

integrated heating Q_1, the peak heating rate q_{p1}, and the peak stagnation pressure p_{p1} for the first pass, for P_0 = 12, 24, and 48 h.

It can be seen that Venus and Earth are generally similar in regard to aerobraking: for a given ρ_p, the decay is a little slower for Venus (smaller velocities), with corresponding smaller peak pressure and heating rate. Venus has also smaller integrated heating, shorter pass duration, smaller "entry" angle, and slower rate of fall of apoapsis and periapsis.

Similarly, for a given ρ_p, Mars (with its lower velocities) has longer decay time, smaller pressures, heating rate, integrated heating, rate of fall of apoapsis and periapsis; but it has greater pass duration and entry angle.

The low gravity of Titan gives it low orbital velocities and a large scale height. Thus, although for a given ρ_p, it

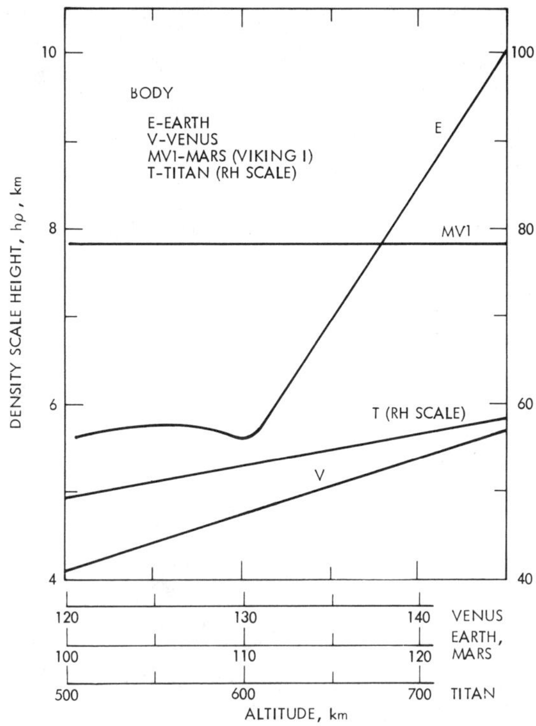

Fig. 2 Density scale height as a function of altitude.

has low pressure, heating rate, and short decay time, it has relatively large values for integrated heating, pass duration, entry angle, and rate of fall of apoapsis and periapsis. Combined with the low potential ΔV loss in aerobraking, this makes Titan a comparatively poor candidate for aerobraking.

A more useful comparison for Venus, Earth, and Mars may be to compare conditions at the VOIR design levels of heating rate and integrated heating (1.5 W/cm^2 and 95 J/cm^2 for which the orbit at Venus decayed in 45 days from an initial 12 h period, at $\rho_p = 3.0 \times 10^{-8}$ kg/m^3. Inspection of Figs. 3 and 4 shows the same levels would be reached for Earth and Mars at ρ_p of 1.9 and 13.5×10^{-8} kg/m^3, respectively (Z_p of 121 and 101 km, respectively), giving decay times of 50 and 10.5 days, respectively, from the same initial orbit.

System Integration/Thermophysics Considerations

In the VOIR study,[5] aerobraking was considered as an option for an existing design of spacecraft and mission.

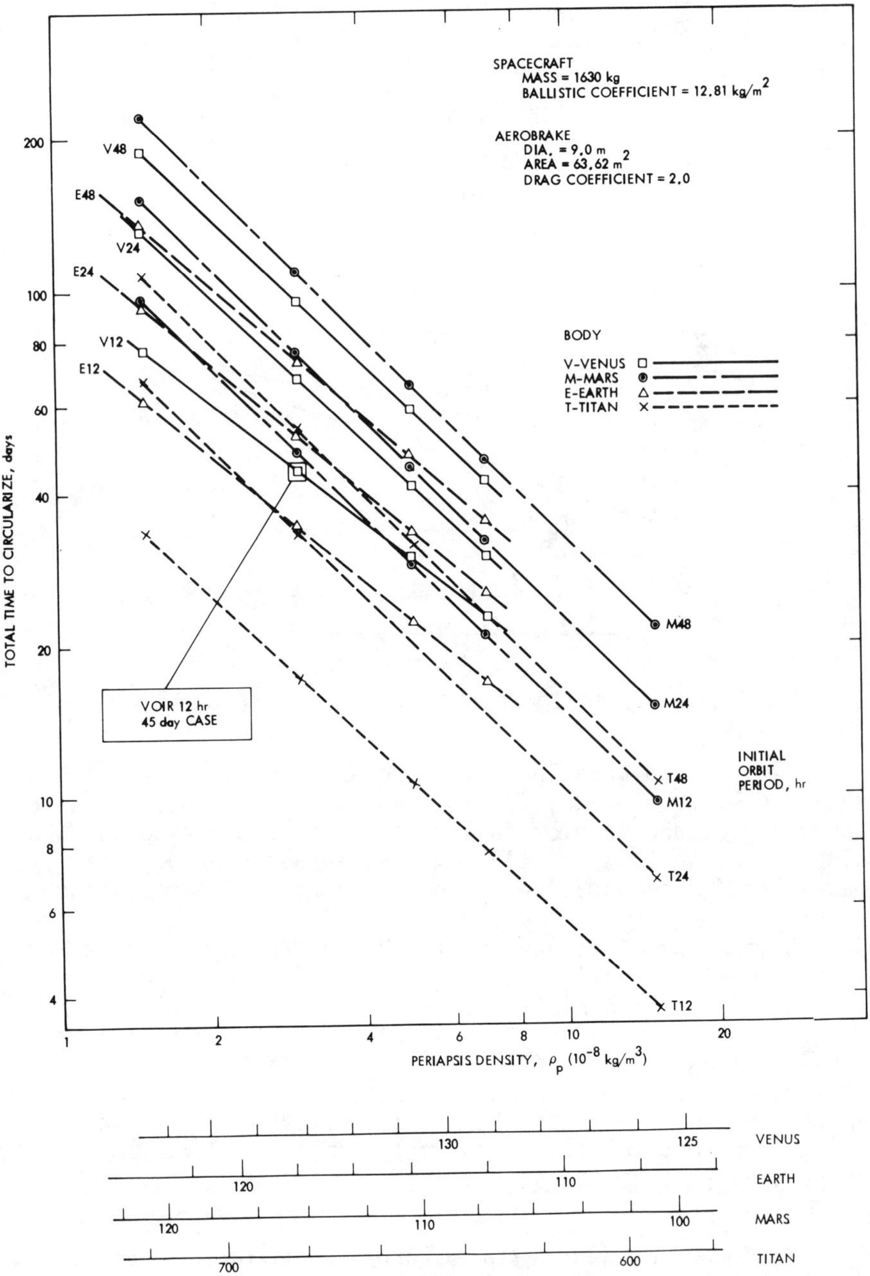

Fig. 3 Circularization time as a function of periapsis density.

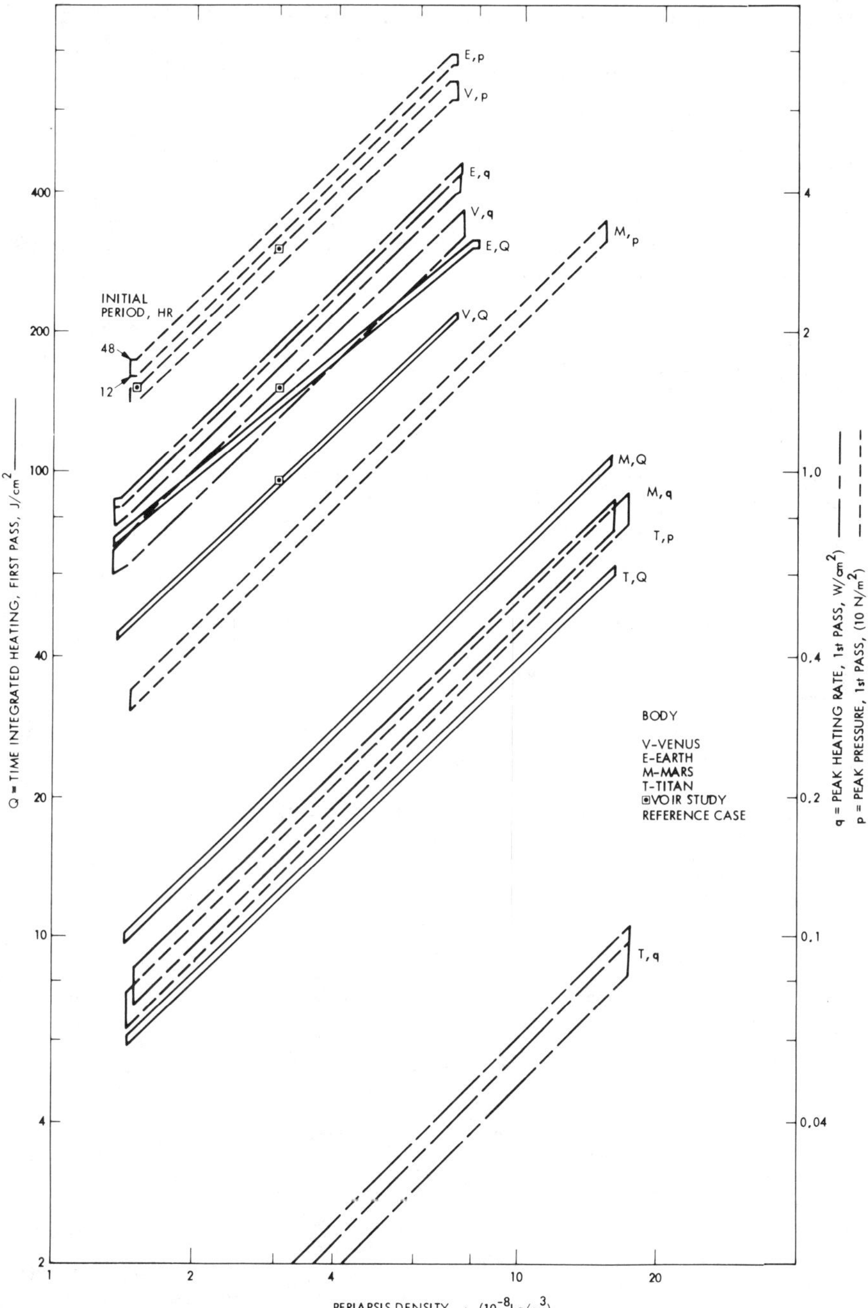

Fig. 4 Pressure and heating as a function of periapsis density (double lines show values for 48 h and 12 h initial period).

Besides the obvious adjustment in the mission profile to allow time for aerobraking (45 days orbit decay), integration of aerobraking has to meet a number of constraints:

1) Thermal shield and drag brake should be added in a "bolt on" manner.

2) Heat load to the spacecraft bus in the aerobraking passes should not exceed the design level (~1 Sol, i.e., the solar radiation level near Earth).

3) Spacecraft bus unrestricted view of space should be retained to facilitate heat radiation by the bus in its thermal control function.

4) Ambient gas density at the spacecraft bus should not exceed a level at which conduction degrades the performance of the thermal blankets (roughly 10^{-4} Torr).

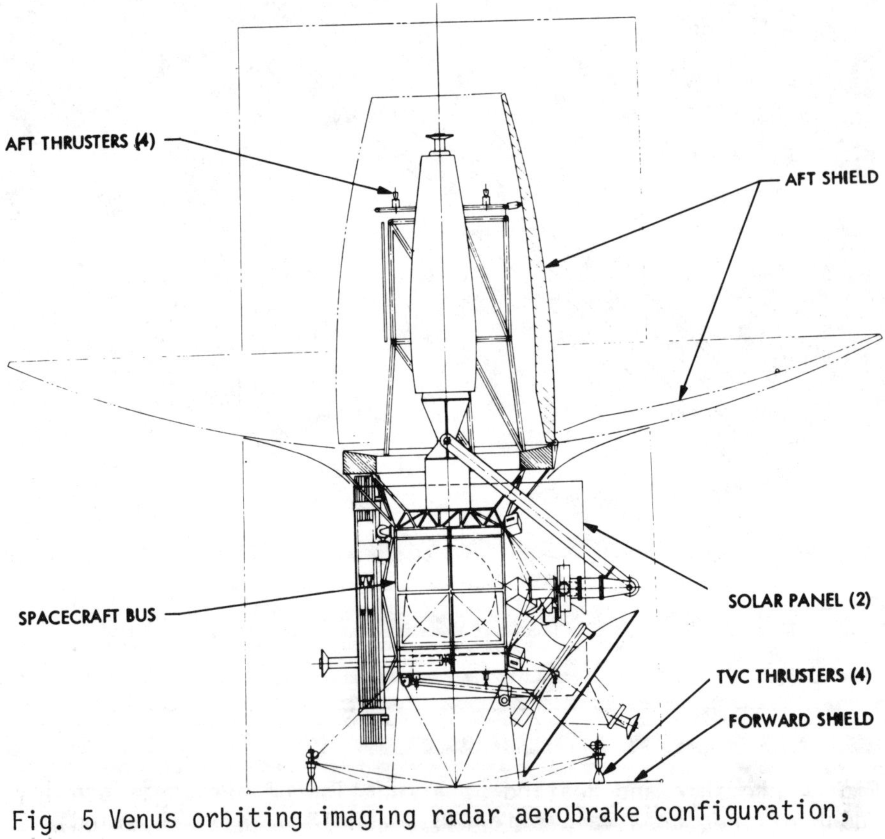

Fig. 5 Venus orbiting imaging radar aerobrake configuration, side view.

The solution adopted to meet these constraints is illustrated in Fig. 5. The two main elements added to the previous nonaerobraking spacecraft design are 1) a fixed forward aeroshield of diameter 4.0 m, about as large as can be fitted in the available launch vehicle space, and 2) a deployable aft drag brake, comprising a fixed central portion of 4.1 m diam with a ribbed skirt deploying to a total diameter of 9.0 m. The radial ribs (12 in all) stiffen the skirt when deployed. The material chosen for the shield and brake is Inconel 718, which has excellent high-temperature properties, is available in foil (down to 1 mil or 25.4 μm), is weldable, and has existing test data on control of optical (thermal radiative) properties.

In aerobraking, the exposed surfaces are heated aerodynamically (front of shield and front outer part of brake). Because these are thin and conduct, the back faces will also be hot; one wishes to minimize the temperature by radiating heat as well as possible (high emittance) away from the spacecraft, but to have low emittance on the parts facing the spacecraft bus. When oxidized, Inconel 718 has high total normal emittance, ranging from 0.6 at 121°C (394 K) to 0.75 at 704°C (977 K). To maintain stress symmetry and prevent curling, both sides are oxidized. Then, the low emittance is achieved on the parts facing the spacecraft by applying and baking a thin gold coating (Hanovia "Liquid Bright Gold" No. 7621). Tests have shown that the coating produces no distortions and has an emittance of $\leq$0.06. Calculations of stress, thermal expansion, creep, and rupture indicate that the drag brake will maintain its integrity. With this arrangement, the maximum temperature reached by any part of the Inconel is estimated to be 704°C (977 K).

One area of concern with regard to the front thermal shield is "wake closure," i.e., expansion of the stream into the area behind the shield. Because the aerodynamic regime is intermediate between free molecular and continuum, accurate calculations of gas density and heating require Monte Carlo type flowfield computations which have not yet been performed. However, studies of wake closure made for a Voyager mission spacecraft plume shield, for a similar stream velocity and a much lower mean free path (and therefore conservative), indicate that the heating rate falls off rapidly with expansion angle ($\sim$1% of the stagnation value at 30 deg expansion). Thus, this conservative result indicates an acceptably low level of heating at the bus location. Similarly, the same calculations (both velocity and density are given) imply that gas density is at an acceptably low level regarding thermal blanket insulation.

Future flowfield computations may indicate that a certain amount of local thermal shielding of selected spacecraft components is required, i.e., those protruding into or located in the region affected by wake closure.

Besides the addition of the shield and the brake, modifications to the VOIR mission to accommodate aerobraking also include:

1) Addition of solar panels behind the aft shield to provide adequate electrical power during aerobraking when the existing panels must be folded because some of them would protrude beyond the shield.

2) Choosing of arrival date and initial orbit at Venus to achieve suitable sun-Earth-Venus geometry for the solar panels to see the sun and for communications with Earth during the aerobraking episode.

3) Increasing capacity of momentum wheels to control and damp fluctuations in angle of attack during the aerobraking passes.

4) Addition of small liquid fuel motors which make it easier to initiate, control, and terminate aerobraking.

With these modifications, which may be by no means optimum in terms of mass, the study of the VOIR mission found that the approach mass at Venus was reduced from about 5200 kg without aerobraking to about 4000 kg with aerobraking.

Conclusions

A comparison of aerobraking to circularize orbits at Venus, Earth, Mars, and Titan indicates that aerobraking offers considerable mass savings over an equivalent mission using propulsive maneuvers. The aerobraking parameters at Venus and Earth are found to be closely similar. For Mars the features are similar, although the potential ΔV saving is less. Because of its low gravity, Titan appears to be a poor candidate for aerobraking.

Studies of the VOIR indicate that the thermophysical and system integration problems can be solved so that aerobraking can be implemented, although perhaps not with maximum efficiency as an add-on to an existing spacecraft design. A worthwhile spacecraft mass reduction of from $\sim$5200 to $\sim$4000 kg at Venus approach was found.

Acknowledgments

The research described in this publication was carried out by the Jet Propulsion Laboratory, California Institute of Technology, under NASA Contract NAS7-100.

References

[1]Hethcoat, J. P., "Aeromaneuvering Orbit Transfer Vehicles for the Space Transport System," Paper A76-46071, 94th Congress of International Astronomy Federation, Anaheim, California, Nov. 1976.

[2]Barbieri, R. W., and Cooley, J. L., "Atmospheric Explorer (AE-C) Adaptive Mission Design and Trajectory Performance," Paper presented at the AAS/AIAA Astrodynamics Conference, Nassau, Bahamas, July 1975.

[3]McRonald, A. D., "Use of Atmospheric Drag to Circularize an Elliptic Spacecraft Orbit at Venus," JPL Internal Document 760-163, March 1, 1977.

[4]McRonald, A. D., and Nock, K. T., "Atmospheric Braking to Circularize an Elliptical Venus Orbit," Paper presented at the AAS/AIAA Astrodynamics Conference, Jackson, Wyoming, Sept. 1977.

[5]French, J. R., "Aerobraking Feasibility: Vol. I. Application to the VOIR Mission, Oct. 30, 1979; Vol II. General Application to Inner Planet Missions,"JPL Internal Document 715-6, Dec. 15, 1979.

[6]"Models of the Venus Atmosphere (1972)," NASA SP-8011, Sept. 1972.

[7]"U.S. Standard Atmosphere, 1976." NASA, Oct. 1976.

[8]"Mars Engineering Models," NASA Document M75-125-3, Nov. 1975.

[9]Seiff, A., and Kirk, D. B., "Structure of the Atmosphere of Mars in Summer at Mid-Latitudes," Journal of Geophysical Research, Vol. 82, No. 28, Sept. 1977, p. 4364.

[10]"The Atmosphere of Titan," NASA SP-340, 1973.

TECHNOLOGY REQUIREMENTS FOR A GENERIC AEROCAPTURE SYSTEM

Manuel I. Cruz*

Jet Propulsion Laboratory, California Institute of Technology, Pasadena, Calif.

Abstract

The technology requirements for the design of a generic aerocapture vehicle system are summarized. These spacecraft have the capability of completely eliminating fuel costly retropropulsion for planetary orbit capture through a single aerodynamically controlled atmospheric braking pass from a hyperbolic trajectory into a near circular orbit. This generic system has application at both the inner and outer planets. Spacecraft design integration, navigation, communications, and aerothermal protection system design problems were assessed in the technology requirements study and are discussed in this paper.

Nomenclature

ACS = attitude control system
B' = normalized mass loss parameter
bps = bit per second
c.g. = center of gravity
CN = cyanogen
ESM = elastomeric silicone material
FBN = forward bent nose
g = gravity
GE = General Electric Company
G_e = entry load in Earth g
GN&C = guidance, navigation, and control
HGA = high gain antenna

Presented as Paper 80-1493 at the AIAA 15th Thermophysics Conference, Snowmass, Colo., July 14-16, 1980.

*Member of Technical Staff, Mission Design Section, Systems Division.

h_w = wall gas enthalpy
I_{sp} = specific impulse
L = vehicle length, meter
L/D = hypersonic lift-to-drag ratio
M_∞ = free stream Mach number
M/C_DA = hypersonic ballistic coefficient, kg/m^2
MMRC = moving mass roll control
MSR = Mars sample return mission
NOMEX = nonmetallic material
$\dot{q}$ = heat flux, W/cm^2
$\dot{q}_{\delta N}$ = heat flux for a given nose bend angle
RCS = reaction control system
RF = radio frequency
RSI = reuseable silicone insulator
RSS = root-sum-square
RTG = radioisotope thermoelectric generator
RTV = room-temperature vulcanizing
SAM = silicone ablator material
SEEGA = solar electric Earth gravity assist
SEPS = solar electric propulsion system
SO2P = Saturn orbiter dual probes mission
STS = space transportation system
SWF = split windward flap
TCM = trajectory correction maneuver
TPS = thermal protection system
V = speed, km/s
V_e = entry speed, km/s
w = area weight density
W = watt
X/R_N = ratio of longitudinal station to nose radius
α = angle of attack, deg
α_T = trim angle of attack, deg
δ = deflection angle
δ_F = flap deflection angle, deg
δ_N = nose bend angle, deg
ΔV = change in velocity, m/s
γ = flight path angle, deg
γ_e = entry flight path angle, deg
ρ = density
σ = standard deviation

Introduction

This paper presents the results of a study[1] whose purposes were to identify technology developments critical to a generic aerocapture vehicle system design and to draw up plans for developing these technologies toward a mission application technology readiness date. The generic set of aerocapture mission applications was studied in an overview sense for the planets from Venus to Uranus. It was decided early in the study that the Uranus aerocapture application would evolve a spacecraft design much different from a spacecraft design for missions to the Venus, Mars, and the Saturn-Titan satellite systems. Of the applications, Mars and Saturn-Titan imposed the most stringent requirements in terms of overall system design.

The aerocapture concept which uses atmospheric braking in a single atmospheric entry pass to place spacecraft in orbit about a planetary body promises not only to more than double science payloads on some planetary spacecraft, but also to significantly decrease interplanetary cruise time.[2] This concept completely eliminates fuel costly retro propulsion. The atmospheric entry is performed using adaptive guidance algorithms which are essentially insensitive to planetary atmospheric density profile uncertainties, aerodynamic performance uncertainties, and navigation and/or planetary ephemerides uncertainties.[3,4] The critical issues associated with the spacecraft design of this concept concern spacecraft integration within aeroshells to meet planetary spacecraft functions and aerothermal design for previously unresearched atmospheric flight regimes and atmospheric chemistry (i.e., Titan's predicted methane-nitrogen atmosphere). The feasibility of this concept was initially studied for a Mars Sample Return (MSR) mission,[5] and the feasibility has now evolved into a viable planetary delivery concept.

Summary

Table 1 is an overall summary of the technical disciplines involved in the study and categorizes the critical technologies identified. A description of each line item and their criticality follows.

In the area of inertial sensors, a vast number of instruments are available which can easily meet both interplanetary and aerocapture atmospheric-flight requirements. It is felt, though, that a single gyro package could not meet both the interplanetary and atmospheric flight requirements. The capture rates are different for both applications. The interplanetary package requires low capture rates for accuracy whereas the atmospheric-flight package requires high capture rates at the expense of accuracy. The tradeoff is the power required.

Attitude control can be simply met using a reaction control system for interplanetary and atmospheric flight. The gas expenditure during atmospheric flight can be relatively large. The addition of an aerodynamic stability augmentation subsystem reduces it and enhances the mission performance.

Structural/payload design of existing spacecraft is not a problem using large aeroshells. In minimizing the aeroshell size for outer planets application, this becomes a mission enabling capability.

Thermal design requires consideration when integrating spacecraft with RTGs. This is particularly true during atmospheric flight when radiator surfaces become ineffective due to a heat load being imposed from aerothermodynamic processes. A number of design solutions exist which, at most, would require a thermal management subsystem.

Table 1 Summary of technical disciplines

Technical discipline	Technology	Technology category
Inertial sensors	None	None
Attitude control	Aerodynamic stability augmentation	Enhancing[a]
Structural design	Aeroshell/spacecraft integrated design	Enabling[a]
Payload design	Aeroshell/spacecraft integrated design	Enabling[a]
Thermal design	None	None
Aerothermal design	Aeroshell design/ flow-field analysis	Enabling
Navigation	Optical navigation processors	Enabling
Communications	Low profile antennas	Enabling
Power	None	None
Radiation	Electronics/RTG exposure	Enabling[a]

[a]Not completely evaluated.

Aerodynamic and aerothermodynamic designs are available for a variety of applications from Mars entry to Jupiter entry. Although the aerocapture GN&C system can tolerate wide dispersions in entry corridor and atmospheric models, the aeroshell must be designed to allow survival in these possible ranges. To increase reliability and enable a planetary application, a system technology study is required to investigate the forcing functions prior to a final design phase.

Entry corridor navigation requires the use of optical navigation for almost all aerocapture applications. Existing optical systems easily meet these requirements. The data-rate requirements, however, are large and can be marginally met with the aerocapture communications subsystems. To enable the use of optical navigation with the aerocapture communications subsystems, some onboard processing will be required to reduce data loads in excess of 20:1.

The communications subsystems of aerocapture require the use of low-profile antennas to reduce the risk of aerocapture by eliminating articulating and large expandable systems. The key technology is to increase the efficiency of low-profile phased array antennas.

Power is not a problem for most aerocapture applications since RTGs are part of the spacecraft system. For solar powered spacecraft, this function is easily met with mosaics of solar cells on the aeroshell.

Radiation from RTGs on an aeroshell packaged spacecraft may be a problem. This is especially true for the electronics which would be exposed to these environments for years.

Discussion

The following paragraphs discuss in detail the technologies and the system integration problems critical to the development of an aerocapture system. These include spacecraft integration, navigation, communications, and aerothermal protection with the latter given the most attention in this paper.

Spacecraft Integration--Delivery Vehicle Selection

Figures 1 and 2 show the entry vehicle geometries that have been investigated for aerocapture. The combination of high volumetric efficiency, low ballistic coefficient, and aerocapture control accuracy led to choosing biconics. Spacecraft geometric requirements are such that they can mostly be integrated within shroudlike cylindrical-conical sections of which biconics are most closely related. To maximize the entry altitude and to reduce the heating load, a low hypersonic ballistic coefficient shape is required. Biconics are at the knee of the ballistic coefficient (M/C_DA) and lift-to-drag (L/D) ratio curve. L/D is required to effect maneuverability and trajectory adaptiveness in the atmosphere. The optimum L/D ratio required for aerocapture control accuracy is 1.0 to 2.0 (the 3 σ accuracy numbers in Fig. 2 are an RSS of a number of error sources which include atmospheric, aerodynamic, navigation, and inertial instrument uncertainties[4]). Biconics have values in this range.

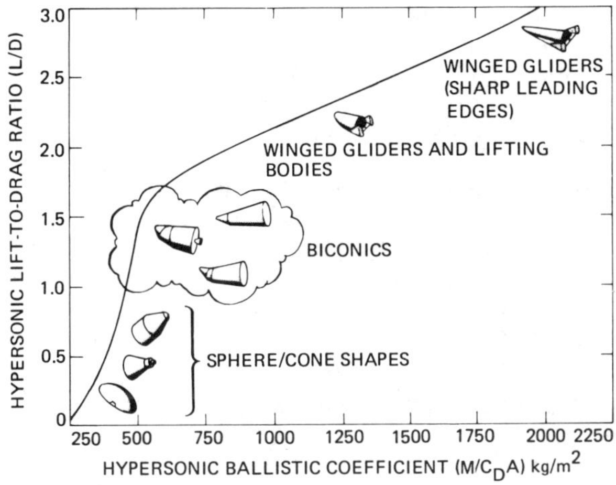

Fig. 1 Potential aerodynamic shapes for a 5000 kg mass.

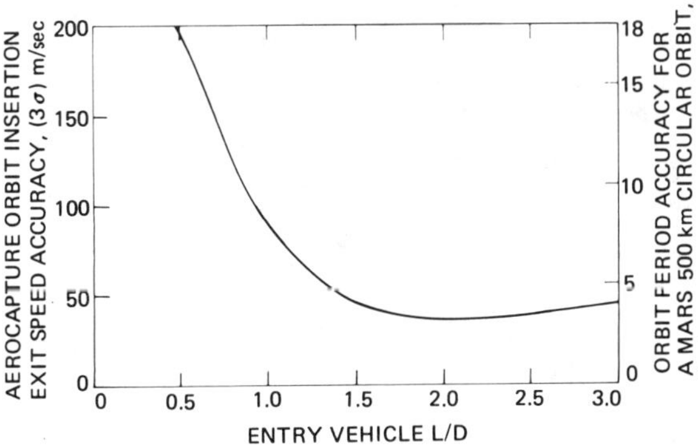

Fig. 2 Aerocapture accuracy as a function of entry vehicle L/D for an entry mass of 5000 kg.

Spacecraft Integration--Delivery Vehicle V-γ Requirements

Figure 3 shows the air relative velocity-flight-path-angle (V-γ) maps for aerocapture application in a generic sense. These translate into navigation and aerothermal protection system design requirements. The more demanding requirements are for central planetary bodies in terms of navigation. Titan, a satellite of Saturn, has a very relaxed requirement due to its physical properties (i.e., density scale height and gravity potential). The more demanding requirements in terms of entry speed are at Uranus and possibly Neptune. Although the navigation requirements are shown to be demanding for Mars and Venus, many missions have been flown to these planetary bodies so that their planetary ephemerides are well known. The entry-speed requirements are well bounded, entry at Mars being the simplest. Entry at Venus is similar to most Earth applications. Also, their atmospheric compositions are well known, since a number of atmospheric probes have been flown to both planetary bodies.

Spacecraft Integration--Saturn Orbiter Dual Probes

Shown in Fig. 4 is a Galileo copy three-axis stabilized spacecraft with two atmospheric entry probes in their interplanetary stored state that could be used for the Saturn Orbiter Dual Probe (SO2P) mission. The science packages are deployed on booms. The power subsystem consists of three Radioisotope Thermoelectric Generators (RTGs) which supply approximately 200 W of power. The main high gain antenna is a furled 5 m antenna. The propulsion tanks shown are required for in-orbit maneuvers and for aerocapture interplanetary flight. The orbiter thruster packages are deployed on booms. The conical volume shown in the forward section would accommodate the Titan and Saturn probes. The Titan probe would be the only one which would require reconfiguration from the Martin Marietta Aerospace design proposal.[6] Table 2 gives a detailed mass breakdown of the Saturn orbiter which would be supported by an aerocapture vehicle. This is exclusive of the aeroshell mass.

A major concern of this configuration is the effect on the science packages from RTG radiation. A very preliminary look at this problem indicates that these packages could be hardened for moderate flight times.

Figures 5 and 6 show the SO2P spacecraft fully deployed. Current mission thoughts consider carrying both probes into orbit with an aerocapture scenario. The Titan probe would be

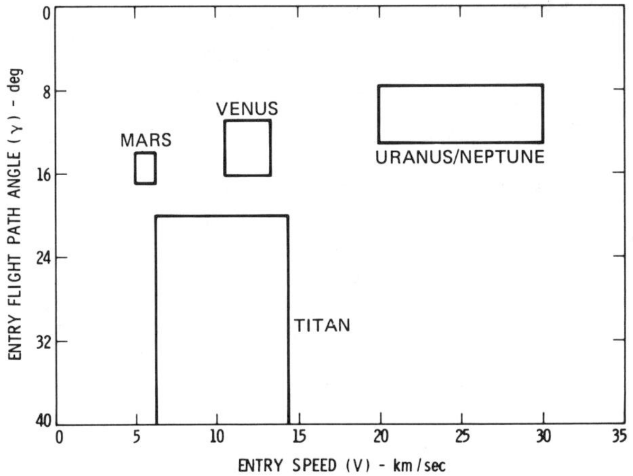

Fig. 3 Air relative planetary entry V - γ requirements for aerocapture.

Table 2 SO2P aerocapture vehicle payload mass breakdown

Item	Mass, kg
Bus, electronics, cabling	300
Adapter trusses	20
Boom outriggers	20
HGA support	6
HGA	32
RTG boom	10
RTGs	96
RTG thermal control system	50
Insulation	5
Scan science boom	7
Scan platform and science	89
Spin science booms	14
Spin platforms and science	52
Thrusters	14
Surface antennas	4
Magnetometer boom and canister	15
Miscellaneous devices, latches, and brackets	20
Probe spin-up table	8
Propellant	310
Orbiter Subtotal	1072

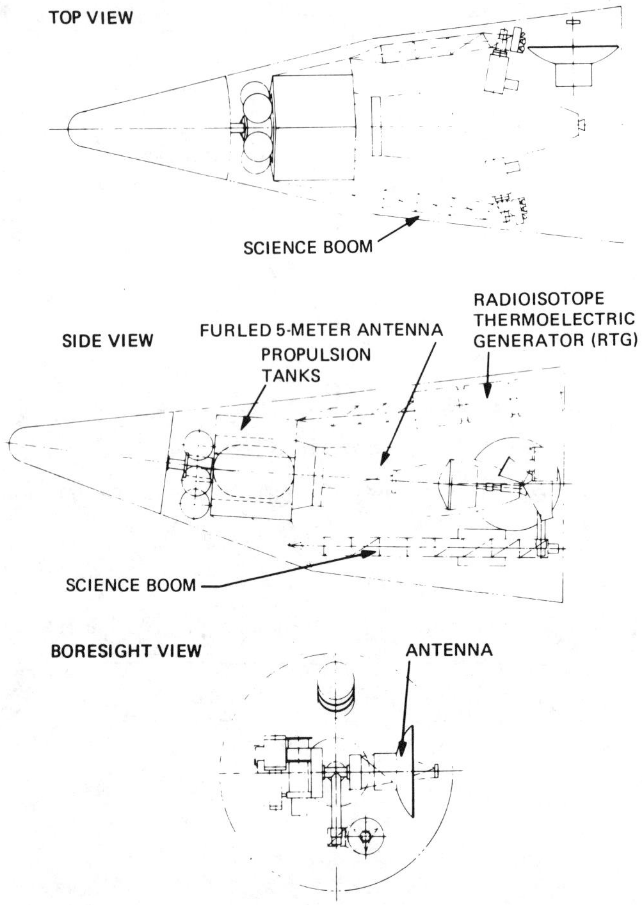

Fig. 4 Galileo-copy three-axis stabilized spacecraft.

deployed first. With the very benign ($V_e \sim 4$ km/s) Titan reentry, the Titan probe could essentially reuse part of the aerocapture aerothermal protection system ($V_e \sim 10$ km/s). The initial aerocapture orbit would be designed to have a low Saturn periapse. This, along with a Titan gravity assist maneuver strategy, would enable the Saturn probe to be deployed into the Saturn atmosphere and an extended orbiter overflight of the entering Saturn probe. Figure 7 shows the nominal initial aerocapture orbit geometry in relation to its hyperbolic approach and Titan orbit.

Spacecraft Integration--Mars Sample Return

Figures 8 and 9 show the Mars sample return spacecraft in its aerocapture configuration. The spacecraft shown consist

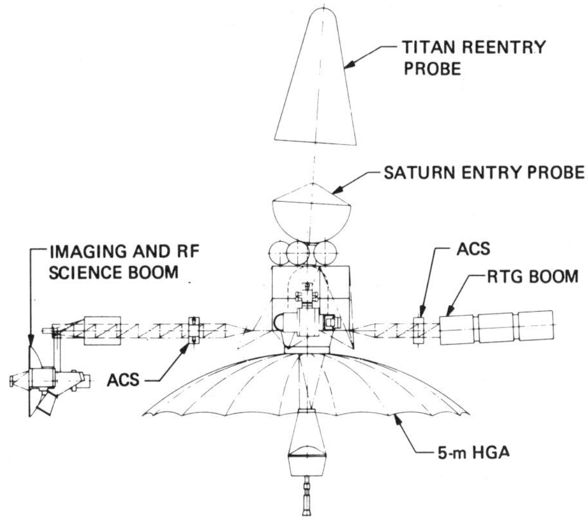

Fig. 5 Fully deployed SO2P spacecraft side view.

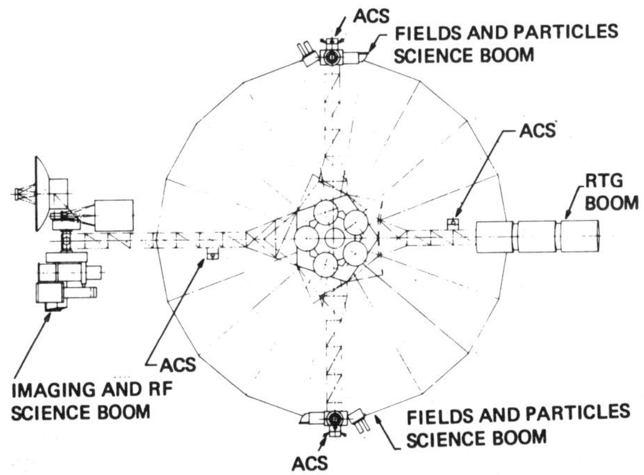

Fig. 6 Fully deployed SO2P spacecraft boresight view.

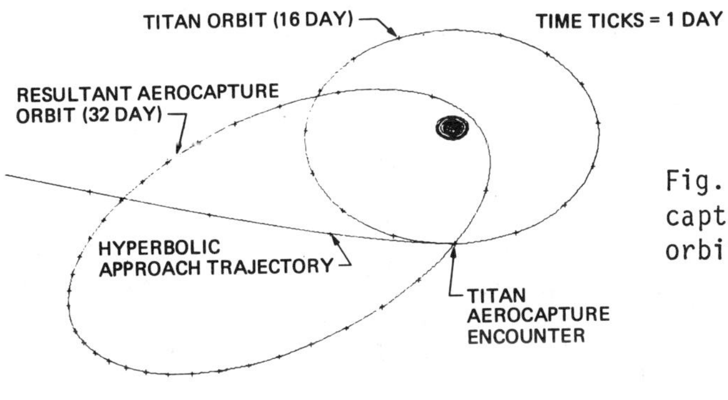

Fig. 7 Titan aero capture for Saturn orbit.

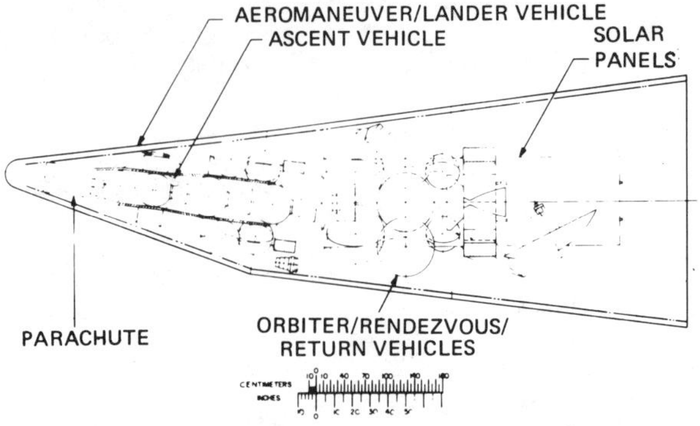

Fig. 8 MSR aerocapture configuration side view.

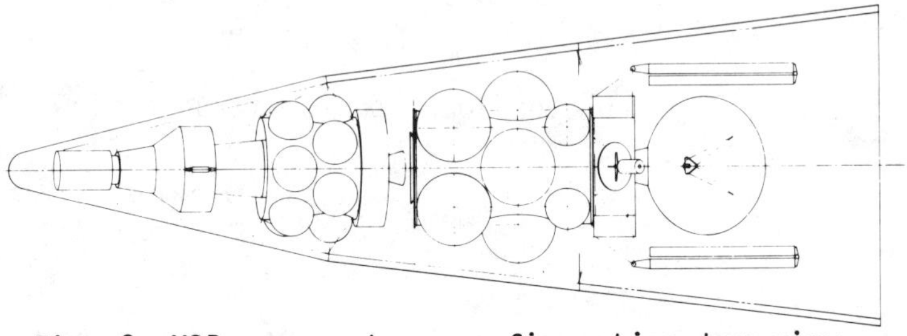

Fig. 9 MSR aerocapture configuration top view.

of re-entry, lander, ascent, orbiter, rendezvous, and return vehicles. These figures to some extent illustrate the large propulsion systems in the aeroshell. These spacecraft are powered by both RTGs and solar arrays. In terms of required payload, it is the most demanding on an aerocapture vehicle. Its total length is 9.9 m with a biconic base diameter of 3.6 m. The re-entry aeromaneuver vehicle shown has the same geometry as the SO2P aerocapture vehicle. In a generic sense, the aerodynamic shape could serve both as a medium and large payload capacity aerocapture system.

Figure 10 shows the fully deployed orbiting spacecraft with the separating Mars re-entry aeromaneuver vehicle. This separation would occur after a number of orbital passes over a variety of possible lander sites. The term re-entry is used to signify that the front end of the aerocapture vehicle is reused for the lander entry system. The biconic configuration for the aeromaneuver vehicle is required to effect the science lander site and accuracy requirements. Table 3 gives a complete breakdown of the spacecraft mass for an MSR aerocapture vehicle. This table, along with Fig. 11, demonstrates the mission enabling capabilities of aerocapture for MSR.

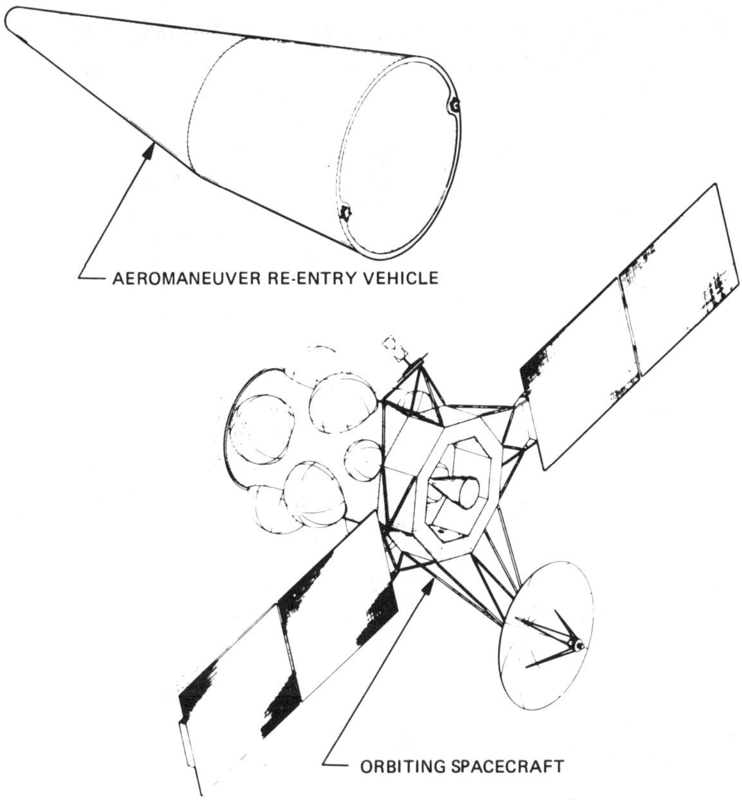

Fig. 10 Fully deployed MSR spacecraft.

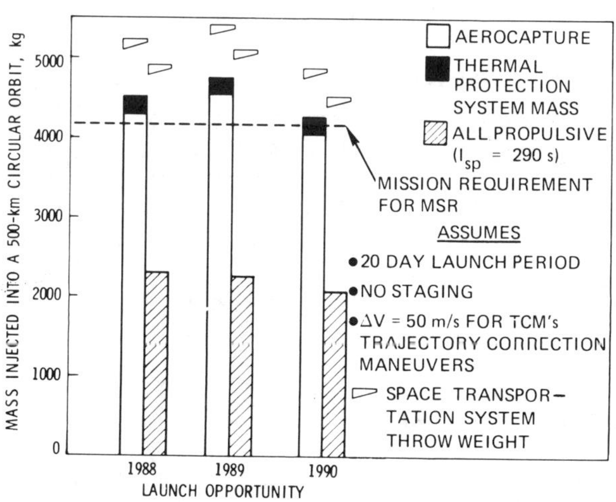

Fig. 11 Aerocapture performance at Mars.

Table 3 Mars sample return spacecraft mass summary

Subsystem	Mass, kg				
	Mars orbiter vehicle	Mars lander vehicle	Mars ascent vehicle	Earth capsule	MSR spacecraft
Structure	235.0	78.0	42.9	8.1	364.0
Telecommunications	52.4	5.4	5.4	3.0	66.2
Power	73.7	26.3	9.8	1.6	111.4
Data handling and control	20.0	6.0	---	1.0	27.0
Attitude and articulation control	45.9	30.2	8.4	---	84.5
Pyrotechnics	4.0	3.5	3.5	0.9	11.9
Cabling	28.0	30.0	8.0	2.4	68.4
Propulsion	1798.0	521.0	600.0	14.5	2933.5
Temperature control	31.4	5.9	4.3	5.8	47.4
Mechanical devices	15.0	11.6	10.2	---	36.8
S/X band antennas	2.8	0.8	0.5	0.3	4.4
Sample acquisition mechanics/sample	---	5.0	---	2.0	7.0
Aeroshell	242.0	203.0	---	---	445.0
Landing subassembly	---	144.3	---	4.9	149.2
Totals	2548.2	1071.0	693.0	45.5	4356.7

Table 4 Aerocapture navigation requirements

Navigation capability	Entry corridor requirement			
	Venus♀ ±1.5 deg	Mars♂ ±1.0 deg	Saturn♄ ±10.0 deg	Uranus♅ ±2.5 deg
Conventional radio navigation	±40.0 deg	±40.0 deg	±15.0 deg	±50.0 deg
Conventional optical navigation	±1.0 deg	±0.5 deg	±4.0 deg	±1.0 deg
Advanced radiometric navigation	±1.0 deg	±3.0 deg	±10.0 deg	±40.0 deg

Aerocapture Planetary Approach Navigation

Aerocapture requires that the planetary approach atmospheric entry vehicle be targeted to a prescribed entry corridor. This entry corridor is dictated by the uncontrolled skip out and limit load boundaries. In this study the criteria for the limit load was the quasi-steady-state aerodynamic load which was limited to 20 Earth g. The uncontrolled skip out boundary was computed using a worse case atmospheric model available to date, for a wide range of ballistic coefficients (about a ±30% variation from an optimum lift-to-drag ratio of 1.5). Table 4 shows what the entry corridor accuracies must be for aerocapture in a generic system. (The numbers also reflect a 1 deg pad away the uncontrolled skip out boundary.) Projected capabilities of conventional radio and optical and advanced radiometric navigation are also shown for each planetary application. In all cases, the table shows that conventional optical navigation with probably some preliminary radio-based estimation will absolutely be required for the aerocapture planetary approach delivery.

The guidance navigation and control (GN&C) system and communications antenna subsystem which will effect both interplanetary and planetary approach command and control functions are shown in Figs. 12 and 13. The TV camera used for approach navigation will be hard mounted on the aeroshell as opposed to pointing with a scan platform. This requires that more pictures than usual be taken to guarantee that the target will be successfully imaged against the star background. The optical data will be transmitted on a realtime basis to Earth for ground computation of planetary approach trajectory correction

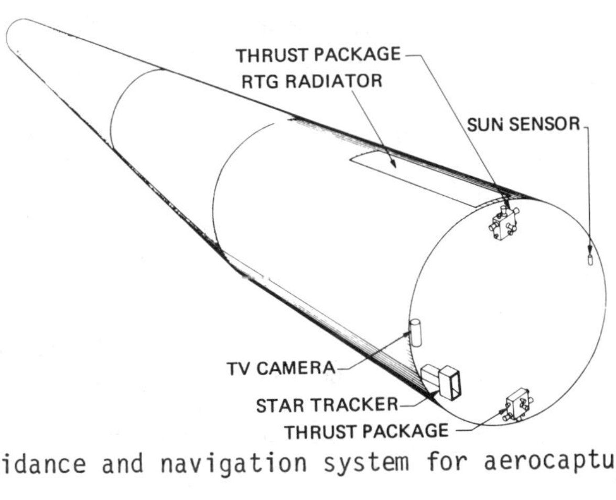

Fig. 12 Guidance and navigation system for aerocapture vehicle.

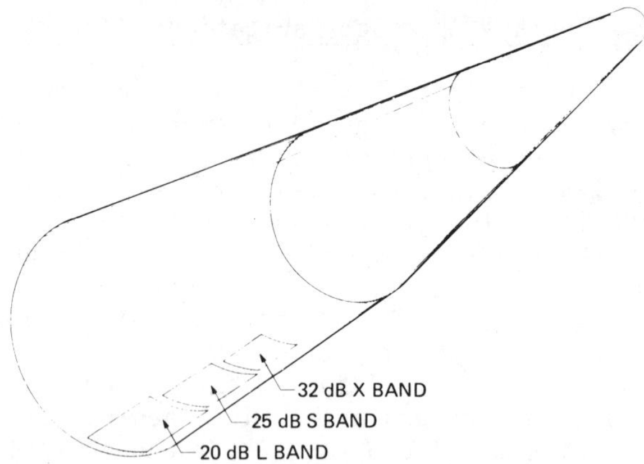

Fig. 13 Aerocapture vehicle antennas.

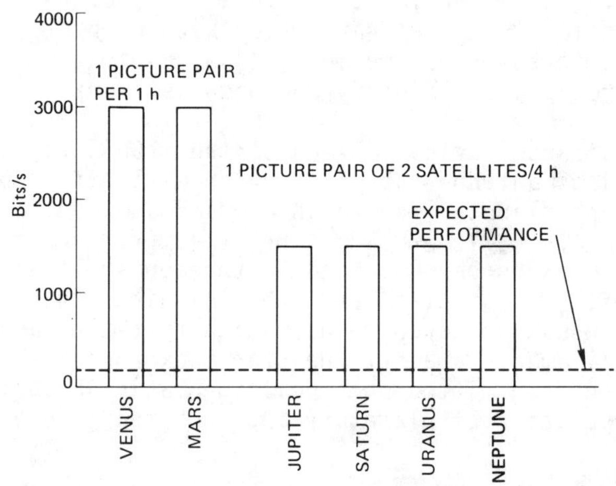

Fig. 14 Data-rate requirements.

maneuvers (TCMs) using low-profile phased array antennas spaced about the aeroshell.

The data rates to meet the navigation function for the range of aerocapture planetary applications are shown in Fig. 14. Figure 15 shows the relationship between the required antenna gain and data rates to be supported as a function of planetary application. The proposed phased array antenna with a technology development to increase efficiency can perform with a gain of 20 to 30 dB at S band and 30 to 35 dB at X band. This infers that the navigation function with this communications subsystem can only be met as far out as Mars. To

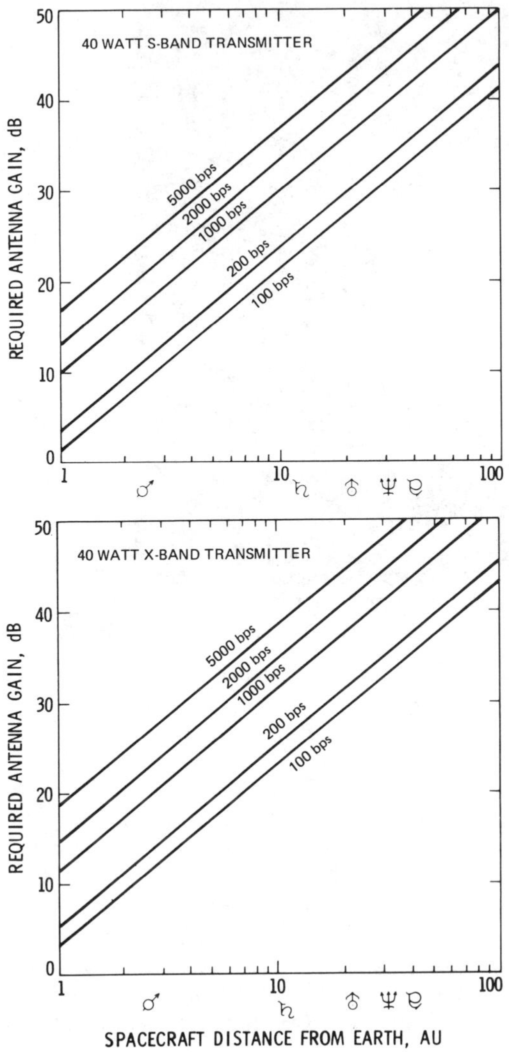

Fig. 15 Required antenna gain.

complement this, a critical technology development will be required which significantly reduces the navigation load to Earth.

Optical navigation uses simultaneous images of the target body and background stars using several exposures of one or more cameras. The basic information sought in each picture is the position of the target center or features on the target surface relative to the star background. A unique determination of the spacecraft trajectory with respect to the target body is then made from a time series of these position ob-

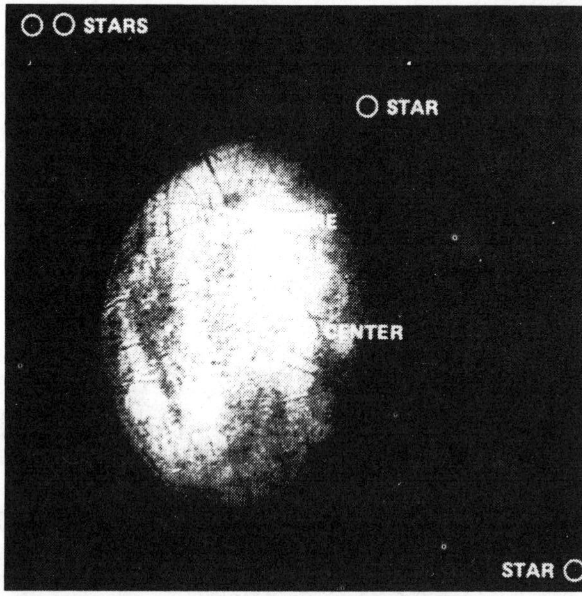

IMAGE LOCATION OF TARGET BODY OR
TARGET BODY FEATURES RELATIVE TO STARS

TARGET BODY = PLANET, SATELLITE, COMET, ASTEROID

Fig. 16 Data content and accuracy.

servations. Figure 16 shows a typical navigation exposure. As may be noted, a large portion of the exposure contains black space which is part of the data transmission. Allowing onboard spacecraft decision making and compressing the data stream can significantly reduce the required data stream for optical navigation. Development of this critical technology for aerocapture would lower the required data rates to within 100 to 200 bps. Using a phased array antenna, the optical navigation requirement could be met as far out as Uranus.

The rationale for proposing the development of high-efficiency, low-profile phased array antennas and optical navigation data compression as opposed to using expendable parabolic high gain antennas is based on the interplanetary trajectory GN&C adaptiveness and the aerocapture vehicle integration with the Space Shuttle Space Transportation Systems. Low-profile phased antennas allow communications redundancy by evenly spacing the antennas about the aeroshell without imposing a large mass penalty as would be the case for heavy parabolic articulating antennas. The low-profile phased antennas allow the aerocapture vehicle to maintain communications with Earth while at the same time maintaining inertial references with respect to target bodies and/or stars in performing both interplanetary and planetary approach navigation. Large parabolic antennas also cannot be readily integrated into the Space Shuttle volume along the aerocapture vehicle aeroshell without complicated articulating mechanisms.

Aerothermal Designs for Aerocapture

As shown earlier in Fig. 3, the application of aerocapture to a planetary system involves a wide range of entry speed and entry flight path angle requirements. This suggests that a single solution generic to all applications cannot be found which will still permit aerocapture to be a mission enabling technology. Some aerocapture applications (i.e., Mars and low-speed Titan entries) result in simple aerothermal protection system design solutions, whereas others may require ambitious aerothermal materials development and highly integrated spacecraft and aeroshell designs (i.e., Uranus and other far outer planets applications) in order to utilize the mission enabling capabilities of aerocapture. Yet other applications, such as a Saturn orbit capture using Titan's atmosphere in the moderate entry speed regime ($V_e \sim 10$ km/s), will require moderate research of the entry problem in the presence of yet unresearched atmospheric compositions (i.e., CH_4, N_2). This R&D is needed to arrive at low weight solutions and to eliminate some conservatism in the aeroshell design due to ignorance of the entry environment.

Figure 17 shows the relationship between the entry heating time and peak heat-transfer rate for a variety of entry vehicle applications. Aerothermal protection material selection is driven by these parameters. The variations of peak turbulent heat-transfer rate and heating time across the Titan aerocapture V-γ map is illustrated along with nominal values for Mars and Uranus. Also illustrated are the heating parameters for several contemporary entry vehicles. Pioneer Venus and Galileo are short time-high heating rate entries where the high performance ablator carbon phenolic is required. The other extreme is the Space Shuttle entry environment with very low peak heating rates, but with very long heating times where reuseable silicone insulator (RSI) is employed as the lightweight heat protection solution. In between these extremes and more in line with the low-speed Titan entries and the nominal Mars entry is the Military Space Satellite reentry vehicle that employs the elastomeric ablator, ESM, and has made numerous successful Earth entries. For the higher Titan entry velocities which significantly decrease the interplanetary cruise time, the peak heating rates and the heating times fall outside the range of current analysis. Therefore,

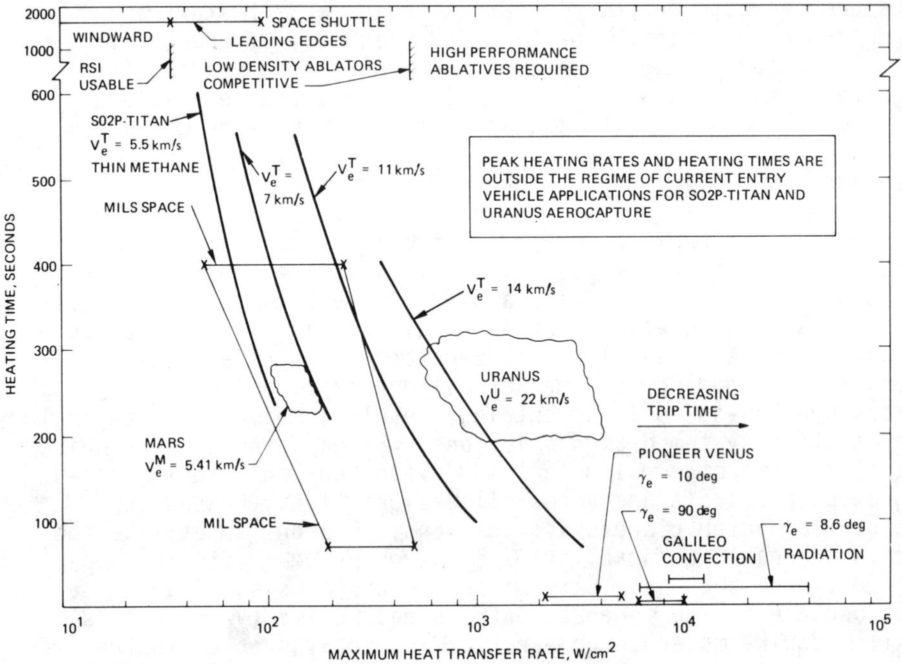

Fig. 17 TPS material selection.

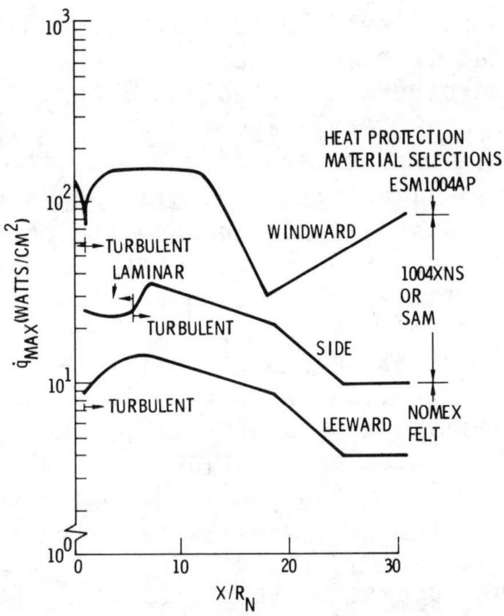

Fig. 18 Peak heat transfer rates of Mars aerocapture vehicle.

plasma arc evaluation tests will be required to assess contemporary ablators for use in these extended heating times. The same would apply to far outer planets aerocapture applications (i.e., Uranus).

Mars aerocapture results in the lowest weight aerothermal protection system solutions using low-density ablators. The entry environment at Mars is well known and was well researched prior to the two Viking landings. MSR aerocapture application required the study of two entry environments, namely aerocapture entry and aeromaneuver re-entry to effect lander operations to select and return Mars surface samples to Earth. The latter requires reusing the aerocapture aerothermal protection system.

Locally on the vehicle, for either the aerocapture and aeromaneuver vehicle, the peak heat-transfer rate usually drives the selection of the aerothermal protection material. Materials such as RSI or reradiative metallic panels have maximum allowable temperatures, whereas low-density ablators such as GE's ESM have higher heat flux thresholds above which char recession is experienced. Most trajectories consist of two important time segments that drive the aerothermal protection system requirements. These are the preheat before pullup (initial entry), and the period after pullup until a time when convective heating has essentially stopped because the velocity has dropped below Mach 8. For aerocapture, the preheat time is short because of the entry flight path angle. For aeromaneuvering, the entry flight path angle can be steepened; hence, preheat time can be shortened. The various aerothermal protection concepts/materials previously described find varying degrees of applicability and reusability for these lifting entry scenarios. The RSI and reradiative metallic cover panels are most attractive for low heat flux areas where peak heat fluxes are limited to 44 and 25 W/cm^2, respectively. A look at Fig. 18 indicates that RSI and metallics would be applicable only to the vehicle side and leeward areas. The low-density charring ablators, such as ESM 1004 and SAM, are more attractive for the higher heat flux areas displayed on the windward and side areas along with the simplicity of application. In the cooler area of the leeward side ($\dot{q}$ < 10 W/cm^2), a lower density material like the RTV coated NOMEX felt being used on the top side of the Space Shuttle is very weight attractive. Figure 19 shows the aerothermal protection system material distribution.

This aerothermal protection system must not only function during the aerocapture phase, but also during the re-entry

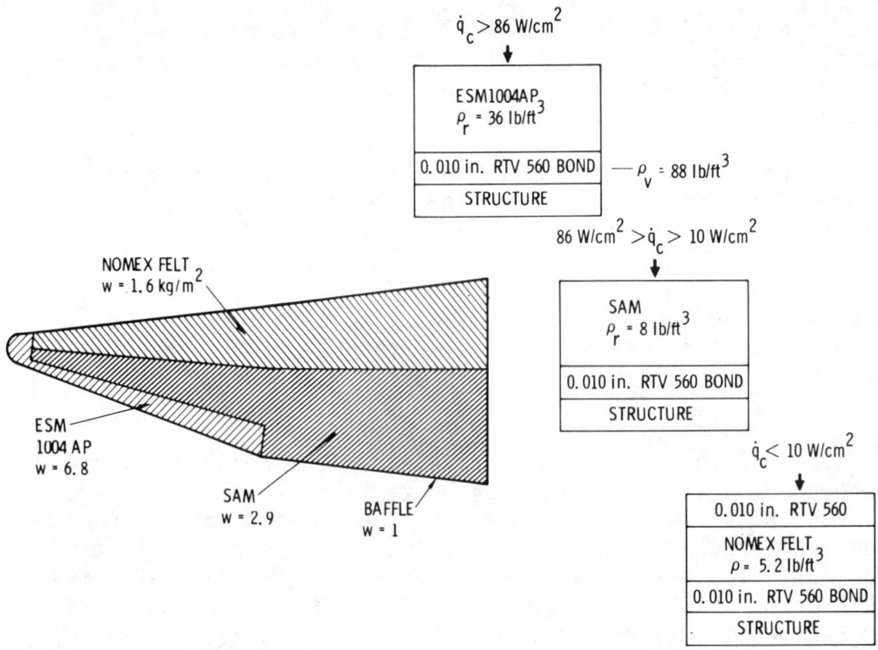

Fig. 19 Aerothermal protection system concepts selected for Mars aerocapture.

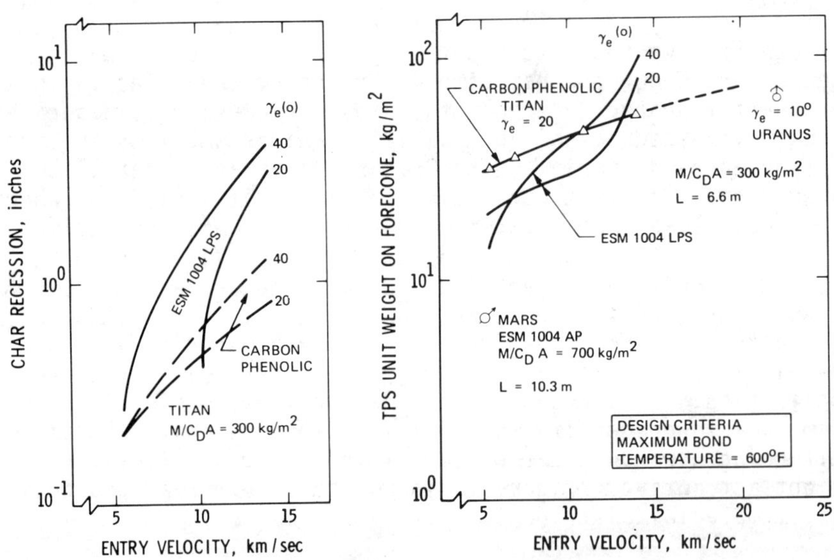

Fig. 20 Titan entry at $V_e \geq 10$ km/s requires carbon phenolic ablator for shape stability.

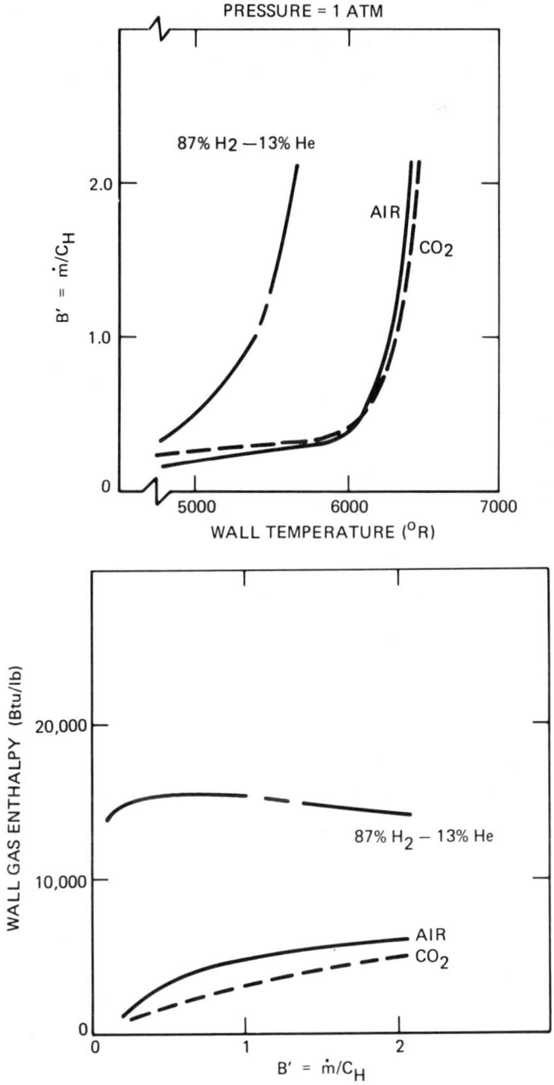

Fig. 21 Atmospheric composition has large effect on carbon ablative performance.

aeromaneuver phase. Low-density ablators, such as ESM and SAM, are expected to be reusable with some small loss of thermal efficiency because of the charring experienced during the aerocapture maneuver. For aeromaneuver, the peak stagnation point heat-transfer rate is a factor of 4 less than aerocapture, the heat load is a factor of 10 less, and the heating time a factor of 6 less. All of these factors add up to a need for significantly less aerothermal protection for the aeromaneuver than aerocapture. It is expected that use of the precharred ablator material for aeromaneuver will result in peak bondline temperatures of much less than 600°F.

Local aerothermal protection weight at the windward forecone location is illustrated in Fig. 20 for aerocapture missions using Titan with reference given to Mars and Uranus. The entry-speed range at Titan is representative of the minimum with corresponding minimum interplanetary cruise times. Entry speeds of 6, 8, and 10 km/s correspond to interplanetary cruise times of 8, 5.7, and 5 yr, respectively. For entry speeds of less than 7 km/s, a low-density elastomeric ablator experiences minimal recession and is very weight attractive. At the higher entry speeds, the elastomeric ablator experiences excessive recession (relative to carbon phenolic) and as a result carbon phenolic provides a lighter weight solution. These results are for the most part for a preliminary aerocapture configuration with an M/C_DA of 300 kg/m^2. As the configuration evolves to M/C_DA greater than 700 kg/m^2, the expected turbulent peak heat-transfer rates will double. This will result in increased surface recession, with carbon phenolic providing the lighter weight solution at entry speeds greater than 10 km/s. New carbon phenolic formulations are available that exhibit both lower density and thermal conductivity, and will provide lighter weight options. Significant development activity has taken place in carbon-carbon shells over a lower-weight fibrous insulation that can provide alternate lightweight options. A concern is the overall material structural response when integrated over a large aerocapture aeroshell.

An additional concern is the effect of atmospheric composition on carbon ablative performance. Ablation material surface reactions with the dissociated boundary layer have been modeled for several carbon-atmosphere combinations. The variation of the normalized mass loss parameter, B', with ablator surface temperature and atmospheric composition is illustrated in Fig. 21. The variation of wall gas enthalpy with B' and atmospheric composition is also illustrated. Wall gas enthalpy, hw, is the primary term in the ablator surface heat balance that controls the carbon surface recession, a higher hw resulting in lower mass loss rates. It is seen that atmospheric composition has a large effect on the carbon ablative performance. (These results are from the Fogaroli-Brant carbon thermochemistry evaluations.) The presence of H_2 and He is highly probable in far outer planets aerocapture application. Other atmospheric compositions, yet unresearched,

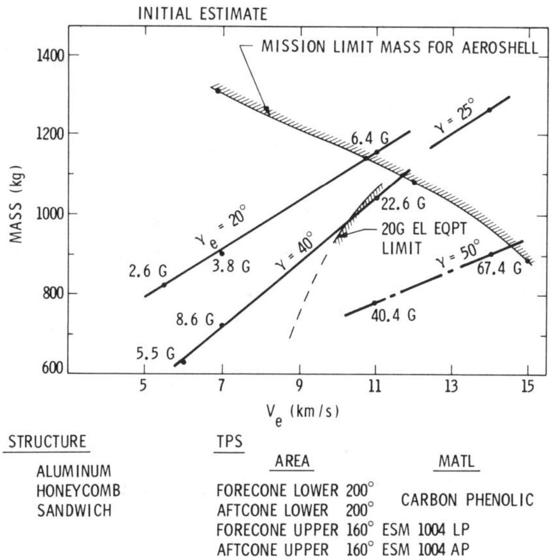

Fig. 22 S02P Titan aerocapture aeroshell mass vs V_e, γ_e, and G_e (initial estimate).

which will be encountered at Titan are CH_4 and N_2. These could result in high radiative heat transfer from CN.

Preliminary aerothermal protection system solutions for Titan aerocapture trajectories are shown in Fig. 22. The aeroshell structure mass estimates are based on an aluminum honeycomb sandwich. The aerothermal protection system consists of carbon phenolic ablator over the fore and aft cone windward areas and ESM 1004 ablator on the leeward side. These mass estimates (assuming ±4 deg navigation capability) are correlated on Fig. 23 to demonstrate the performance of aerocapture for an S02P mission. This figure shows that aerocapture can perform the S02P mission with a 5-yr flight time and a positive mass margin using the Solar Electric Earth Gravity Assist (SEEGA) interplanetary transfer option. The all-propulsive SEEGA mission requires an 8-yr mission time period prior to beginning intensive Saturn science, with an approach mass requirement of 3100 kg which places the mission in the red in terms of mass margin.

Aerocapture Atmospheric Entry Attitude Control

Figure 24 shows atmospheric entry attitude control system weight estimates for a Titan aerocapture. The high reaction control system (RCS) mass represents a simple dual thrust on-

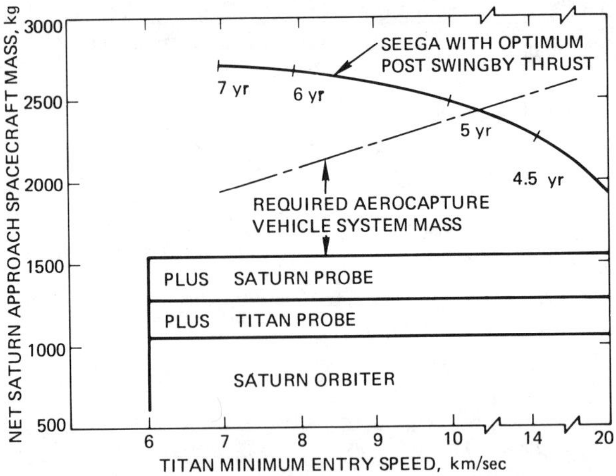

Fig. 23 Saturn orbiter and probes mission performance with aerocapture/SEPS.

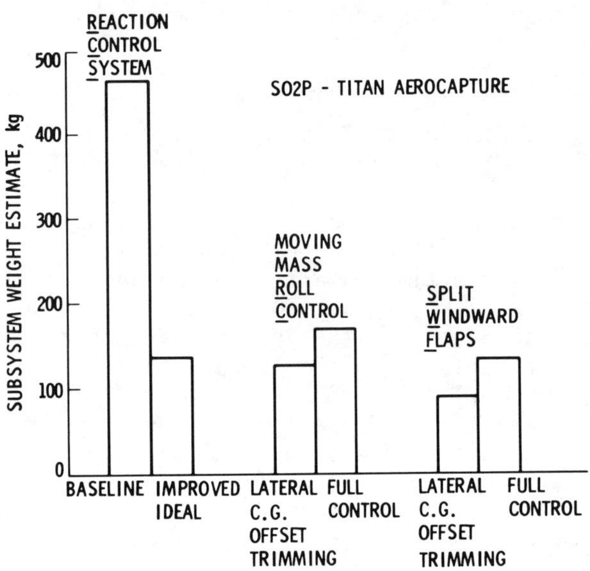

Fig. 24 Three control options are available.

off system. The low weight represents an idealized RCS system with a 2 deg deadband. The moving mass roll control (MMRC) and the split windward flap (SWF) masses include some RCS fuel (12 kg) for pitch and yaw damping. The full control SWF uses no RCS fuel. The MMRC, requiring moving masses of 100 to 200 kg, would be difficult to implement in a generic system due to variations in payload volume. The system was sized for a

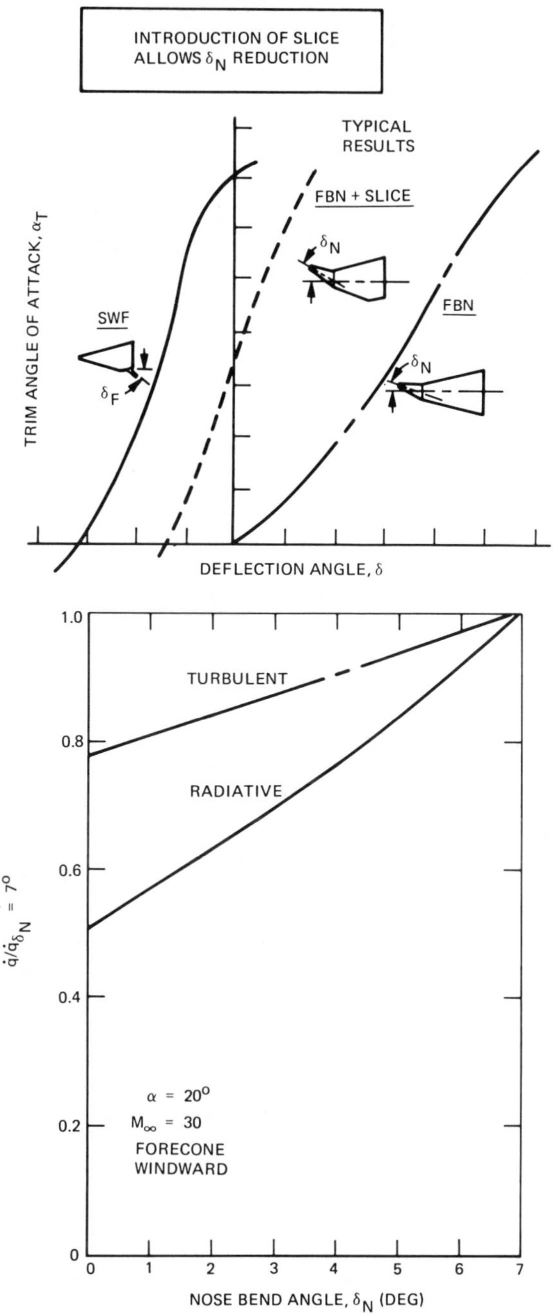

Fig. 25 Configuration tailoring can reduce heat transfer to forecone.

1.6 cm lateral roll center of gravity offset. Both RCS and MMRC masses are very sensitive to changes in c.g. offset numbers while the SWF mass remains relatively insensitive.

Usage of SWF for control of the aerocapture vehicle provides an additional number of basic vehicle flowfield interaction problems plus a possible opportunity to reduce the thermal protection weight on the forecone by reducing the nose bend angle. Two major options exist for installation of the flaps. They can be installed as trailing flaps, or on a slice of the vehicle aft end. The slice option produces a larger trim angle of attack for a given nose bend angle. The variation of trim angle of attack with nose bend and flap deflection angles is illustrated in Fig. 25. Typically, a bent nose configuration exhibits a much larger level of pitch stability than a plain cone with flaps; a bent nose with flaps configuration falls in between. The reduction in radiative and turbulent heat-transfer rate to the windward meridian area of the forecone as the nose bend angle is decreased is also illustrated. At the higher entry velocities where radiative heat transfer is large, this potential reduction is significant.

Conclusion

The research and technology developments which are critical to the design of a generic aerocapture vehicle system are few. These few primarily deal with the communications and navigation of the spacecraft, aerothermal research, and subsystem integration, all of which are highly interdependent. The aerothermal and spacecraft design depend in turn on the spacecraft navigation capability and delivery entry speed. It is expected that a modest research and technology effort will allow the design and application of this mission enabling concept to future planetary missions.

Acknowledgment

The research described in this paper was carried out by the Jet Propulsion Laboratory, California Institute of Technology, under NASA Contract No. NAS7-100. The main contributors to this study were Albert G. Brejcha, Thomas C. Duxbury, Jordan Ellis, John Hultberg, Richard H. Key, Donald W. Noon, Louis Keeler and Edward Rinderle of JPL; and Raphael Armento, Dwight Florence, Robert Hassett, and George Reehl of General Electric Company Reentry Systems Division; and James N. Moss of Langley Research Center.

References

[1]Cruz, M. I., Duxbury, T. C., Brejcha, A. G., Florence, D., and Moss, J., "Generic Aerocapture Technology Requirements Review," JPL presentation to NASA Office of Aeronautics and Space Technology, Document 725-26, 18 June 1980.

[2]French, James R., and Cruz, Manuel I., "Aerobraking and Aerocapture for Planetary Missions," Astronautics & Aeronautics, Vol. 18, No. 2, Feb. 1980, pp. 48-71.

[3]Cruz, M. I., "The Aerocapture Vehicle Mission Design Concept," AIAA Paper 79-0893, NASA/AIAA Conference on Advanced Technology for Future Space Systems, NASA Langley Research Center, Hampton, Va, May 8, 1979.

[4]Cruz, M. I., "Aerocapture Vehicle Mission Design Concepts for the Inner and Outer Planets," AIAA Paper 79-115, AAS/AIAA Astrodynamics Specialist Conference, Provincetown, Mass., June 25, 1979.

[5]Cruz, M. I., Armento, R. F., and Giles, W. H., "Aerocapture - A System Design for Planetary Exploration," Paper 79-160, 30th International Astronautical Federation Congress, Munich, FRG, Sept. 16, 1979.

[6]Castro, A. J., "Titan Probe Technology Assessment and Technology Development Plan Study," Martin Marrietta Corp., Denver Division, Report No. TPT-MA-02-3, July, 1980.

Author Index for Volume 77